UNITED STATES–JAPAN SEMINAR ON
HOST–GUEST CHEMISTRY

Advances in Inclusion Science

VOLUME 6

United States–Japan Seminar on Host–Guest Chemistry

Proceedings of the U.S.–Japan Seminar on Host–Guest Chemistry, Miami, Florida, U.S.A, 2–6 November 1987

Edited by

GEORGE W. GOKEL
University of Miami, Florida, U.S.A.

and

KENJI KOGA
University of Tokyo, Japan

Reprinted from
Journal of Inclusion Phenomena and Molecular Recognition in Chemistry,
Vol. 7, Nos. 1 and 2 (1989)

Kluwer Academic Publishers
Dordrecht / Boston / London

Library of Congress Cataloging in Publication Data

United States-Japan Seminar on Host-Guest Chemistry
    (1987 : Miami, Fla.)
    United States-Japan Seminar on Host-Guest Chemistry.

    (Advances in inclusion science)
    Includes index.
    1. Electron donor-acceptor complexes--Congresses.
I. Gokel, George W., 1946-   . II. Koga, Kenji,
1938-   . III. Title. IV. Series.
QD474.U55 1987   541.2'2   89-11064
ISBN-13:978-94-010-6925-0   e-ISBN-13:978-94-009-0969-4
DOI: 10.1007/978-94-009-0969-4

Published by Kluwer Academic Publishers,
P.O. Box 17, 3300 AA Dordrecht, The Netherlands.

Kluwer Academic Publishers incorporates
the publishing programmes of
D. Reidel, Martinus Nijhoff, Dr W. Junk and MTP Press.

Sold and distributed in the U.S.A. and Canada
by Kluwer Academic Publishers,
101 Philip Drive, Norwell, MA 02061, U.S.A.

In all other countries, sold and distributed
by Kluwer Academic Publishers Group,
P.O. Box 322, 3300 AH Dordrecht, The Netherlands.

Table of Contents

Journal of Inclusion Phenomena and Molecular Recognition in Chemistry 7 (1989), 1.
© 1989 *by Kluwer Academic Publishers.*

Editorial

The year 1987 was especially notable in that the Nobel Prize was awarded to three of the pioneers of the field. Unfortunately, 1987 was also the year in which we must sadly mark the passing of two great men of science, Professor Iwao Tabushi and Professor James J. Christensen. It is to these two men that the contributions from the United States-Japan Joint Seminar are dedicated.

The seminar was held 2–6 November 1987 at the Coconut Grove Hotel, Miami, Florida. The organization of the event was directed by Professor George W. Gokel and Professor Kenji Koga, and we are grateful to them for assembling this outstanding set of manuscripts.

JERRY L. ATWOOD
J. ERIC DAVIES

Journal of Inclusion Phenomena and Molecular Recognition in Chemistry 7 (1989), 3–6.
© 1989 *by Kluwer Academic Publishers.*

Iwao Tabushi (1933–1987)

My only close association with Iwao Tabushi being limited to a period of about two weeks in Kyoto when I, together with my wife, visited Kyoto on his invitation, I cannot lay claim to a long standing personal intimacy with him. Nor for that matter can I claim to be expert over the wide range of chemical behavior which he explored in his research so diligently and productively. Still, even in the short time that I interacted with him, this altogether remarkable man impressed me so strongly by his qualities as a human being, and by the way that he addressed his professional interests that I feel moved to add my own testimony to that made by others in commemorating him.

As I look back on our altogether too brief an association, the impression left with me that comes through most strongly is his generosity. He gave generously of his time and his energies, from what seemed an inexhaustible reserve of vitality, in sharing professional interests, and in seeing even to the details of accommodation and travel arrangements for what appeared to be a steady stream of visitors to his laboratories. This generosity extended also to social arrangements which invariably included the accompanying individuals and involved on occasion his colleagues, usually some or all of his coworkers, sometimes his wife, Sakiko, and almost always himself. I continue to marvel at his tact and restraint, when even at a late evening hour his enjoyment of the occasion never seemed to flag, knowing as I did that after parting with the last lingering guest, he would return to his laboratory to the work that he enjoyed so keenly.

So great was his enjoyment of his work and his enthusiasm for new ideas and new results that it is altogether likely that the late hours in the laboratory in the company of some of his coworkers were refreshing. At any rate, when I would see him the next day, even after we had spent a long evening together, there was no diminution in his clarity of thought, nor of his enthusiasm as our discussions turned to his research interests. This is not to say that he lacked interest in the work of his visitors – they were given ample opportunity to present their results and views in public seminars. Besides, though I was aware of Tabushi's reputation prior to my visit to Kyoto, I was well informed only about a small part of his work, and it was clearly in my interest to acquaint myself with a wider range of his activities.

As much as anyone I have known Tabushi was the complete chemist. His involvement in research was not limited by the traditional subject boundaries: both his interests and capabilities transcended them. Nor was he limited by a concern with whether his research was basic or applied; he tried to understand Nature and equally he tried to put his knowledge to use. His training was that of a physical organic chemist, and he had a firm grasp of the physico-chemical principles that underlie that discipline. At the same time, he had an imposing command of descriptive chemistry, organic and inorganic. His multifacetted approach to chemistry is illustrated, though inadequately, by the following excerpts taken from rather recent papers selected from the approximately 200 he published.

"A comprehensive model of the inclusion process of α-cyclodextrin is presented herein. Van der Waals interaction energy, Allinger's conformation energy, solvation energy of apolar solute in water, hydrogen bond energy of water molecules in the cavity of α-cyclodextrin hexahydrate, and all other possible energies were taken into account for the calculation of free-energy change by complexing an apolar guest molecule by α-cyclodextrin. Motional freedoms of all the particles relevant to the inclusion phenomenon were taken into consideration." ('Approach to the Aspects of Driving Force of Inclusion by α-Cyclodextrin', I. Tabushi, Y. Kiyosuke, T. Sugimoto and K. Yamamura, *J. Am. Chem. Soc.* (1978), **100**, 916.)

"Picket-fence porphyrin (TpivPP)-iron-N-methylimidazole-O_2 complex is used as an artificial P-450, and the decomposition rates are investigated in details in the presence of HCl and H_2-colloidal platinum supported on poly(vinylpyrrolidone) with or without addition of benzoic anhydride. From the decay rates of the oxy complex followed by electronic spectrum under a variety of conditions, pseudo-first-order rate (with the complex) constants are obtained." ('Kinetics and Mechanism of Reductive Dioxygen Activation Catalyzed by P-450 Model System. Iron Picket

Fence as a Catalytic Center', I. Tabushi, M. Kodera, and M. Yokoyama, *J. Am. Chem. Soc.* (1985), **107**, 4466.)

"Concept of coupling between oxidation and generation of pH gradient across membrane is presented by use of artificial liposome modified with electron transport catalysts, cyt c_3 or C_4V^{++}. The presented coupling mechanism based on facilitated 'down-hill' electron flow, electroneutrality preservation and permeability control is confirmed by independent and direct measurements." ('Basic Principle of Coupling between Oxidation and pH Gradient Generation. Artificial Liposome Digesting H_2', I. Tabushi and T. Nishiya, Tetrahedron Letters, (1982), **23**, 2661.)

"Molecular design of uranyl specific ligands is reviewed. Basic principles to characterise the unique nature of uranyl complexes are scrutinized from ligand field calculations, crystallographic analyses and stability constants of various complexes. The most typical characteristic of uranyl coordination is a planar hexacoordination by anionic ligands giving rise to the formation of the 'ate-complex.' Several macrocycles were synthesized by the appropriate combination of these principles. Approaches to recovery of uranium from sea water are briefly reviewed. Organic chelating resins are the most promising candidates for the practical adsorbent". ('Molecular Design of Specific Uranophiles', I. Tabushi and Y. Kobuke, *Israel J. Chem.*, (1985), **25**, 217.)

The excerpts also serve to indicate what the major themes in Tabushi's research were toward the end of his career. His work on cyclodextrins was begun in 1976. The motivation was to gain an understanding of molecular recognition processes and, armed with such understanding, to mimic – in this particular instance – hydrolytic enzymatic processes. Even his early work in modifying cyclodextrins so as to increase catalytic activity and selectivity attracted world-wide attention, and his position as a pioneer in this active and important field of research is secure.

Tabushi's interest in biomimetic processes extended also to oxidation-reduction enzymes. His success in this field can be gauged from these quotations taken from an article published in *Japan Chemical Week* (April 17, 1986) under the heading 'Efficient Proc. for Oxygen Activation Pioneered'.

"A research group led by Professor I. Tabushi of Kyoto University has successfully developed a new process to activate oxygen using an artificial system modeled on cytochrome P450 enzyme. This process is more than 1000 and $5 \sim 6$ times more efficient than conventional methods and natural enzyme, respectively...

"Their study is 'ultimately' aimed at pioneering artificial livers. The liver biochemically decomposes poisonous substances and synthesizes biomaterials. Cytochrome P450 is involved in such biochemical processes.

"The said study has paved the way for realizing an artificial system which resembles a natural enzymatic reaction process and which is needed for conditioning the existence of biomaterials. It is expected to be applied to the industrial-use oxidation process in which benzene and paraffin are transformed into phenol and alcohol, respectively."

It should not be thought that Tabushi's interest in the biological processes was restricted to invention. Invention grew out of understanding, and understanding grew out of research, at the most basic level. An important such interest was the study of electron transport and attendant ion transport through artificial membranes, where again pioneering and definitive experiments were performed in his laboratories. I make specific mention of this interest because it provided the greatest overlap between his research activities and my own, and this overlap accounts for his sponsorship of my visit to his laboratories.

On an earlier occasion I wrote this about Tabushi:

"Few match his passionate devotion to research and his intensity of effort. He is an inspiring leader of a large group that contains a high proportion of gifted individuals. Their quality accounts in part for the productivity of the laboratory. It is however mainly to be credited to Professor Tabushi himself. He is very imaginative and ingenious. He is quick to master a new subject area, and to find in it opportunities which others have missed. Ideas are soon translated into action in his laboratories, and his entire research group is driven by Tabushi's restless energy, where he sets an example by his own remarkable diligence."

I mourn the loss of a gifted scientist, and the early termination of what promised to be for me a warm and rewarding friendship.

H. TAUBE

Journal of Inclusion Phenomena and Molecular Recognition in Chemistry 7 (1989), 7–17.

New Molecular Shapes for Recognition and Catalysis

JULIUS REBEK, JR
Department of Chemistry, University of Pittsburgh, Pittsburgh, Pennsylvania 15260, U.S.A.

(Received: 1 February 1988)

Abstract. Progress in molecular recognition is reviewed with special emphasis on the advantages offered by molecular clefts. These new structures are rapidly assembled from readily accessible starting materials and feature functional groups that converge on smaller species that present complementary surfaces. The sizes and shapes of the clefts are controlled by the use of appropriate spacer elements. The selective binding of acids, amines, amino acids, metal ions, heterocyclic compounds and nucleosides is described. Their special applicability to problems involving concerted catalysis is also introduced.

Key words. Recognition, hydrogen bonding, molecular clefts, convergent functionality, concerted catalysis, stereoelectronic effects.

1. Introduction

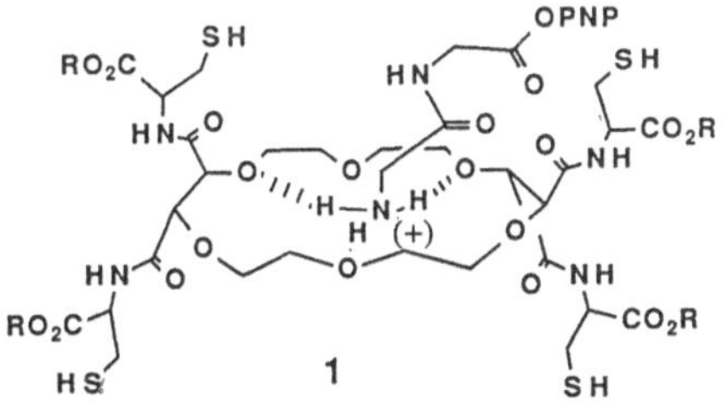

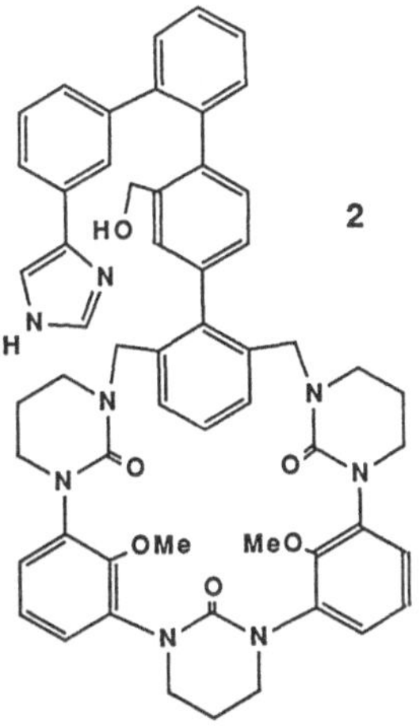

With the award of the 1987 Nobel Prize in chemistry to Cram, Lehn and Pedersen, model systems for bioorganic chemistry have become respectable. In the not-too-distant past, biochemical phenomena such as catalysis, recognition and transport were believed to be unique properties of macromolecules. Recent successes in imitating such phenomena using much smaller compounds has shown that chemical behaviour can be engineered into simple molecules. Crown ethers, for example, are satisfactory models for allosteric cooperativity [1] and they have gained much popularity as enzyme models [2]. In particular, the crown ether **1** described by Lehn [2c] and the macrocyclic cavitand **2** described by Cram [3] have been successful in showing large rate enhancements in their reactions with p-NO$_2$-phenylester derivatives. The cyclodextrin systems explored by Bender and Breslow [4] have shown similar rate enhancements for the hydrolysis of other esters such as **3** and **4**.

2. Structural and Synthetic Considerations

These models stress the use of binding to enhance reaction rates, and the advantage of readily available binding forces provided by these systems is the primary reason for their widespread use. The difficulty in attaching catalytically useful functionality to the parent macrocyclics is their principal disadvantage. For example, Lehn's system **1** requires four functional arms to provide a nucleophile near the bound ester, and a heroic synthetic effort is required in the Cram protease model **2**. Even so, only two of the three functional groups of the catalytic triad are in position to react with the substrate. Similar problems face cyclophane-based structures [5]. The use of such model systems to demonstrate concerted catalysis is even more difficult [6]. We have turned to new molecular shapes to overcome these disadvantages, and while development of macrocylic structures continues unabated, a few recent reports [7] suggest our views are gaining some acceptance.

The premise is that well-placed functionality can create a unique micro-environment for rate enhancements, and we are applying our recent discoveries [8] concerning molecular clefts to such situations. Specifically, we are developing systems in which functional groups converge (as in **5**) on substrates bound within. This structural strategy permits functional groups to act on several sides of a substrate molecule simultaneously and thereby enhance both recognition and catalysis.

Our departure from the classical macrocyclic shapes was made possible by access to the unusual triacid **6**, first described by Kemp [9], in which three equatorial methyl groups force the carboxyl functions to a triaxial arrangement. The resulting U-shaped relationship that exists between any two carboxyl functions provides an opportunity to reverse the direction of bonds within a given molecule. For example, condensation of **6** with most primary amines gives imides, but with diamines derived from aromatics (e.g. *m*-xylidinediamine) an unusual C-shaped structure **7** results. The *ortho* substituents serve to prevent rotation about the C_{aryl}—N_{imide} bond, and enforce the convergent conformation of two carboxyl groups as shown. Indeed, **7** shows all the spectroscopic earmarks of a conventional hydrogen-bonded dimer of carboxylic acids, but is unable to dissociate into the monomer.

Convergent functionality

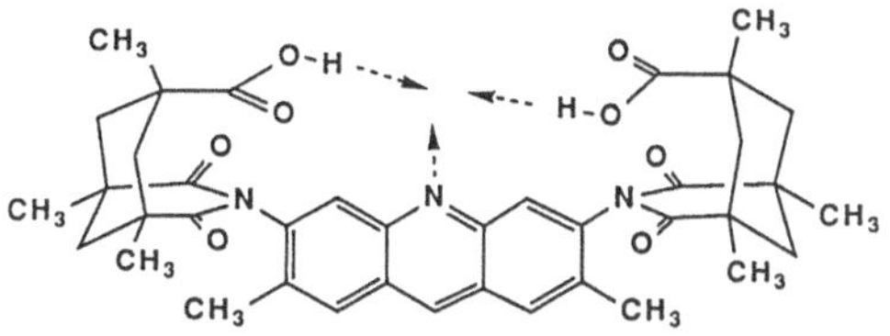

5

Scheme 1

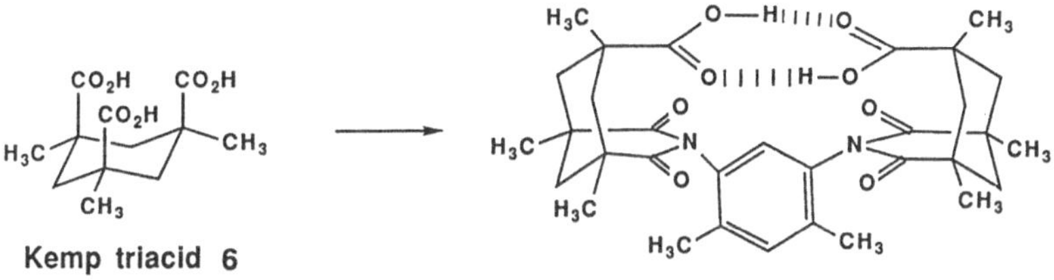

Kemp triacid 6

7

Using larger aromatic systems as spacers the distances between the acid functions can be increased (Scheme 2). The diacid derived from acridine yellow gives a system (**8**) in which ~ 8.5 Å separates the opposing carboxyl oxygens, while the naphthalene spacer **9** provides a distance of 5.6 Å. Other suitable diamines are also commercially available; e.g. 2,7-diaminoacridine gives **10** in which rotation permits several conformations [10]. The diaminofluorene derivative **11** features slightly different angles and distances; these subtle changes are reflected in its altered binding selectivities.

Scheme 2

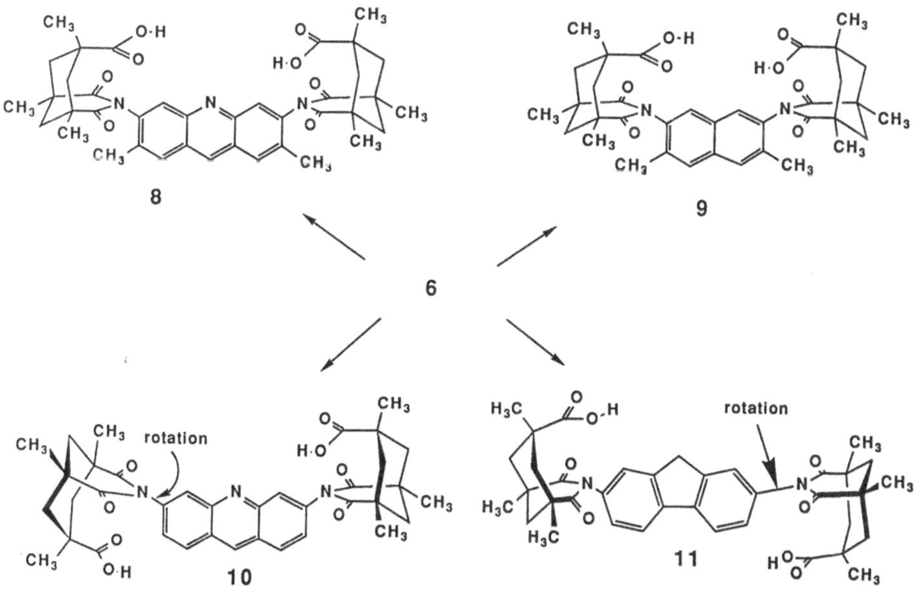

8

9

6

10

11

These systems all permit conformations in which two carboxyl groups converge, but the condensation of **6** with primary amines is sufficiently general that it may be used as an architectural cliche. A number of sizes and shapes for model molecular receptors can be constructed with a minimal synthetic investment. For example, with tren, a new triacid **12** is obtained and four convergent carboxyls e.g. **13** may be prepared through Lindsey's [11] porphyrin synthesis, using the appropriate aldehyde (Scheme 3).

Scheme 3

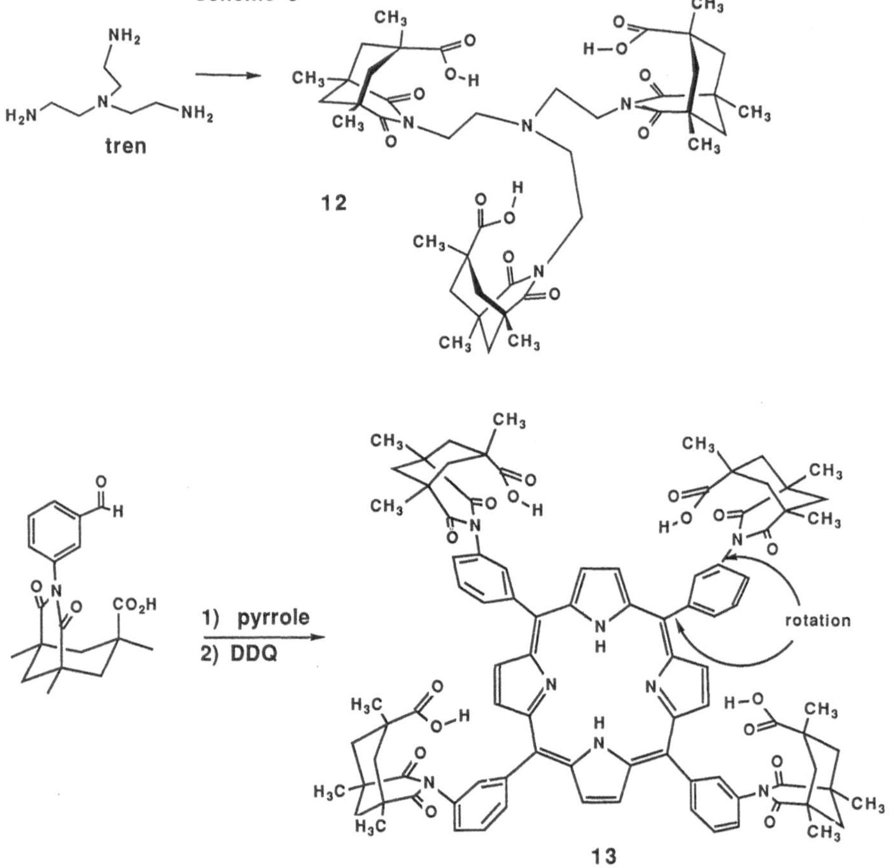

12

13

3. Complexation of Complementary Structures

The model receptors **7–13** provide a range of sizes and shapes with which the rules for binding selectivity to smaller molecules can be laid out. Substrates with complementary basicity are ideal and we have studied the complexation of heterocyclic diamines extensively. The acridine **13** binds tenaciously to pyrazine, DABCO and other heterocycles that bridge the gap between the carboxyl functions [12]. With imidazole and its derivatives 2 : 1 complexes are formed; both the acid and base character of the heterocycle is expressed in the binding event (Scheme 4).

Scheme 4

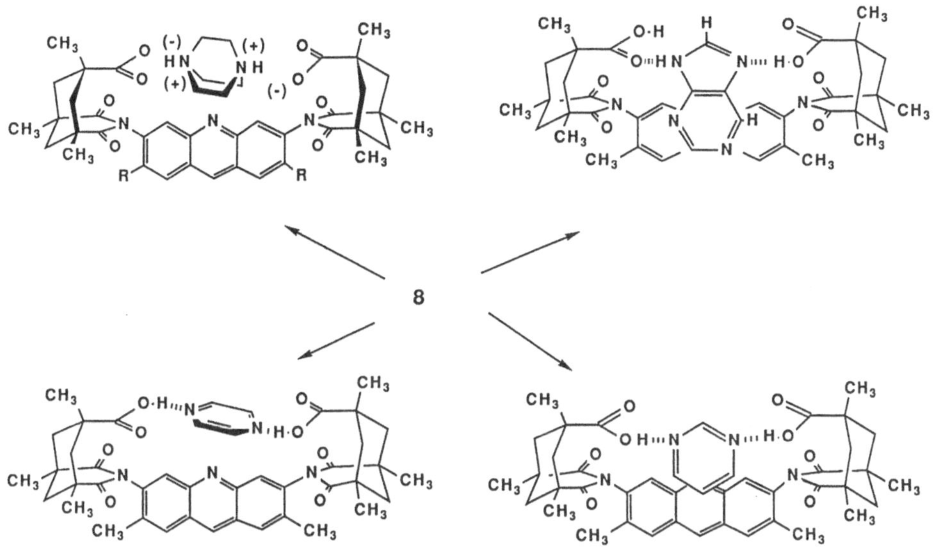

8

Dicarboxylic diacids offer complementary functionality in the form of hydrogen bonding that gives rise to the dimerization of most carboxylic acids in non-competing media. The acridine spaces of **8** and **10** accommodate oxalic and malonic acids but reject longer diacids such as succinic or glutaric acid [13]. The fluorene spacer of **11** shows affinity for glutaric and camphoric acids but eschews the smaller diacids (Scheme 5).

Scheme 5

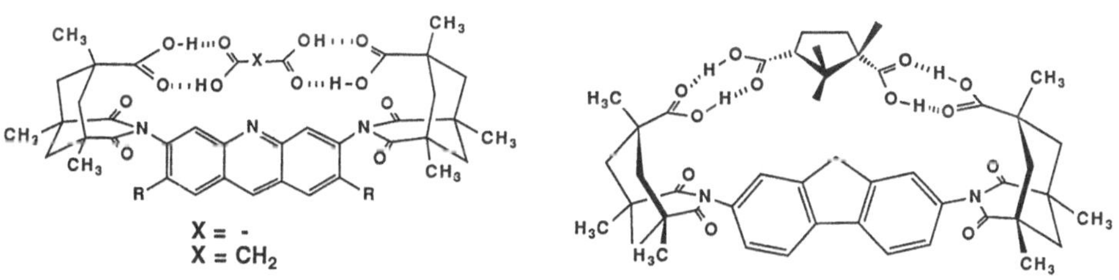

X = -
X = CH₂

The chelation of small molecules described above may be extended to metal ions. The convergence of the carboxyls within the molecular clefts provides a microenvironment ideal for divalent metals. A special structural feature of the new ligands involves stereoelectronic effects at carboxyl oxygen. Classical chelates such as EDTA present the less *anti* lone pairs to the metal ion, but the new structures offer the more basic *syn* lone pairs. Metals such as Ca⁺⁺ and Mg⁺⁺ are tightly bound and readily transported across liquid membranes (Scheme 6). In addition, the mode of binding within the new ligands is exclusively *trans*, a feature which is likely to lead to altered reactivity of the bound metal ions as catalysts.

Scheme 6

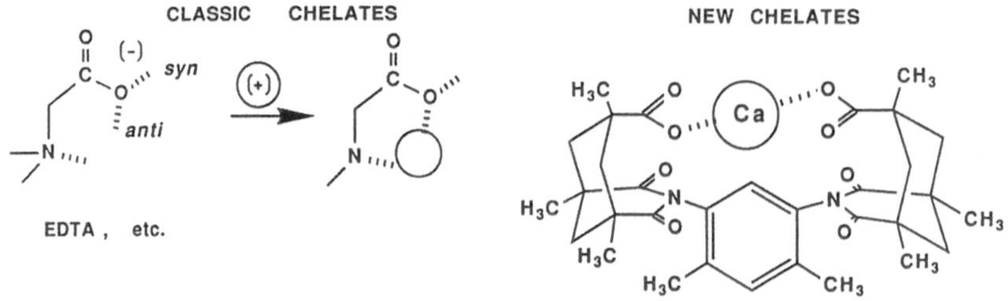

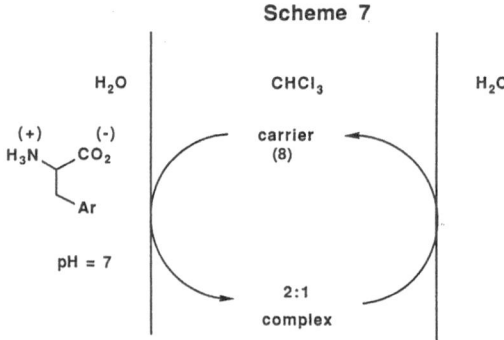

Binding of certain amino acids, as neutral (zwitterionic) substrates within the cleft of the acridine substrates has also been observed [14]. For such complexes, an additional element of recognition exists: stacking interactions between the acridine surface and the aromatic side chains of tryptophan, phenylalanine and tyrosine are detected by NMR methods. This specific contact – which is also seen with other β-phenethylamines [15] – results in selective transport of the β-aryl amino acids across liquid membranes with these carriers as 2 : 1 complexes (Scheme 7).

Scheme 7

H_2O	$CHCl_3$	H_2O

(+)   (-)
H_3N CO_2

Ar

carrier
(8)

pH = 7

2:1
complex

Other neutral heterocyclics can be bound within these molecular receptors provided that appropriate matching of functionality is arranged. For example, the acridine **8** can form cooperative hydrogen bonds with cyclic diamides such as primidone **14**, whereas its diamide derivative **15** provides an ideal microenvironment for diketopiperazines [16] (Scheme 8). A variety of heterocycles of differing size, shape and functionality were used to delineate the 'promiscuity' of the synthetic molecular clefts.

Scheme 8

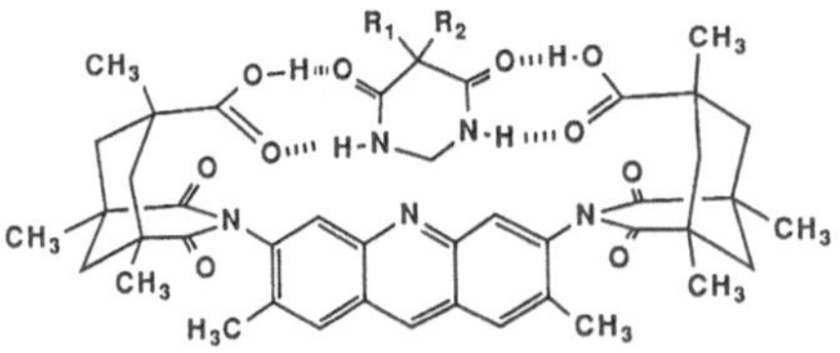

14

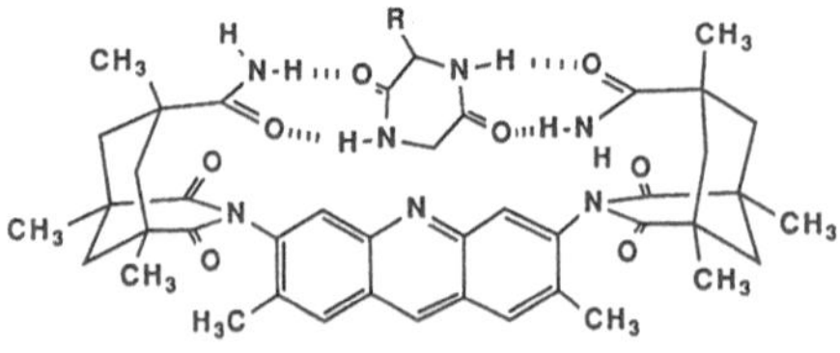

15

4. Nucleic Acids

Another recent target for practitioners of molecular recognition involves nucleic acid components. For this, the Kemp triacid is used to prepare simple imides having aromatic surfaces near and parallel to the plane of the imide function. This arrangement of functionality presents a surface complementary to adenine derivatives; base-pairing and aryl stacking forces converge from perpendicular directions to provide a complementary surface for adenine [17]. Watson-Crick (16), Hoogsteen (17), and even bifurcated hydrogen bonding (18) is detected by NMR methods [8]. Simultaneous binding in both case-pairing senses can be achieved by the bisimide (19), a molecular chelate for adenine derivatives (Scheme 9).

Scheme 9

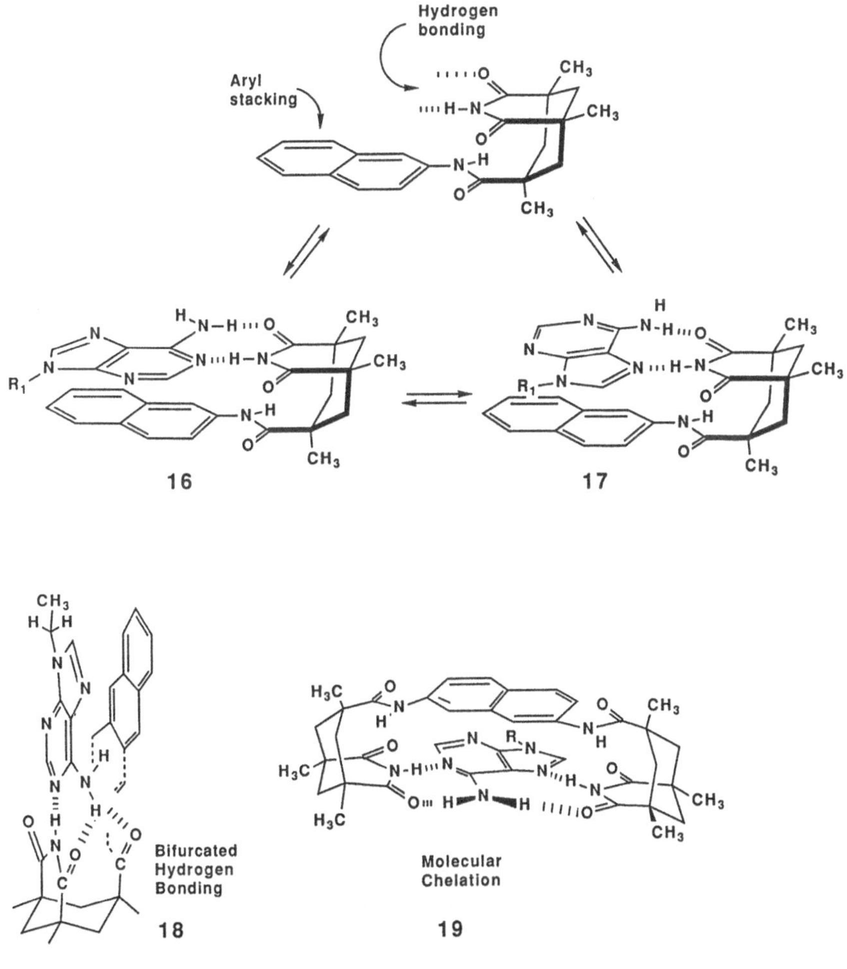

16

17

18

Bifurcated Hydrogen Bonding

19

Molecular Chelation

5. Catalytic Applications

A unique advantage of the new structures relevant to their significance as enzyme models is the possibility of exploring stereoelectronic effects at carboxyl oxygen. Unlike the situation with acyl carbon, reaction trajectories and lone pair orientation at carboxyl oxygen have been difficult to assess. Gandour [19] has pointed out that in most enzymes where the carboxylate appears at the active site, the more basic *syn* lone pair is involved in catalysis. (Scheme 10).

Scheme 10

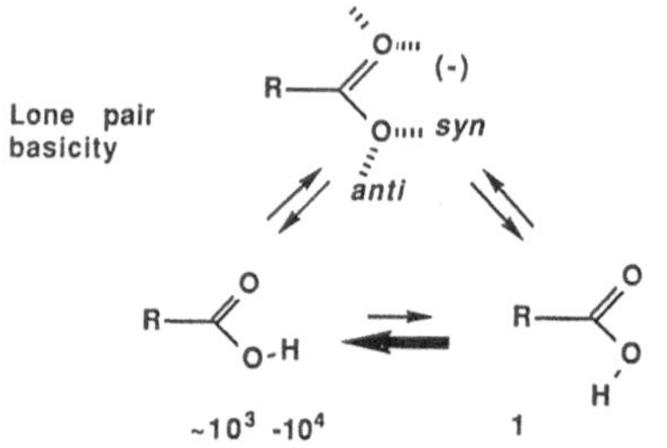

In previous models for lysozyme or the serine proteases this orientation effect has been neglected. The Loudon [20] structure **20** involves the less basic *anti* lone pair for stabilizing the developing carbonium ion, while the Fife [21] case **21** involves both the less basic lone pair and the *anti* form of the carboxylic acid. The Bruice [22] system **22** is typical of serine protease models. These may be contrasted with the convergent arrangement of the diacids in lysozyme **23**, or the involvement of the Asp *syn* lone pair in the serine proteases **24**.

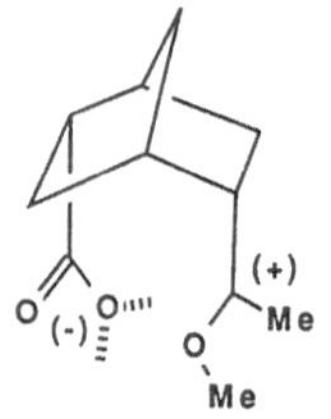

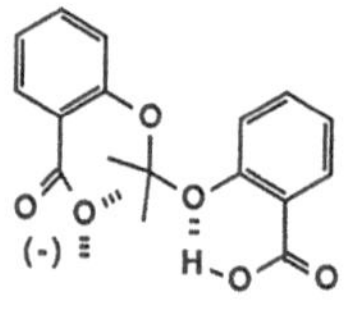

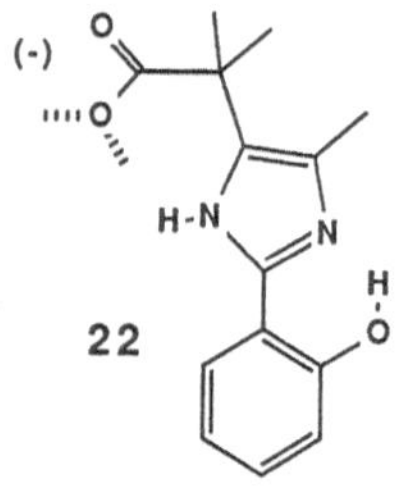

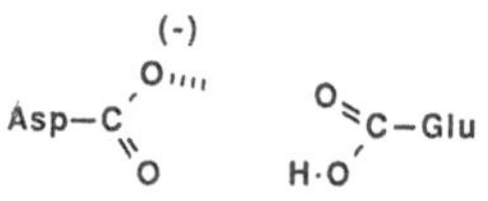

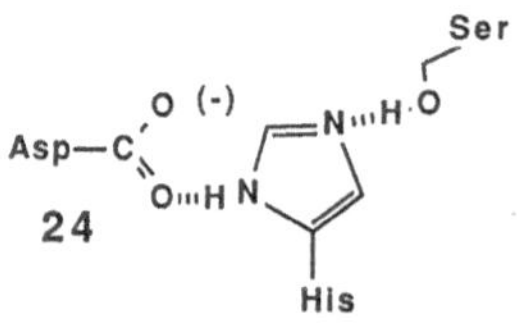

The molecular clefts are able to orient carboxyl groups in a specified direction and we have observed unusual reactivity with hemiacetals that fit within its confines [23]. The convergent diacid **8** causes the rapid dissociation of the dimer of glycolaldehyde (Scheme 11). Initial complexation followed by catalysis of hemiacetal cleavage is the most probable scenario for this reaction structures **25** and **26** offer two possibilities. Pyridone or simple carboxylic acids, much admired for their ability to catalyze glucose mutarotation [24], are ineffective in this dissociation reaction.

Scheme 11

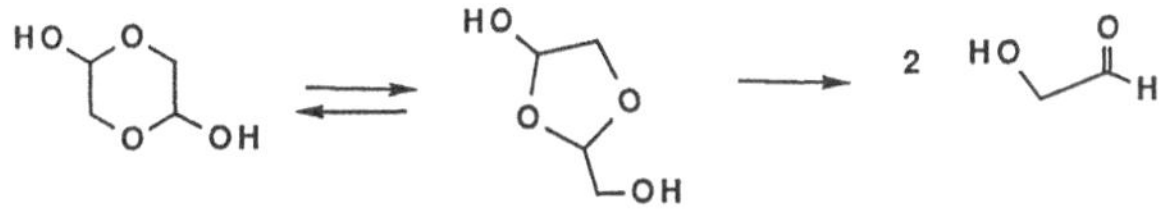

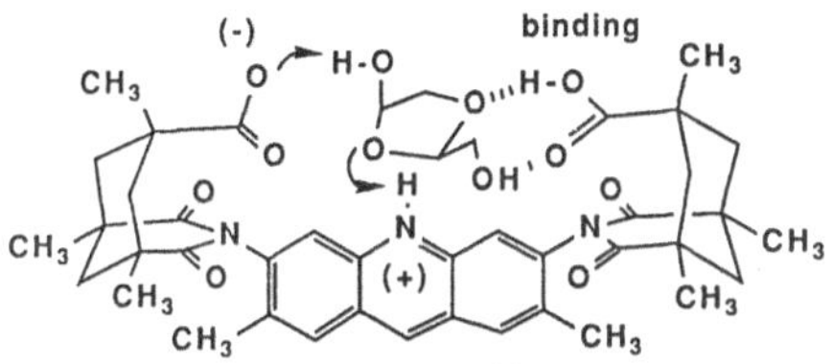

general acid-base catalysis

25

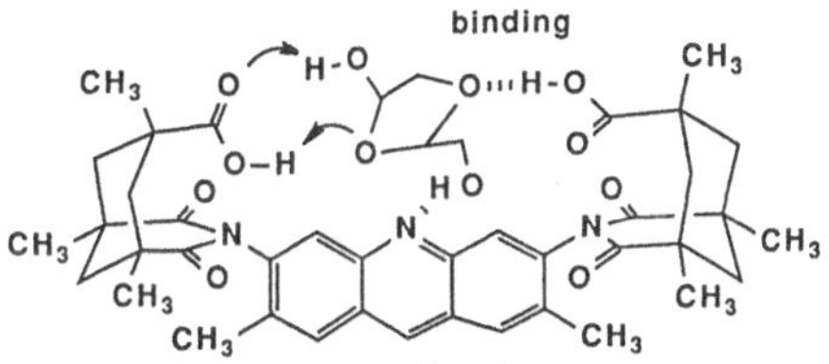

conventional catalysis

26

Another promising system involves the concerted acid/base catalysis of enolization of ketones. Stereoelectronic considerations [25] indicate that an optimal catalyst for this reaction requires that acid and base components converge from perpendicular directions on the ketone (Scheme 12). The structure **27** (a glycine derivative of Kemp's triacid) exhibits considerable activity in the enolization of phenylacetone. It should be possible to engineer additional points of contact between ketone (substrate) and diacid (catalyst) to enhance the enolization process, and we are working toward this goal.

Scheme 12

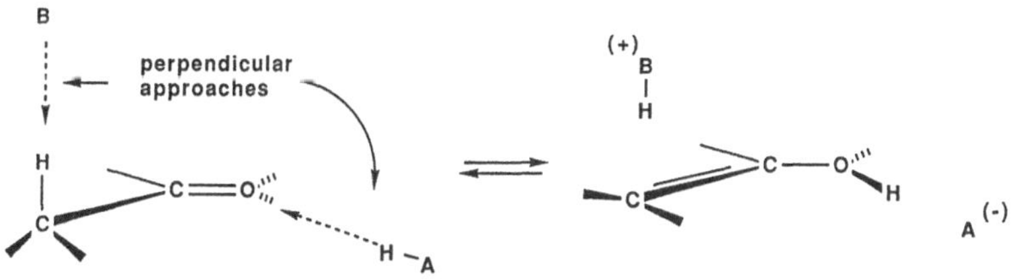

27

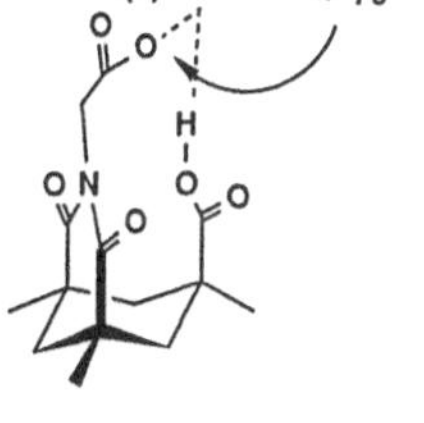

It is possible to 'line' the sides of the cleft with a number of functional groups. For example, derivatization of the acridine diacid **8** with histidino [26] gives **28** (Scheme 13). This structure represents the first model system in which the appropriate lone pair of the carboxylate is directed toward an imidazolium ion. We have also prepared the corresponding amide **29**; its reactivity will be compared to that of **28** in the same way as the mutant chymotrypsin was recently compared with the wild-type enzyme [27].

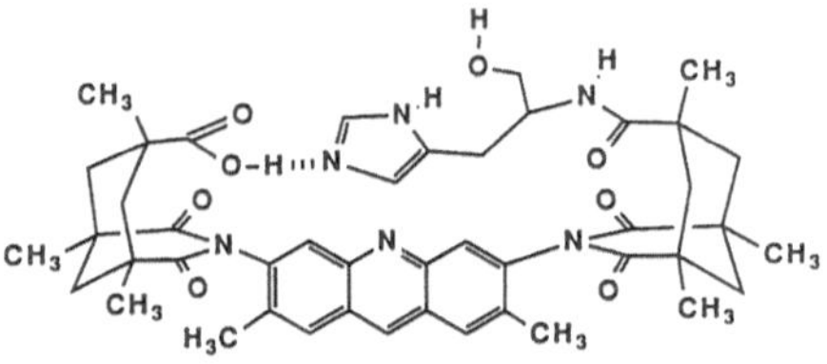

28

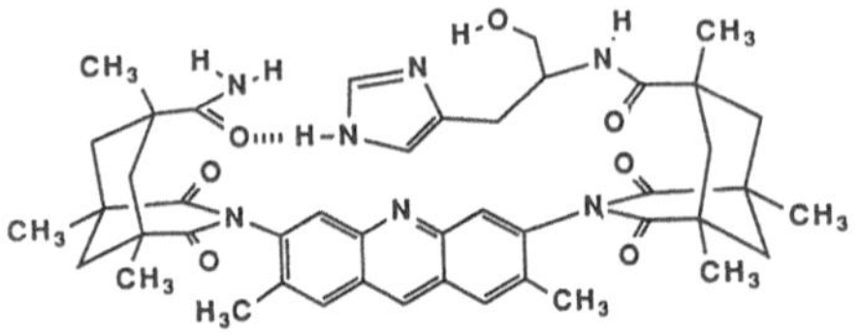

29

6. Future Directions

Perhaps the greatest advantage of the new systems is a consequence of their rigidity. This provides an opportunity to place both electrophilic and nucleophilic centers on a single substrate. This notion was first suggested by Swain [28] as ideal for concerted catalysis (Scheme 14), but flexible structures are unable to prevent intramolecular reactions of the catalytic acid-base pairs. We are pursuing these notions with suitable spacers and functional groups and will report on developments in due course.

Scheme 14

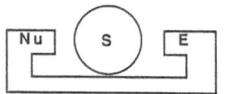

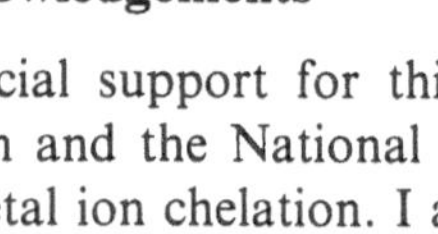

Substrate between nucleophilic
and electrophilic centers

Acknowledgements

Financial support for this research was provided by the National Institutes of Health and the National Science Foundation. Year Laboratories supported work on metal ion chelation. I am most grateful for the outstanding contributions of my co-workers, whose names may be found in the literature citations.

References

1. J. Rebek, Jr., T. Costello, L. Marshall, R. Wattley, R. C. Gadwood and K. Onan: *J. Am. Chem. Soc.* **107**, 7481 (1985).
2. (a) J.-M. Lehn: *Science* **227**, 846 (1985); (b) D. J. Cram: *Ibid.* **219**, 1177 (1983); (c) J.-M. Lehn and C. Sirlin: *J. Chem. Soc., Chem. Commun.* 949 (1978).

3. D. J. Cram, P.Y.-S. Lam and S. P. Ho: *J. Am. Chem. Soc.* **108**, 839 (1986).
4. G. Trainor and R. Breslow: *ibid.* **103**, 154 (1981); V. T. D'Souza and M. L. Bender: *Acc. Chem. Res.* **20**, 146 (1987).
5. H. Stetter and E.-E. Roos: *Chem. Ber.* **88**, 1390, 1985; K. Odashima, A. Itai, Y. Iitaka, and K. Koga: *J. Am. Chem. Soc.* **102**, 2504 (1980); S. P. Miller, and H. W. Whitlock, Jr.: *ibid.* **106**, 1492 (1984); J. Winkler, E. Coutouli-Argyropopoulou, R. Leppkes and R. Breslow: *ibid.* **105**, 7198 (1983), F. Diederich and D. Griebel: *ibid.* **106**, 8037 (1984); C. D. Gutsche: *Acc. Chem. Res.* **16**, 161 (1983).
6. See for example, R. Breslow, P. Bouy and C. L. Hersh: *J. Am. Chem. Soc.* **102**, 2115 (1980) and references therein.
7. C. S. Wilcox, L. M. Greer and V. Lynch: *J. Am. Chem. Soc.* **109**, 1865 (1987); T. R. Kelly and M. P. Maguire: *ibid.* **109**, 6549 (1987).
8. J. Rebek, Jr.: *Science* **235**, 1478 (1987); J. Rebek, Jr., L. Marshall, R. Wolak, K. Parris, M. Killoran, B. Askew, D. Nemeth and N. Islam: *J. Am. Chem. Soc.* **107**, 7476 (1985).
9. D. S. Kemp and K. S. Petrakis: *J. Org. Chem.* **46**, 5140 (1981).
10. J. Rebek, Jr., B. Askew, M. Killoran, D. Nemeth and F.-T. Lin: *J. Am. Chem. Soc.* **109**, 2426 (1987).
11. J. S. Lindsey, I. C. Schreiman, H. C. Hsu, P. C. Kearney and A. M. Marguerattaz: *J. Org. Chem.* **52**, 827 (1987).
12. J. Rebek, Jr. and D. Nemeth: *J. Am. Chem. Soc.* **108**, 5637 (1986).
13. J. Rebek, Jr., D. Nemeth, P. Ballester and F.-T. Lin: *ibid.* **109**, 3474 (1987).
14. J. Rebek, Jr., B. Askew, D. Nemeth and K. Parris: *ibid.* **109**, 2432 (1987).
15. J. Rebek, Jr., B. Askew, P. Ballester and A. Costero: *ibid.* **109**, 6866 (1987).
16. J. Rebek, Jr., B. Askew, N. Islan, M. Killoran, D. Nemeth and R. Wolak: *ibid.* **107**, 6736 (1985).
17. J. Rebek, Jr., B. Askew, P. Ballester, C. Buhr, S. Jones, D. Nemeth and K. Williams: *ibid.* **109**, 5033 (1987). For another system using macrocyclic structures see A. D. Hamilton, and D. Van Engen: *ibid.* **109**, 5035 (1987).
18. J. Rebek, Jr., B. Askew, P. Ballester, C. Buhr, A. Costero, S. Jones and K. Williams: *ibid.* **109**, 6866 (1987).
19. R. Gandour: *Bioorg. Chem.* **10**, 169 (1981).
20. D. E. Ryono and G. M. Loudon: *J. Am. Chem. Soc.* **98**, 1889 (1976); G. M. Luodon and D. E. Ryono: *ibid.* **19**, 1900 (1976).
21. T. H. Fife and T. J. Przystas: *J. Am. Chem. Soc.* **102**, 292 (1980).
22. G. A. Rogers and T. C. Bruice: *ibid.* **96**, 2473 (1974).
23. J. Wolfe, D. Nemeth, A. Costero and J. Rebek: *ibid.* **110**, 983 (1988).
24. (a) C. G. Swain and J. F. Brown: *ibid.* **74**, 2534 (1952); (b) K. A. Engdahl, H. Birehed, O. Bohman, U. Obenius, and P. Ahlberg: *Chimica Scripa* **18**, 176 (1981); (c) P. R. Rony and R. O. Nef: *J. Am. Chem. Soc.* **95**, 2896 (1973) and earlier work by these authors. For other studies of the dissociation, see R. P. Bell, and J. P. H. Hirst: *J. Chem. Soc.* 1777 (1939), P. W. Wertz, J. C. Garver, and L. Anderson: *J. Am. Chem. Soc.* **103**, 3916 (1981) and references therein.
25. (a) E. J. Corey and R. A. Sneen: *J. Am. Chem. Soc.* **78**, 6279 (1956); (b) B. Capon, A. K. Siddhanta, and C. Zucco; *J. Org. Chem.* **50**, 3580 (1985); (c) B. Capon and A. K. Siddhanta; *ibid.* **49**, 255 (1984); (d) P. Deslongchamps: *Stereoelectronic Effects in Organic Chemistry* (Pergamon Press, Oxford) Chapter 2 (1983); E. S. Hand and W. P. Jencks: *J. Am. Chem. Soc.* **97**, 6221 (1975).
26. Reported at the 191st ACS meeting, New York City, 1986.
27. C. S. Craik, S. Roczniak, C. Largman and W. J. Rutter: *Science* **237**, 909 (1987).
28. C. G. Swain and J. F. Brown, Jr: *J. Am. Chem. Soc.* **74**, 2538 (1952).

Journal of Inclusion Phenomena and Molecular Recognition in Chemistry 7 (1989), 19–26.

The Preparation of Ribose-Free 'Artificial Dinucleotides': A New Approach to Molecular Recognition?

JONATHAN L. SESSLER*, DARREN MAGDA, and JEFF HUGDAHL
Department of Chemistry, University of Texas, Austin, Texas 78712, U.S.A.

(Received: 1 February 1988)

Abstract. The synthesis of the cytosine-guanine linked dimer **1**, and its symmetric guanine-guanine analogue **2** are described. Compound **1** is the first homogeneous ribose-free system to be prepared and characterized in which molecular recognition by complementary guanine-cytosine base-pairing might be possible. It is thus prototypical of a new and potentially large class of 'artificial oligonucleotides'.

Key words. Artificial dinucleotide, molecular recognition, cytosine-guanine dimer.

1. Introduction

Recently there has been tremendous interest in reproducing, in simple model compounds, the substrate specificity and catalytic efficiency of enzymatic systems and a number of complicated approaches involving crown ethers [1, 2], cryptands [3], calixarenes [4], spherands [5], clefts [6, 7], cyclophanes [8], and cyclodextrins [9, 10] are currently being pursued. Little effort, however, is being devoted to developing template-type catalysts inspired by DNA replication or ribosomal syntheses [11]. Nonetheless, complementary purine-pyrimidine base-pairing could offer an attractive new approach to molecular recognition [11–12] and might provide a unique means of controlling chemical reactivity. Our interest in this area derives from our long-standing fascination with the conceptual simplicity and chemical efficiency of DNA replication: we remain curious as to just which factors are responsible for self-replication at the molecular level. In particular, we have been intrigued by the question of whether it might be possible to construct artificial self-replicating systems from simple ribose and phosphate-free purine and pyrimidine precursors [13]. This is clearly a highly challenging goal; in this paper we describe the results of initial synthetic studies which might lead to its eventual realization.

A generalized representation of a possible template-based self-replicating system is shown in Scheme 1. Here □ and ▥ could represent complementary base-pairs and —— a rigid spacer group. From the point of strategic planning, several steps can thus be identified as crucial to the success of the overall catalytic sequence. Firstly, it is necessary to prepare covalently linked base-pairs, such as **a**, which are joined in such a way that they could serve as the initial dimeric template. These must be capable of binding by base-pairing the appropriate complementary monomeric substrates (e.g. **b** and **c**), without interference from competitive 'narcissistic' interactions involving just the original template **a**. Secondly, bond formation between the

---

* Author for correspondence.

Scheme 1

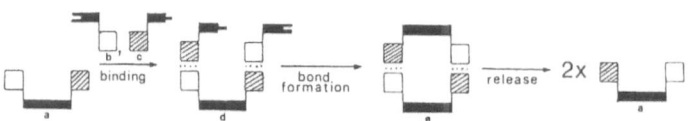

cobound substrates in **d** must be effected (to give **e**). For optimum results, this bond formation step should involve a reaction which is intrinsically exothermic but which would be slow in the absence of the neighboring group effect imposed by the template. Finally, once bond formation is complete, the newly formed dimer must be released from the original template (to give two equivalents of **a**). Clearly, if a *catalytic* self-replicating system is to be constructed, product release (from **e**) must occur at a rate such that a new generation of monomeric substrates may bind to the templates and thus enter into the cycle. For noncatalytic systems, these stringent kinetic and/or thermodynamic restrictions would be relaxed. In this case, changes in solvent or other conditions could be used to 'pry' the newly formed dimer from the original template.

The postulated self-replicating sequence outlined above is obviously predicated on a number of assumptions. The first, and most fundamental, of these is that base-pairing interactions may be used as controlling elements in simple biomimetic systems. Clearly, until this basic point is established, it makes little sense to discuss subsequent issues such as whether the thermodynamics and kinetics of bond formation and substrate release can be controlled as desired. Unfortunately, this first basic premise has yet to be established: although several generalized biomimetic hydrogen bonding model systems have been reported recently [12, 14, 15], only one involved possible base-pairing interactions [12]. In fact, at present, we are not aware of any covalently linked purine and pyrimidine het-erodimers which might be capable of carrying out selective Watson-Crick base-pairing (for various nonbinding dimeric systems see refs. 16–19). Indeed, it is not clear at present that such systems can be made and, if prepared, whether they will be capable of effecting molecular recognition by base-pairing. As our immediate objective, we therefore sought to prepare simple ribose and phosphate-free 'artificial dinucleotides' in which complementary guanine-cytosine base-pairing might be possible. In this paper we report our initial results in this area; we describe the synthesis of the cytosine-guanine dimer **1** and its symmetric guanine-guanine analogue **2**.

1

2

2. Design Considerations

In approaching the problem of developing simple acyclic 'artificial dinucleotides' we were confronted with the choice of which particular base-pair to work with: adenine/thymine or guanine/cytosine. We selected the latter pair, reasoning that the three hydrogen bonds involved in their pairing would lead to stronger interactions in an artificially constructed binding site (Figure 1). The equilibrium constant for guanine/cytosine base-pairing is $3.7\,M^{-1}$ at $32°$ in DMSO [20]. Under these conditions hydrogen bonding interactions are considered to be dominant. (This should be contrasted to the situation in water where stacking predominates and base-pairing interactions are weak [21].) Since nucleoside bases (especially guanine) are notoriously insoluble in most organic solvents, we could anticipate DMSO to be the solvent of choice of initial enzyme models.

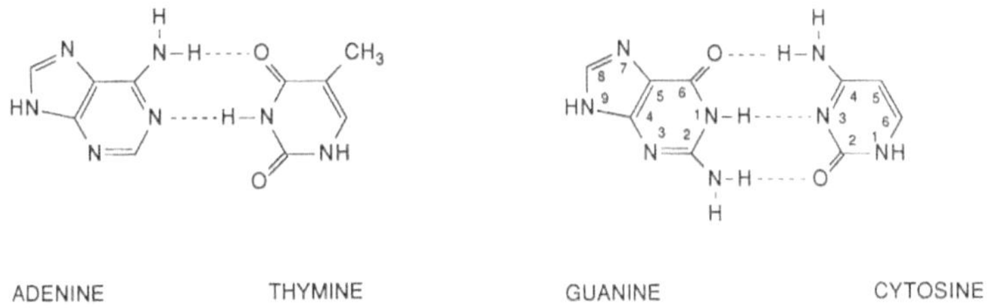

Fig. 1. Watson-Crick base-pairing interactions for common nucleic acid heterocycles.

In a linked dimeric molecule where multiple base-pairs are free to form, the strength of intermolecular interactions should be larger than in a simple cytosine/guanine pair. This assumption is valid provided that *intramolecular* base-pairing does not occur. The system must therefore be designed in such a way as to preclude these interactions. An examination of CPK models suggests that steric constraints would prevent undesirable internal hydrogen bonding in covalently linked guanine-cytosine dimers, such as **1**, in which the two bases are spanned by a five atom tether. Of course, in order to retain the ability to carry out *intermolecular* base-pairing, both the tether itself and its site of attachment to the bases must be chosen so that Watson-Crick hydrogen bonding can still occur. The obvious points of attachment are thus N^1 for cytosine and either N^9, C^8, or N^7 for guanine. Although N^9 substitution is found in DNA, for a variety of synthetic considerations, we chose to prepare N^7-substituted guanine-containing systems. CPK models suggest that no difficulties should be engendered by this choice: N^7-substituted guanines should be able to recognize N^1-substituted acyclic cytosines as complementary substrates.

3. Synthesis

In order to facilitate entry into a wide variety of artificial oligonucleotide systems, electrophilic synthons **3** and **4** were developed. Both compounds have their exo-cyclic amino group protected as the benzamide. This not only protects this functionality, but also improves the solubility of the heterocycles in organic solvents. Moreover, because both incorporate a reactive electrophilic functionality, we felt that they could easily be elaborated to give five atom bridging systems by reaction with a variety of monatomic nucleophiles (e.g. sulfide anion, amines, etc.).

The 2-hydroxyethyl substituted forms of both bases (**5** and **6**) were readily obtained following literature procedures [18, 22]. The synthesis of 7-(2-hydroxy-ethyl)guanine **5** called for the alkylation of guanosine (**7**) with ethylene oxide. We found this reaction especially attractive owing to the absence of the regioisomeric by-products such as those that result from the direct alkylation of guanine [23].

We found that the amino-protected guanine **8** could be synthesized by two methods (Scheme 2) [24]. One involved the selective deprotection of the dibenzoyl derivative **9** [25], obtained from **5** by treatment with benzoic anhydride in pyridine. Alternatively, N^2-benzoylguanosine (**10**) could be alkylated with ethylene oxide and subsequently depurinated by heating in water. The latter procedure was found to be more convenient.

Scheme 2

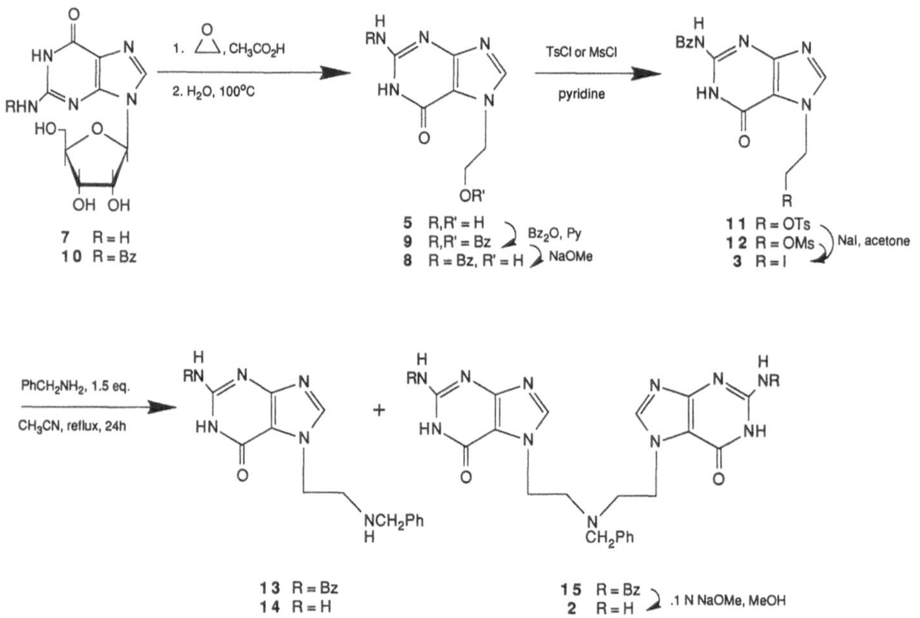

Once the protected species **8** was in hand, the next step involved activating the hydroxyl substituent. Our first attempt resulted in its conversion to the tosylate **11** by reaction with *p*-toluenesulfonyl chloride in pyridine. Compound **11** proved to be inadequately reactive for our purposes. The methanesulfonyl derivative **12** was therefore prepared from **8** using similar conditions. The mesylate was found to be more reactive than the tosylate, giving the desired reaction with benzylamine in acetonitrile, although somewhat slowly. Compounds **11** and **12** were therefore converted to the iodide analogue **3** using sodium iodide in refluxing acetone. Reaction of **3** with benzylamine produced three compounds **13–15**; the relative yields depended on the stoichiometry of the reaction. If a large excess of benzyl-amine is employed, both the secondary amine **13** and its deprotected analogue **14**

were formed. The latter presumably resulted from the nucleophilic attack by benzylamine at the benzamide carbonyl. Interestingly, the monomeric precursor is much less susceptible to debenzoylation under these conditions. When 1.5 eq. of benzylamine was used, the protected secondary amine **13** was obtained in ca. 50% yield as the major product. The protected dimer **15** was also isolated in 10% yield as a minor byproduct. The dimer was deprotected by treating with sodium methoxide/methanol under reflux, to produce **2** in nearly quantitative yield.

The cytosine analogues were produced in a manner similar to the guanine derivatives, with a few minor differences (Scheme 3). For instance, 1-(2-hydrox-yethyl)cytosine **6** was produced by reaction of cytosine (**16**) with ethylene carbonate [22]. Also, the selective deprotection of dibenzoylated 1-(2-hydroxyethyl)cytosine **17** failed under a variety of conditions, contrary to our expectations [26]. Fortunately, the desired monoprotected compound **18** was easily separated from **17** after the reaction of benzoic anhydride and **6**. Finally, reaction of **18** with *p*-toluenesulfonyl chloride/pyridine at room temperature led to the chloride **19**, presumably via displacement by chloride of the intermediate tosylate. This suggests that the cytosine electrophiles are in general more reactive than the guanine analogues. Nevertheless, the chloride was converted to the iodide **4** after initial attempts at substitution failed. (It is conceivable that **4** could be reacted with benzylamine to produce a protected dimer **20**, which could be further elaborated to afford **21**, a *bis*-cytosine dimer analogous to **2**; this chemistry is being explored at the moment.)

Scheme 3

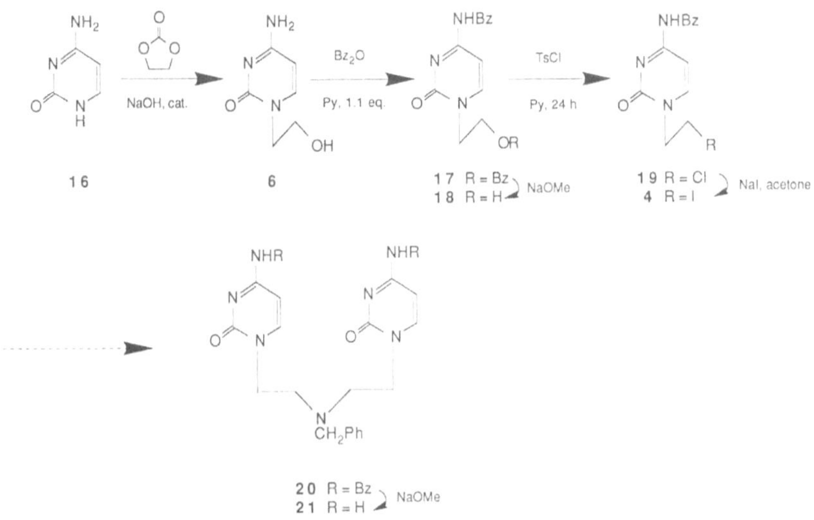

Compound **4** was found to react readily with the secondary amine **13** in refluxing acetonitrile to produce the fully protected, 'mixed dimer' **22** (Scheme 4). The monobenzoyl species **23** could be generated by deprotection of the cytosine moiety with sodium methoxide/methanol at room temperature, while one hour at reflux was required to achieve the fully deprotected 'dinucleotide' **1**.

Scheme 4

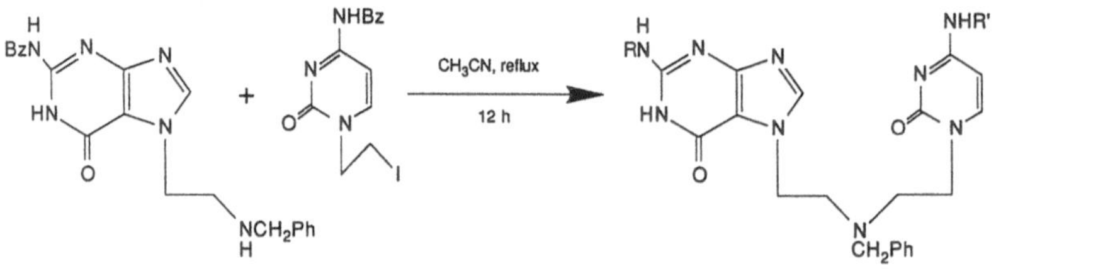

CH₃CN, reflux
12 h
NCH₂Ph
H
NHBz
NHR'
CH₂Ph
0.1 M NaOMe
MeOH, reflux, 1.5 h

13 4

22 R, R' = Bz
23 R = Bz, R' = H
1 R, R' = H

The ^{1}H NMR spectra of dimers **1** and **2** are clearly different (c.f. Table I). Not only are the expected changes due to differences in chemical structure observed, but also slight perturbations exist for the acidic protons of guanine. (The cytosine amino protons resonate in the same region of the spectrum as the phenyl protons (at ca. δ 7.00), and hence are not well resolved.) For instance, as compared to the homodimer **2**, the lactam and amino protons of **1** are shifted downfield by 36 and 18 Hz, respectively. Shifts of similar magnitude were reported for the purine N^1-H and C^2-NH_2 protons in a dimeric Watson-Crick type complex formed from cytidine and guanosine in DMSO [20]. This suggests that the heterodimer **1** may be undergoing intermolecular hydrogen bonding to form, by biomimetic molecular recognition, a supramolecular complex (Figure 2). We are currently exploring this intriguing possibility.

Table I. Selected ^{1}H NMR data for 'artificial dinucleotides' and synthetic precursors[a]

Compound	Purines				Pyrimidines		
	C8-H	C2-NH_2	C2-NHBz	N1-H	C5-H	C6-H	C4-NHBz
18					7.27	8.02	11.13
19					7.32	8.18	11.23
4					8.15	8.89	12.38
8	8.09		12.33	11.85			
11	8.14		12.21	11.98			
12	8.24		12.41	11.89			
3	8.23		12.40	11.88			
13	8.14		3.38	3.38			
14	7.89	6.13		b			
15	7.96		12.22	11.81			
2	7.73	6.05		10.62			
22	8.11		12.10	11.18	7.22	7.53	12.10
23	8.03		12.05	b	5.62	7.28	
1	7.78	6.10		10.72	5.57	7.20	

[a] DMSO-d_6; 360 MHz

[b] Peaks too broad to be assigned chemical shift values.

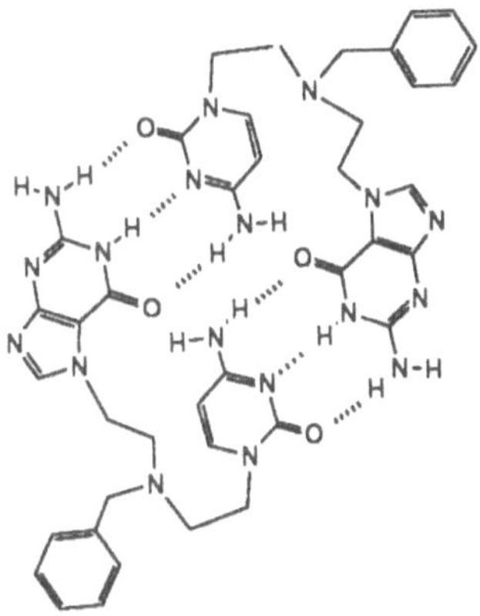

Fig. 2. Schematic representation of possible supramolecular complex which might be formed from the 'artificial dinucleotide' 1.

4. Conclusions

We have described here the synthesis of the first rationally designed 'artificial dinucleotide', 1, in which self-recognition by Watson-Crick hydrogen bonding might be possible. The synthetic strategy employed involved the preparation of several activated 'tailed' guanine and cytosine precursors and their subsequent reaction with benzylamine. It appears likely that this procedure will prove general and that it will be possible to couple these reactive synthons with a variety of other nucleophiles. In this way it should prove possible to prepare a variety of novel structures including other dimeric systems and unprecedented higher 'artificial oligonucleotides'. Studies with these should allow us to probe further important factors, such as hydrogen bonding and stacking, which influence base-pairing in both simple systems and naturally occurring oligonucleotides.

Acknowledgements

This work was supported by the National Science Foundation (PYI Award 1986), The Camille and Henry Dreyfus Foundation (New Faculty Award 1984), and the Procter and Gamble Co.

References and Notes

1. R. M. Izatt, J. S. Bradshaw, S. A. Nielsen, J. D. Lamb, J. J. Christensen, and D. Sen: *Chem. Rev.* **85**, 271 (1985).

2. F. Vögtle and H. Sieger: *Complexation of Uncharged Molecules and Anions by Crown-Type Host Molecules* (Host Guest Complex Chemistry-Macrocycles, Eds. F. Vögtle, E. Weber), pp. 319–373. Springer Verlag (1985).

3. (a) P. G. Potvin and J. M. Lehn: *Design of Cation and Anion Receptors, Catalysts and Carriers* (Progress in Macrocyclic Chemistry, v. 3, Eds. R. M. Izatt, J. J. Christensen), pp. 167–239. John Wiley and Sons (1987); (b) J. M. Lehn: *Science* (*Washington D.C.*) **227**, 849 (1985).

4. (a) C. D. Gutsche: *Calixarenes and the Art of Molecular Basket Making* (Progress in Macrocyclic Chemistry, v. 3, Eds. R. M. Izatt, J. J. Christensen), pp. 93–165. John Wiley and Sons (1987); (b) C. D. Gutsche: *Top. Curr. Chem.* **123**, 1 (1984).

5. (a) D. J. Cram: *CHEMTECH* **17**, 120 (1987); (b) D. J. Cram: *Acc. Chem. Res.* **98**, 1041 (1984).
6. (a) J. Rebek, Jr.: *Science (Washington D.C.)* **235**, 1478 (1987); (b) J. Rebek, Jr.: *Acc. Chem. Res.* **98**, 258 (1986).
7. C. S. Wilcox, L. M. Greer, and V. Lynch: *J. Am. Chem. Soc.* **109**, 1865 (1987).
8. Y. Murakami: *Top. Curr. Chem.* **115**, 107 (1983).
9. I. Tabushi: *Pure Appl. Chem.* **58**, 1529 (1986).
10. R. Breslow: *Acc. Chem Res.* **13**, 170 (1980).
11. Catalytic systems based on polymerized purines and pyrimidines have been reported: (a) T. Inoue and L. E. Orgel: *Science (Washington D.C.)* **219**, 859 (1983); R. Naylor and P. T. Gilham: *Biochemistry* **5**, 2722 (1966).
12. M. S. Kim and G. Gokel: *J. Chem. Soc., Chem. Commun.*, 1686 (1987).
13. Self-replicating systems based on hexanucleotides have, however, been reported: G. v. Kiedrowski: *Angew. Chem., Int. Ed. Engl.* **25**, 932 (1986).
14. (a) J. Rebek, Jr. and D. Nemeth: *J. Am. Chem. Soc.* **108**, 5637 (1986); (b) J. Rebek, Jr., B. Askew, P. Ballester, C. Buhr, S. Jones, D. Nemeth, and K. Williams: *J. Am. Chem. Soc.* **109**, 5033 (1987); (c) J. Rebek, Jr., B. Askew, P. Ballester, C. Buhr, A. Costero, S. Jones, D. Nemeth, and K. Williams: *J. Am. Chem. Soc.* **109**, 6866 (1987).
15. A. D. Hamilton and D. Van Engen: *J. Am. Chem. Soc.* **109**, 5035 (1987).
16. A. F. Maggio, M. Lucas, J. L. Barascut, A. Pompon, and J. L. Imbach: *Nouv. J. Chim.* **10**, 643 (1986).
17. P. Brookes and P. D. Lawley: *Biochem. J.* **77**, 478 (1960).
18. P. Brookes and P. D. Lawley: *J. Chem. Soc.* **1961**, 3923 (1961).
19. C.-W. Chen and H. W. Whitlock, Jr.: *J. Am. Chem. Soc.* **100**, 4921 (1978).
20. R. A. Newmark and C. R. Canter: *J. Am. Chem. Soc.* **90**, 5010 (1968).
21. (a) M. P. Schweizer, S. I. Chan, and P. O. P. Ts'o: *J. Am. Chem. Soc.* **87**, 5241 (1965); (b) A. D. Broom, M. P. Schweizer, and P. O. P. Ts'o: *J. Am. Chem. Soc.* **89**, 3612 (1967); (c) M. P. Schweizer, A. D. Broom, P. O. P. Ts'o, and D. P. Hollis: *J. Am. Chem. Soc.* **90**, 1042 (1968); (d) S. R. Jadkunas, C. R. Cantor, and I. Tinoco Jr.: *Biochemistry* **7**, 3164 (1968).
22. N. Ueda, K. Kondo, M. Kono, K. Takemoto, and M. Imoto: *Die Makromolekulare Chemie* **120**, 13 (1968).
23. K. K. Ogilvie, S. L. Beaucage, and M. F. Gillen: *Tetrahedron Lett.* **35**, 3203 (1978).
24. All new compounds gave spectral and C. I. mass spectrometric data in accord with the assigned structures.
25. J. L. Sessler, D. J. Magda, V. Lynch, D. I. Bernstein, and G. M. Schiff: *Nucleosides and Nucleotides*, in press.
26. H. Khorana, A. F. Turner, and J. P. Vizsolyi: *J. Am. Chem. Soc.* **83**, 686 (1961).

Journal of Inclusion Phenomena and Molecular Recognition in Chemistry 7 (1989), 27–38.

Nucleotide Recognition by Macrocyclic Receptors

ANDREW D. HAMILTON,* ALEX MUEHLDORF, SUK-KYU CHANG,**
NALIN PANT, SHYAMAPROSAD GOSWAMI, and DONNA VAN ENGEN
Department of Chemistry, Princeton University, Princeton, NJ 08544, U.S.A.

(Received: 1 February 1988)

Abstract. It is shown that complementary positioning of recognition sites (particularly hydrogen bonding, stacking and hydrophobic groups) into a macrocyclic structure can lead to very strong and specific complexation of uncharged organic molecules.

Key words. Molecular recognition, nucleotide, thymine, guanine, barbiturate.

1. Introduction

The recognition and binding of nucleotide substrates by proteins is at the heart of gene expression and metabolic control. The possible future design of 'synthetic repressor' molecules that might artificially activate or control genes depends on an understanding of the key features of nucleotide recognition. The most important recognition features on each nucleotide base are the hydrogen bonding groups at its periphery (Figure 1). These form the basis of the Watson-Crick hydrogen-bonding

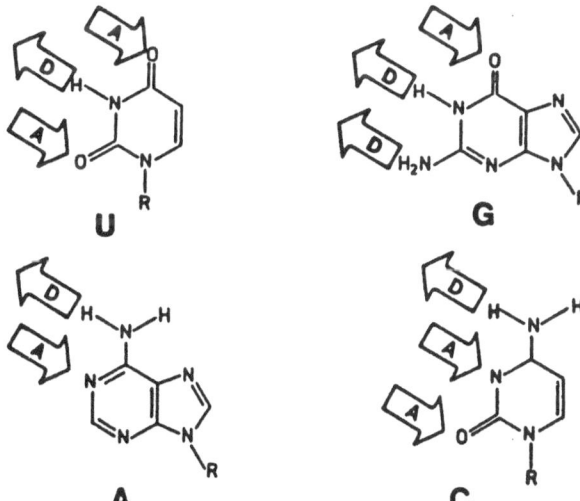

Fig. 1.

* Address correspondence to this author at Department of Chemistry, University of Pittsburgh, Pittsburgh, PA.
** On leave from Chung-Ang University, Seoul, Korea.

scheme and the double helical structure of DNA [1]. However, additional recognition can occur perpendicularly to the hydrogen bonding groups by means of a stacking interaction to the plane of the aromatic base. It is significant that several key nucleotide-binding proteins use both hydrogen bonding and π–π stacking forces to provide strong and selective complexation [1]. For example, ribonuclease T_1 binds guanine via two hydrogen bonds between the peptide backbone and the O(6) and NH(1) groups of guanine *plus* a stacking interaction (at 3.4 Å) between a tyrosine residue (Tyr-45) and the purine plane (Figure 2) [2]. The electron density map [2] further suggests that Tyr-45 swings from an 'unbound' conformation into a stacking position on guanine binding, exemplifying an induced fit recognition process [3].

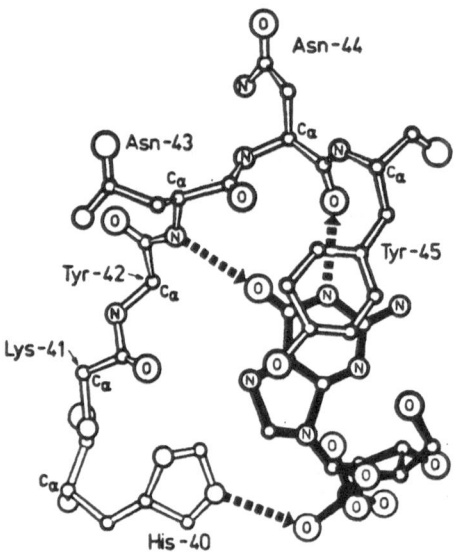

Fig. 2.   X-ray structure of ribonuclease-T_1 active site.

In designing a new class of artificial receptors for nucleotides we sought to incorporate both recognition features. Our strategy (Figure 3) was to link, within a macrocycle, a group capable of stacking with the nucleotide base to one complementary to its hydrogen bonding periphery. Eventually several of these monomeric hosts might be linked to form a receptor for specific sequences of nucleotide bases.

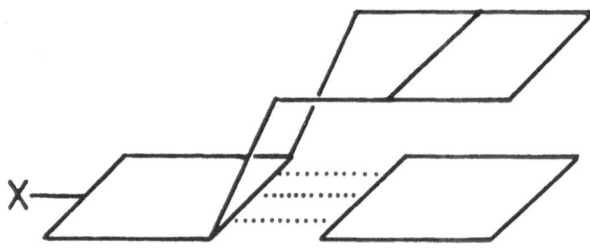

Fig. 3.

2. Thymine Recognition

The first receptor studied [4] was that for thymine and was based on the triple hydrogen-bonding complementarity between 2,6-diaminopyridines and the imide group of thymine [5] (Figure 4). The stacking unit was derived from 2,7-dihydroxy-naphthalene, a component known for its intercalating properties in several DNA-binding drugs (e.g. neocarzinostatin) [6]. The two groups were linked into a macrocycle (3) via the synthesis shown in Figure 5. Reaction of 2,7-dihydroxynaph-thalene with an ethyl bromoalkanoate (acetone, K₂CO₃) gave diester (1) which was converted into its diacid chloride 2 (a, HCl-acetone; b, (COCl)₂) and reacted, under high dilution conditions with 2,6-diaminopyridine to give 3. The yield of the final macrocyclization step varied from 20–26% according to ring size.

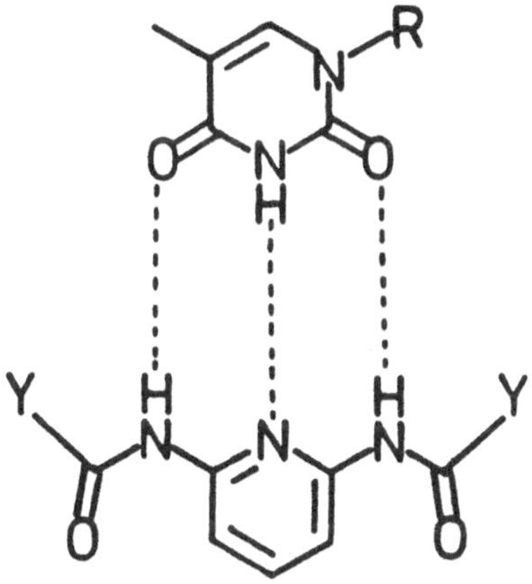

Fig. 4.

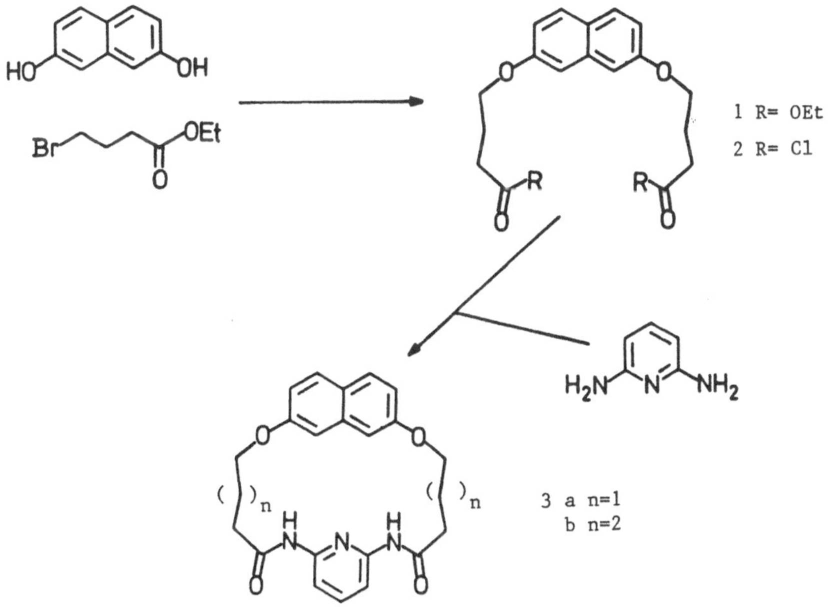

1 R= OEt

2 R= Cl

3 a n=1
  b n=2

Fig. 5.

The open structure of **3a** (as opposed to an intramolecularly stacked conformation) was confirmed by X-ray crystallography [4] which showed the naphthalene
poised at a 127.5° angle to the pyridine ring (Figures 6 and 8a). The thymine
recognition properties of **3** were studied using ^{1}H NMR and X-ray crystallography.
Treatment of a CDCl$_3$ solution of **3a** with one equivalent of *N*-butylthymine (**4**)
caused several characteristic changes in the ^{1}H NMR of both host and guest. The
NH proton resonances on both **3a** and **4** are shifted downfield by 2.25 and 2.6 ppm
respectively, confirming the formation of a triple hydrogen bond complex. In
addition, upfield shifts (0.19, 0.29 and 0.24 ppm) are seen in the ring-proton,
-methyl and -N-methylene resonances of the thymine substrate. These are consistent
with the approach of the naphthalene to the thymine plane and the influence of its
ring current on the nearby protons on the substrate (the terminal alkyl CH$_3$ on **4**
experiences no upfield shift). The X-ray structure of the complex (Figure 7)
confirms the ditopic nature of the two recognition sites. Three hydrogen bonds are
formed between the pyridine and thymine rings at distances of N—O, 2.87, 2.99 and
N—N, 3.06 Å. The naphthalene ring lies directly above the substrate at an angle of
14° and closest inter-plane contact of 3.37 Å. The close contact as well as the ring
current induced shifts in the ^{1}H NMR confirm that there is a strong correlation
between solution and solid state structures. Figures 8a and b show side views of
both free and bound forms of **3a** and clearly demonstrate a 'molecular hinge-like
motion of the macrocycle. On substrate binding, the naphthalene swings through an
arc of 34.1° to within Van der Waals distance (3.4 Å) of the thymine. This induced

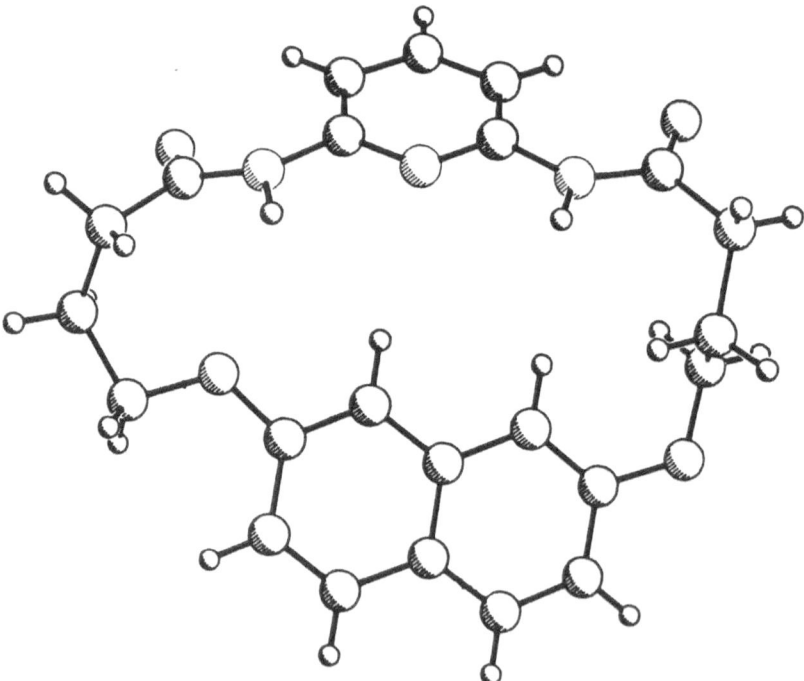

Fig. 6.

C(3)
C(2)
C(4)
O(22)
C(1)
C(5)
O(6)
C(22)
N(3)
C(6)
C(21)
N(1)
N(2)
C(7)
C(8)
C(20)
O(18)
C(25)
C(23)
O(10)
C(9)
C(19)
C(17)
C(24)
O(26)
C(11)
O(29)
N(4)
C(29)
C(26)
C(16)
C(12)
C(15)
C(14)
C(13)
N(5)
C(27)
C(31)
C(28)
C(30)
C(32)
C(33)
C(34)

Fig. 7.

a
b
4

Fig. 8.

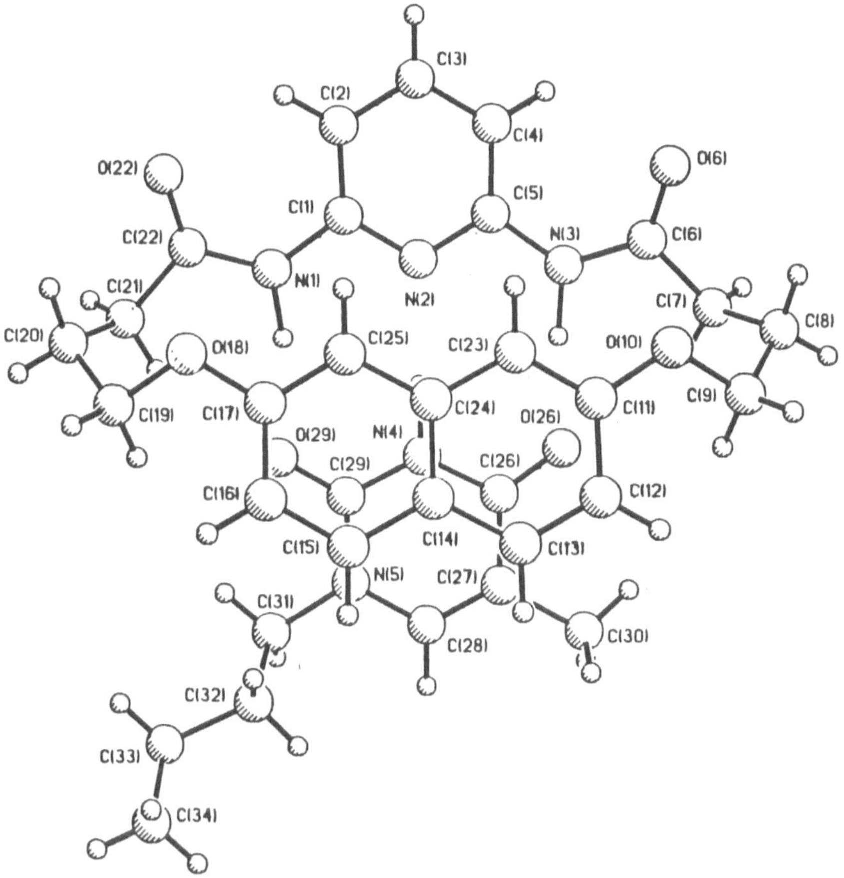

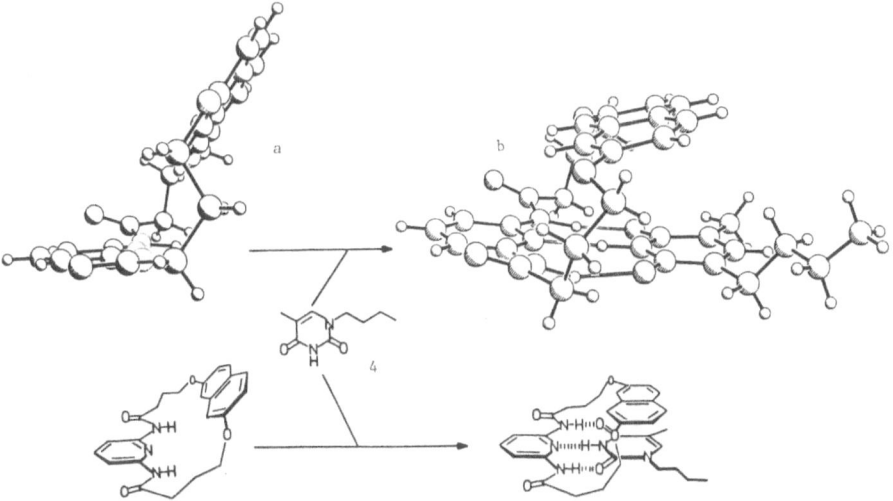

fit-like behavior mimics that of the tyrosine residue of ribonuclease T_1 which stacks with the guanine substrate.

The ditopic nature of the binding of thymine by **3a** is further reflected in its association constant. These were measured by monitoring the chemical shift of different protons in **3a** as a function of N-butylthymine concentration and then analyzing the titration data by means of a Foster-Fife analysis [7]. Both **3a** ($K_s = 290$ M^{-1}) and the larger **3b** ($K_s = 251$ M^{-1}) show an approximately threefold increase in association constant for binding to N-butylthymine compared to simple 2,6-dibutramidopyridine ($K_s = 90$ M^{-1}) which lacks the stacking component. This corresponds to a contribution of ~ 0.75 kcal/mol from the naphthalene–thymine interaction to the overall binding free energy for **3a**:**4** (~ 3.5 kcal/mol). Similar binding enhancements have been seen in related ditopic receptors for adenine [8].

3. Guanine Recognition

The hydrogen bonding component of the nucleotide receptor can be readily varied in order to modify its substrate selectivity. For example, we have recently [9] prepared a ditopic (hydrogen bonding and stacking) receptor for guanine based on the triple hydrogen bond complementarity between 7-amino-1,8-naphthyridines and the peripheral hydrogen bonding groups of guanine (Figure 9). The necessary 2-substituted-7-amino-1,8-naphthyridine was synthesized by the route outlined in Figure 10. Condensation of 2,6-diaminopyridine and ethyl acetoacetate in H_3PO_4 gave aminonaphthyridone **5** which after acetylation with acetic anhydride and chlorination with $POCl_3$ gave **6** [10]. Reaction of **6** with sodium 2-hydroxyethoxide afforded the aminoalcohol **7** which was then cyclized under high dilution conditions with diacid chloride **2** to form the macrocyclic receptor **8** in 20% yield.

Once again nucleotide base binding was conveniently followed using ^{1}H NMR. Addition of one equivalent of 2′,3′,5′-tri-O-pentanoylguanosine **9** to a CDCl$_3$ solution of **8** causes downfield shifts in the NH resonance of **8** and the NH and NH$_2$ resonances of **9** (1.36, 0.31 and 0.25 ppm, respectively), consistent with their forming a triple hydrogen bond complex (as in Figure 9). In addition, all of the naphthalene proton resonances are shifted upfield due to the close approach of the naphthalene to the bound guanine. These shifts are greater for the naphthalene-5, -6

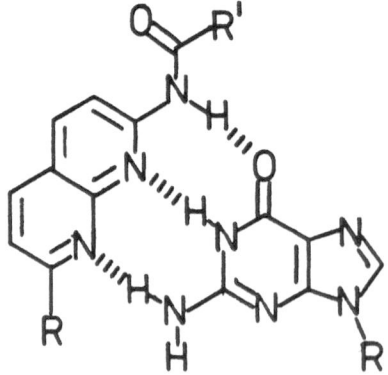

Fig. 9.   **9**: R = 2′, 3′, 5′-tri-O-pentanoylribose. **10**: R = OEt, R′ = CH$_3$.

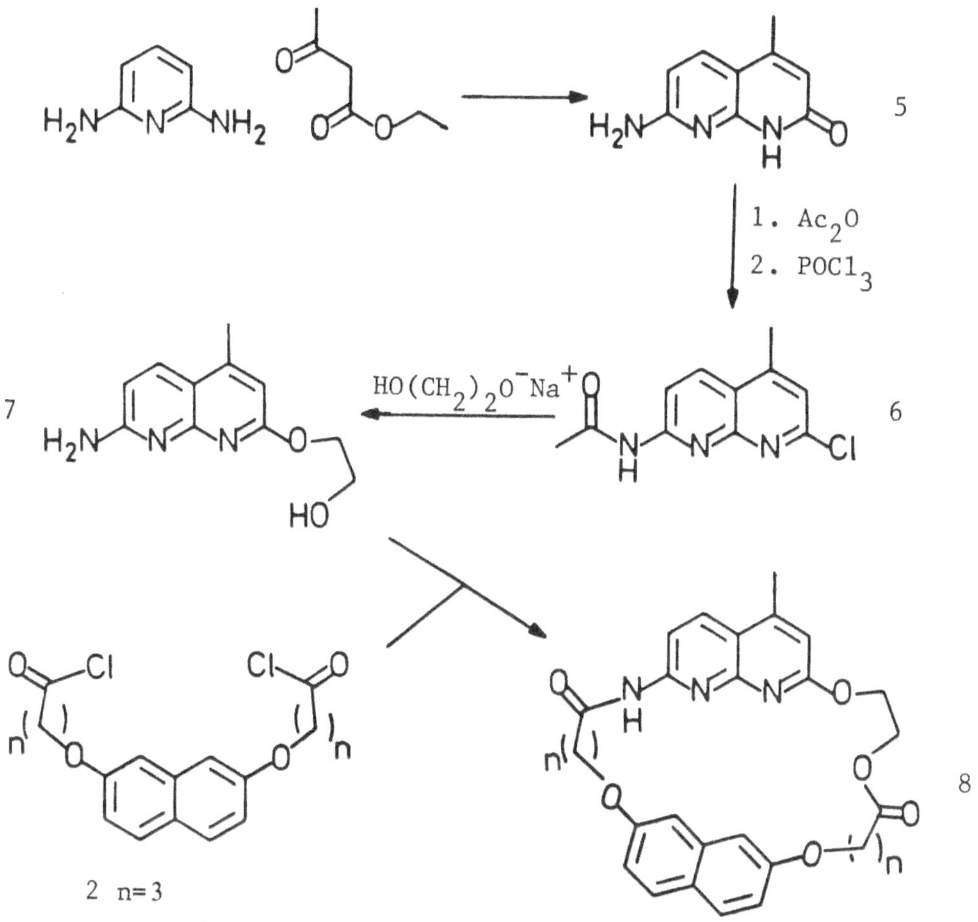

Fig. 10.

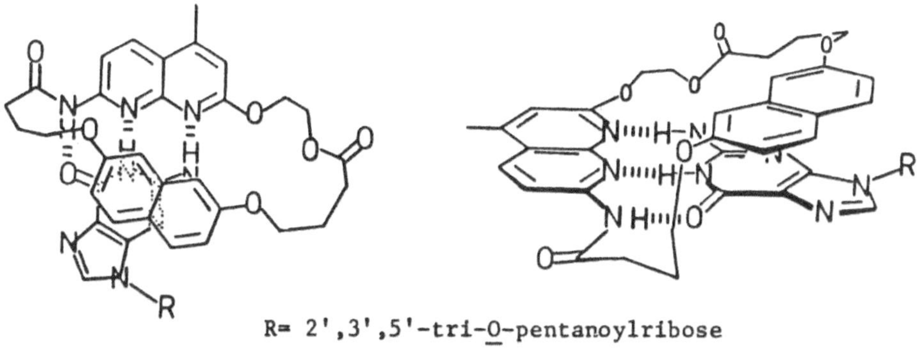

R= 2',3',5'-tri-O-pentanoylribose

Fig. 11.

and -8 protons (0.15, 0.14 and 0.18 ppm) than for those at the -4,3 and -1 positions (0.1, 0.09, and 0.11 ppm), suggesting a skewed orientation of the naphthalene relative to the guanine (Figure 11). In this structure for the complex $8:9$ (Figure 11) the -5, -6 and -8 protons on the naphthalene would lie closer to the purine and naphthyridine ring currents and would, thus, experience greater upfield shifts. As in the thymine case, the stacking interaction causes a substantial increase in the association constant between 9 and naphthyridine macrocycle 8 ($K_s = 530$ M^{-1}) compared to acyclic naphthyridine 10 ($K_s = 126$ M^{-1}). The contribution of guanine–naphthalene stacking to the overall binding energy between $8:9$ ($\sim$ 4 kcal/mol) is approximately 1 kcal mol^{-1}, in good agreement with the thymine case.

4. Barbiturate Recognition

Our success in incorporating hydrogen-bonding groups into macrocyclic receptors prompted us to extend the approach to other biologically significant substrates. This and other work [8, 11, 12] suggested that incorporation of several inwardly-facing hydrogen bonding groups into a cleft or cavity would lead to strong binding to substrates with complementary shape and hydrogen bonding characteristics. We focused our attention on the barbiturate family of drugs (Figure 12) which are widely used as sedatives [13] and as anticonvulsants [14]. Molecular modelling studies suggested that incorporation of two 2,6-diaminopyridine units into a macrocyle would allow the complexation of all six accessible hydrogen-bonding groups (four CO lone pairs and two NHs) in 5,5-disubstituted barbiturates (Figure 13). Appropriate choice of spacer Y may further allow a secondary recognition of the substituents in the 5,5-positions. For example, a diphenylmethane unit (as Y) should accommodate the 5,5-ethyl groups of barbital 12.

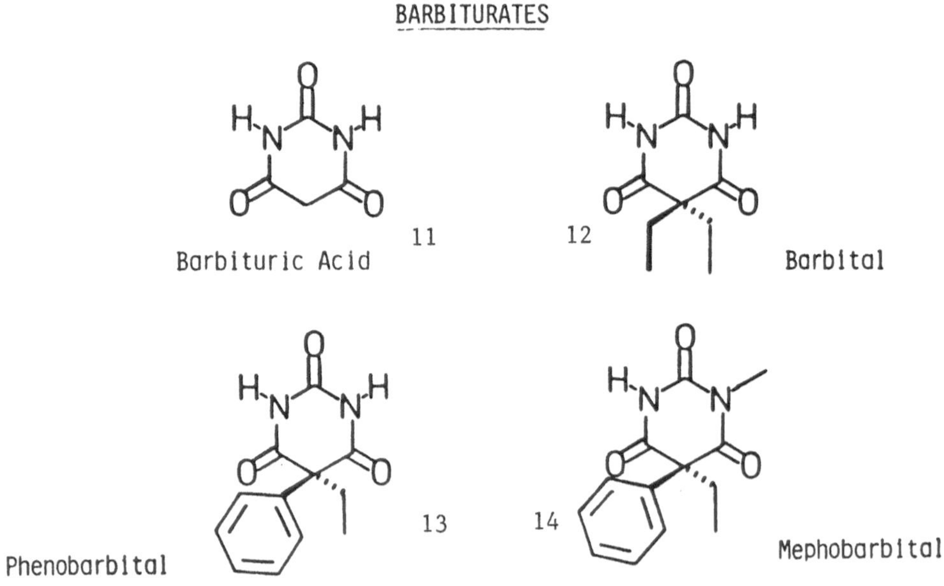

BARBITURATES

Fig. 12.

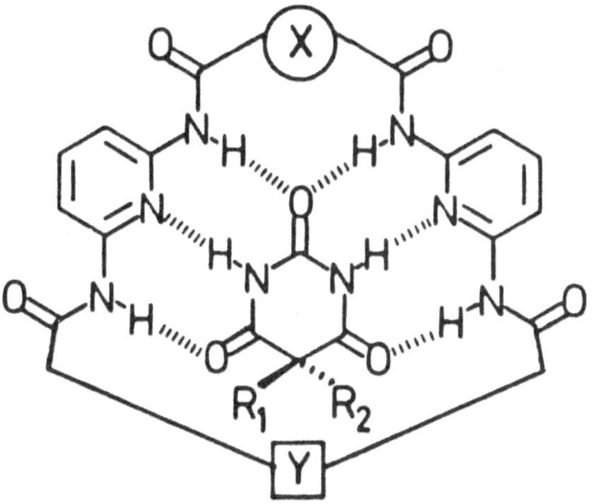

Fig. 13.

The first barbiturate receptors **18, 19, 20,** were prepared by standard high dilution methods from diamine **15** and acid chlorides **16** and **17** [15] (Figure 14). Their open conformation was confirmed by an X-ray structure of **19** which shows a tetrahydrofuran molecule occupying the central cavity (Figure 15). There is a high degree of preorganization in this structure which requires very small conformational changes to achieve the proposed hexahydrogen bonded complex. Complex formation was followed by ^{1}H NMR. Addition of one equivalent of barbital **12** to a CDCl$_3$ solution of **18** caused large downfield shifts of the amide resonances of **18** (1.65 and 1.63 ppm) and the imide resonance of **12** (4.38 ppm) indicating the formation of a

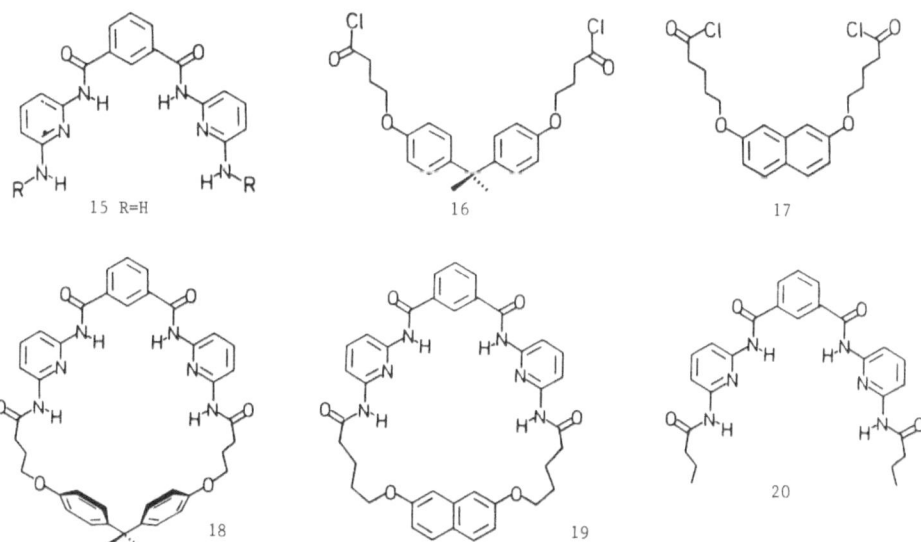

Fig. 14.

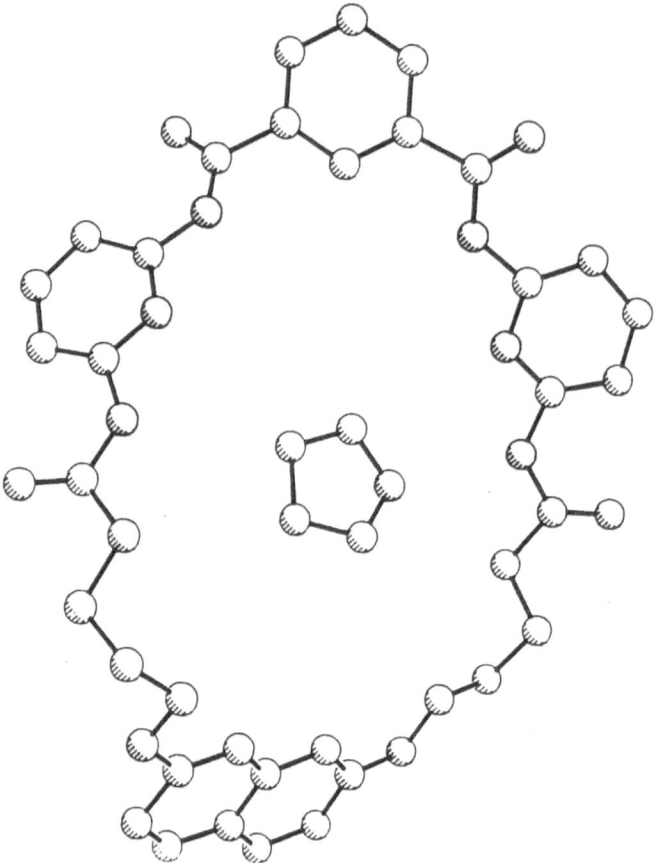

Fig. 15.

hexahydrogen bonded complex as shown in Figure 16. Also, the CH_2 and CH_3 resonances of the barbital ethyl groups were shifted upfield (0.25 and 0.23 ppm) confirming their proximity to the diphenylmethane cleft in Figure 16. A large downfield shift (0.4 ppm) is seen in the isophthaloyl-2-proton resonance and CPK models suggest that in the complex this proton lies in the deshielding region of the barbital-2-carbonyl group. In addition the isophthaloyl resonances in uncomplexed **18** are broad due to the conformational mobility of the macrocycle. In the complex (Figure 16) the motion of the isophthaloyl group is restricted and its ¹H resonances sharpen. Essentially similar NMR changes are seen with the other receptors, **19** and **20**, in their interactions with non-N-alkylated barbiturates.

Association constants were determined from ¹H NMR titration data using either Foster-Fife [7] or nonlinear least squares analysis and are collected in Table I. The strongest complex ($K_s = 1.37 \times 10^6 \, M^{-1}$) is formed between barbital **12** and diphenylmethane receptor **18**. This result is expected due to the strong complementarity in both shape and hydrogen bonding specificity that exists between **12** and **18** (Figure 16). Alkylation of one barbiturate N-atom (as in mephobarbital **14**) essentially removes three H-bonding groups from participating in complexation and

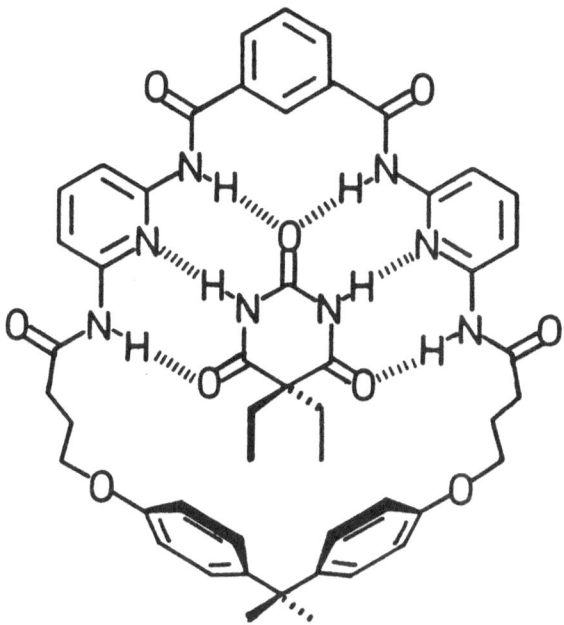

Fig. 16.

results in a 1000 fold decrease in binding to **18**. When the inwardly-pointing binding site is no longer enforced by a macrocyclic superstructure (as in acyclic **20**) association to barbital **12** diminishes by almost 100-fold. Finally, incorporation into the barbiturate-5 position of a bulky substituent that cannot fit neatly into the receptor cavity (as with phenobarbital **13** and **18**) causes a 10-fold reduction in the binding constant. A degree of secondary recognition of the 5, 5-substituents in the barbiturate can also be seen. While **18** binds barbital **12** more strongly than phenobarbital **13**, receptor **19**, which contains a flat naphthalene moiety, binds **13** nearly twice as strongly as **12**.

In summary, we have shown that complementary positioning of recognition sites (particularly hydrogen bonding, stacking and hydrophobic groups) into a macro-cyclic structure can lead to very strong and specific complexation of uncharged organic molecules.

Table I. Association constants for the receptor–barbiturate interaction.

Receptor	Barbiturate	K_s, M^{-1} (25°C, CDCl$_3$)
20	Barbital (**12**)	2.08×10^4
18	Mephobarbital (**14**)	6.80×10^2
19	Phenobarbital (**13**)	2.80×10^5
18	Phenobarbital (**13**)	1.97×10^5
19	Barbital (**12**)	1.35×10^5
18	Barbital (**12**)	1.37×10^6

Acknowledgements

We thank KOSEF, Korea for a fellowship to S.K.C. and the National Institutes of Health for financial support of this work.

References

1. W. Saenger: in *Principles of Nucleic Acid Structure*, Springer-Verlag, New York, 1984, p. 105.
2. U. Heinemann and W. Saenger: *Nature* **299**, 27 (1982).
3. A. Fersht: in *Enzyme Structure and Function*, Freeman, Reading, 1977, p. 262.
4. A. D. Hamilton and D. Van Engen: *J. Am. Chem. Soc.* 1987, **109**, 5035.
5. B. Feibush, A. Figueroa, R. Charles, K. D. Onan, P. Feibush, and B. Kargar: *J. Am. Chem. Soc.* **108**, 3310 (1986).
6. K. Edo, M. Mizugaki, Y. Koide, H. Seto, K. Furihata, N. Otake, and N. Ishida: *Tetrahedron Lett.* **26**, 331 (1985).
7. R. Foster and C. A. Fife: *Prog. Nucl. Magn. Reson. Spectrosc.* **4**, 1 (1969).
8. J. Rebek Jr., B. Askew, P. Ballester, C. Buhr, S. Jones, D. Nemeth, and K. Williams: *J. Am. Chem. Soc.* **109**, 5033 (1987).
9. A. D. Hamilton and N. Pant: submitted for publication.
10. E. V. Brown: *J. Org. Chem.* **30**, 1607 (1965).
11. J. Rebek Jr.: *Science* **235**, 1478 (1987).
12. M. Newcomb, S. S. Moore, and D. J. Cram: *J. Am. Chem. Soc.* **99**, 6405 (1977); V. M. L. J. Aarts, C. J. V. Staveren, P. D. J. Grootenhuis, J. V. Eerden, L. Kruise, S. Harkema, and D. N. Reinhoudt: *J. Am. Chem. Soc.* **108**, 5035 (1986); R. E. Sheridan and H. W. Whitlock Jr.: *J. Am. Chem. Soc.* **108**, 7120 (1986).
13. J. A. Vida: in *Burger's Medicinal Chemistry*, Wolff, M. E., Ed., Wiley-Interscience, New York, 787 (1981).
14. E. I. Isaacson and J. N. Delgado: in *Burger's Medicinal Chemistry*, Wolff, M. E. Ed., Wiley-Interscience, New York, 829 (1981).
15. S. K. Chang and A. D. Hamilton: *J. Am. Chem. Soc.*, in press.

Journal of Inclusion Phenomena and Molecular Recognition in Chemistry 7 (1989), 39–51.

Molecular Recognition in Carnitine Acyltransferases

RICHARD D. GANDOUR
Department of Chemistry, Louisiana State University, Baton Rouge, LA 70803-1804, U.S.A.

(Received: 1 February 1988)

Abstract. We are designing and synthesizing rigid guests to probe the topography of the carnitine acyltransferases, regulatory enzymes in lipid metabolism. Our designs are based on structural studies of substrates and possible molecular mechanisms of enzymatic activity. Recent X-ray, ^{1}H NMR, and force-field computational studies on carnitine and acetylcarnitine, coupled with the known stereospecificity for activity in carnitine acyltransferases, have led us to propose a molecular mechanism for acyl transfer in these enzymes. The 'folded' conformation of an acylcarnitine is most populated and should be preferred for binding to these enzymes, because, in this conformation, the acyloxy is the most sterically accessible. There are four key recognition sites on the enzymes: I, carboxylate; II, trimethylammonium; III, coenzyme A; IV, acyl. Sites, I, II and III serve as the three loci required to create a chiral environment on the enzymes for carnitine. An addition-elimination reaction involving the formation of a tetrahedral intermediate is suggested as the mechanism for O-to-S acyl transfer. This proposed tetrahedral intermediate is chiral and the enzymes should prefer the *R* configuration at this center. Based on this proposal, conformationally rigid tetrahedral-intermediate analogues have been designed, synthesized and assayed. Morpholinium and 2-hydroxymorpholinium derivatives inhibit carnitine acetyltransferase and palmitoyltransferase. Because of rigidity at their two chiral centers, these inhibitors serve as probes of molecular topography of recognition sites, I, II, and IV.

Key words. Enzyme, inhibitor, force field calculations.

1. Molecular Recognition

Molecular recognition refers to all interactions between two molecules or two remote pieces of the same molecule. A time-dependent phenomenon, it includes both stable and unstable complexes. Host–guest chemistry, which covers stable complexes from ions–ionophores, substrates–enzymes, drugs–receptors, antigens–antibodies, is a major chemical research area with substantial practical applications in catalysis, separation science, medicine, and material science. Recognition by a host requires varying degrees of flexibility. An implicit assumption is that a perfectly constructed cavity will have ultraselectivity.

1.1. FLEXIBILITY OF HOST

The structural flexibility of a crown ether and its tendency to adopt a conformation appropriate to its environment have been recognized since the first structures appeared [1]. Ligand flexibility is necessary for complexation by cryptands [2]. Complexation of ions by these flexible ligands largely depends on neutralizing the charge on the metal, which requires an appropriate number of donors in an appropriate topography. Cram and Trueblood [3] have enunciated the principle that guests organize the hosts and that preorganization of the host improves binding. Gokel [4] has proposed the idea that cavity sizes of crowns are adjustable and has investigated the dynamics of binding by retaining a flexible component in his captivating lariat ethers. We have subsequently demonstrated [5] that the guest determines the cavity size of the host, depending on the number and identity of donors in the macrocycle. The carbon framework in these macrocycles primarily maintains the connectivity relationships among the donors rather than imposing a rigid conformational or steric bias on the system.

1.2. TOPOGRAPHICAL MAPPING OF FLEXIBLE HOSTS WITH RIGID GUESTS

When the guests are rigid ions, then the host simply must accommodate. In applying Cram and Trueblood's hypothesis [3] to the design of guests, we conclude that rigid guests can probe the structure of unknown hosts, an idea that has its origins in medicinal chemistry [6]. We are using this concept of a rigid guest's organizing a flexible host (induced-fit model [7]) to design enzyme inhibitors in order to map the topographies of catalytic centers in enzymes. Mapping topographies of the active sites of enzymes is a formidable task, which we are undertaking by preparing conformationally rigid analogs of reaction intermediates proposed for carnitine-acyltransferase catalyzed reactions. These analogs have groups anchored to a rigid molecular framework in a well-defined stereochemistry. Their inhibitory ability depends on their complementing the topography of the catalytic center.

2. Carnitine

Carnitine, the biological carrier-molecule of fatty acids destined for transport into and oxidation by mitochondria [8], is required for efficient metabolism of long-chain fatty acids [9–11]. Carnitine transports fatty acyl groups across mitochondrial membranes after accepting the acyl group from an acyl CoA in a reaction, Eq. 1, catalyzed by carnitine palmitoyltransferase (CPT).

$$\tag{1}$$

After being transported across the inner mitochondrial membrane, acyl carnitine donates the acyl group to an endogenous CoA molecule in the mitochondrial matrix. After fatty-acid oxidation, the acetyl CoA transfers the acetyl group to carnitine in a reaction catalyzed by carnitine acetyltransferase (CAT). Acetylcarnitine is transported out of the mitochondrial matrix and donates the acetyl group to exogenous CoA.

3. Reaction-Intermediate Analogs

Wolfenden [12] has pioneered the development of transition-state analog inhibitors of enzymes. The idea is that enzymes bind transition structures or reaction intermediates more tightly than reactants or products (Figure 1). Molecules that resemble the structures of transition states or reaction intermediates but are unreactive will bind strongly to the enzyme. Our goal is to design a conformationally rigid, reaction-intermediate analog inhibitor in order to map the topographies of the catalytic centers of CAT and CPT.

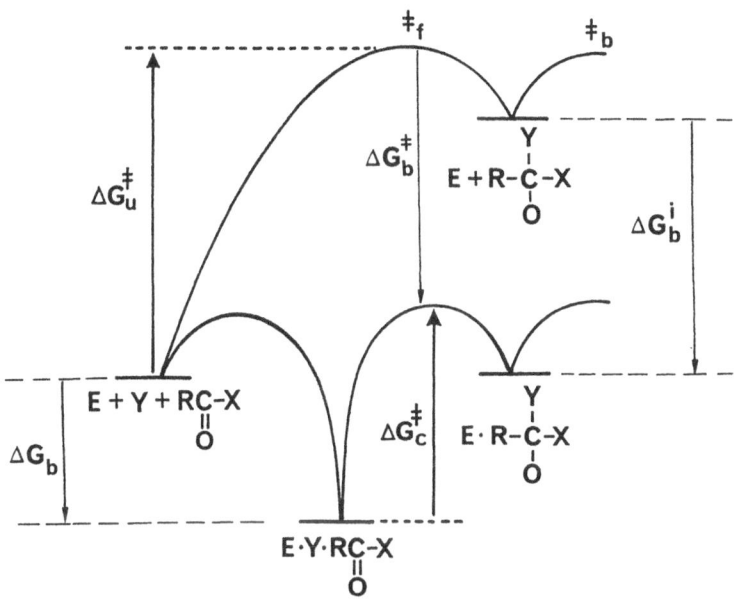

Fig. 1. Free-energy diagram illustrating the energetic advantage of binding transition structures and reaction intermediates. For a complete discussion of the ideas from which this diagram is derived, see the excellent review by Schowen [13].

We have proposed [14] a mechanism (Eq. 2) for acetyl transfer in CAT involving a tetrahedral intermediate, which contains both carnitine and coenzyme A. We assume that a similar mechanism operates in CPT.

(2)

The catalytic centers of both CPT and CAT contain a carnitine recognition site and a CoA recognition site, as well as an alkyl recognition site that is juxtaposed to the other two sites (Figure 2). The carnitine recognition site has two recognition points, one for carboxylate, the other for trimethylammonium. Together with the CoA site, these two recognition points create the stereoselectivity observed for the reaction. The alkyl group is probably not detached from its recognition site during the transfer, hence creating an additional chirality for the active site (i.e., the tetrahedral intermediate). This chirality is recognized during the transfer and must be mimicked by a reaction-intermediate analog.

To design a rigid analog, we must know which of the nine possible conformations of carnitine is (are) preferred on the surface of the enzyme. This question is still unanswered but we have addressed the question of the preferred conformation of carnitine and acetylcarnitine in other states of matter. We have used single crystal X-ray for the solid state, NMR for the solution state, and *ab-initio*-enhanced MM2 for the computational state.

4. Structural Studies

4.1. SINGLE CRYSTAL X-RAY ANALYSIS

We have determined [14] the crystal structures of the zwitterions of carnitine and acetylcarnitine, which are similar to those of hydrochloride salts. Because the zwitterion is the physiologically active form, we need to know if there are any

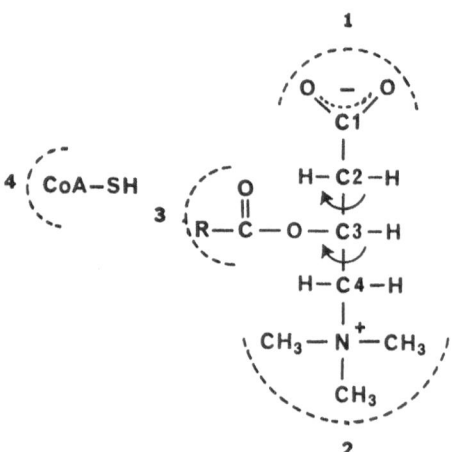

Fig. 2. Coenzyme A and acylcarnitine illustrating the possible recognition sites. The carnitine site has two recognition points: (1) carboxylate and (2) trimethylammonium. The acyl recognition site (3) is juxtaposed to the carnitine and coenzyme A site (4). Carnitine has flexibility about the C2—C3 and C3—C4 bonds. The preferred conformations about these bonds must be known in order to design rigid inhibitors.

changes in conformation arising from ionization of the carboxyl. In the solid state, the conformations do not change on ionization. Carnitine has a different conformation to acetylcarnitine. The conformation about C3—C4 is similar in both but the conformation about C2—C3 changes from *anti*(*a*) to *g*⁻. Murray, Reed, and Roche [15] have labeled the conformation of carnitine as 'extended' and acetylcarnitine as 'folded' (Figure 3).

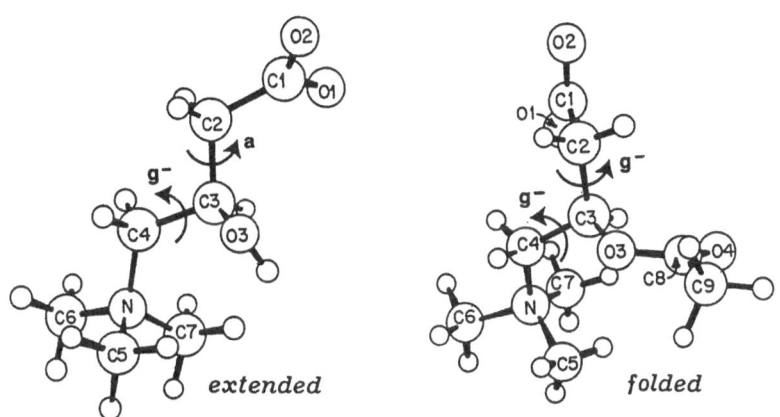

Fig. 3. Crystal structures of the zwitterions of carnitine (left) and acetylcarnitine (right) from ref. [14]. Carnitine is in the 'extended' conformation (*a*, *g*⁻) and acetylcarnitine is in the 'folded' conformation (*g*⁻, *g*⁻).

4.2. COMPUTATIONAL STUDIES

We have computed the relative energies of the conformations of carnitine and acetylcarnitine [16]. Atomic point charges from a single-point *ab initio* (3–21G basis set) calculations of the zwitterions using the crystal-structure geometries form the electrostatic force field. The total charges on the three polar fragments of carnitine and acetylcarnitine are shown below.

The MM2 results show that electrostatic energy accounts for enhancement of the proportion of folded conformer. Attraction between carboxylato and quaternary ammonium groups increases for both carnitine and acetylcarnitine. The greater enhancement for acetylcarnitine than carnitine results from relief of both electrostatic and steric repulsion between carboxylato and acetoxyl groups.

Because we wanted to determine the relative energetics of conformations in the physiological state, we explored the effect of dielectric. As we are dealing with charged structures the effects are quite dramatic. The dielectric has only small effects on populations down to a value of about $\varepsilon = 40$, below which electrostatics primarily determine the energy of the zwitterions.

Figure 4 shows computational results for a dielectric of water ($\varepsilon = 80$). As in the X-ray study, 'folded' was favored for acetylcarnitine and 'extended' for carnitine. Previous semi-empirical studies [15] on carnitine's conformation suggested a strong electrostatic attraction between the carboxylate and trimethylammonium. The difference between those calculations and ours is that the semi-empirical methods calculate the energy of an isolated molecule and thus resemble the gas phase. Because the dielectric constant is part of the force field in our MM2 calculations, we can simulate the solution state.

4.3. ^{1}H NMR

Our computational studies were done in concert with an ^{1}H NMR study of the conformations in deuterium oxide. The problem of how to determine the conformation of flexible molecules in solution has challenged researchers for decades and will continue to do so. The problem is even more difficult for flexible molecules that are either quite polar or charged, especially if these molecules interact with the solvent. The techniques of NMR spectroscopy and computational chemistry,

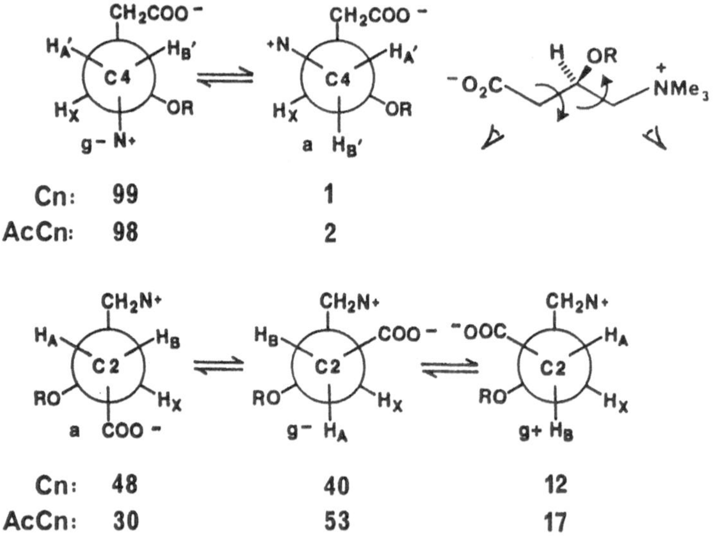

Fig. 4. Proportions of conformations of carnitine (Cn) and acetylcarnitine (AcCn) determined by MM2 computation from ref. [16]. Top line of Newman projections have C4 in front and C3 at the back. g^- and a refer to the torsion angle between C4—N$^+$ and C3—OR. Bottom line of Newman projections have C2 in front and C3 at the back. a, g^-, and g^+ refer to the torsion angle between C2—COO$^-$ and C3—CH$_2$N$^+$.

especially when used in tandem have resulted in considerable progress in conformational analysis.

My student, Dr. William J. Colucci, with the help of Dr. Steven Jungk, has
developed an equation (Eq. 3) for determining conformations about C—C bonds.
His equation [17] like those of Pachler [18] and of Altona [19] is a modified
Karplus expression [20]. Rather than using electronegativities to account for
substituent effects, Colucci's equation employs empirically derived substituent constants. The ΔS_i term is a group substituent effect. The equation is a classical
free-energy relationship such as the Taft equation.

$$^3J(\text{HCCH}) = A + B \cos \theta + C \cos 2\theta + \sum_{i=1}^{4} \Delta S_i \cos \theta \cos \phi \, \text{HX}_i \qquad (3)$$

The question is: how do we get the values for ΔS_i? We have measured or taken
from the literature coupling constants for monosubstituted ethanes. Because all
conformations are populated due to free rotation, an average coupling constant is
measured. Integrating the equation gives the simple result that the average coupling constant for a monosubstituted ethane is simply the average coupling constant for ethane, A, minus 0.25 times ΔS_i. A, B, and C were obtained from
calculations on ethane [21].

We measured the average coupling constants for the appropriate ethyl compounds at the same pH as we measured the spectra for carnitine and acetylcarnitine [16]. This illustrates the simplicity of Colucci's equation because the values
of ΔS_i are determined in the medium of choice and thus any solvent effects on
the substituent effects are accounted for. The results of the ^{1}H NMR study are
shown in Figure 5. 'Folded' is favored for acetylcarnitine and 'extended' for
carnitine.

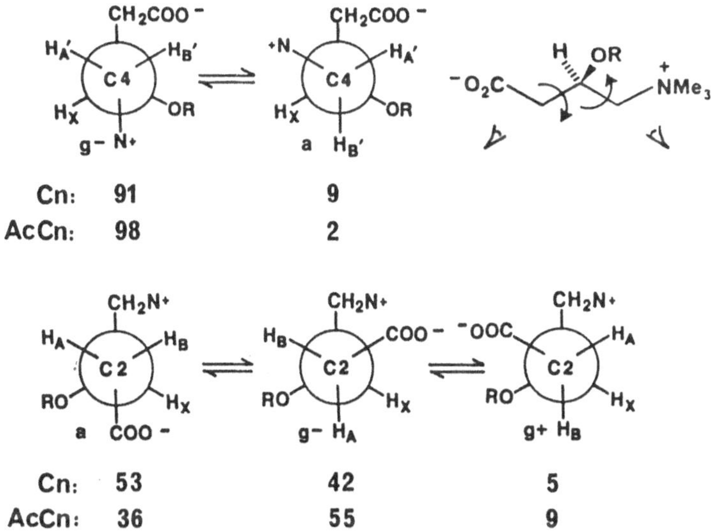

Fig. 5. Proportions of conformations of carnitine and acetylcarnitine determined in solution by
^{1}H NMR, from ref. [16]. Notation is the same as in Figure 4.

In summary, our conformational analysis of carnitine and acetylcarnitine shows that carnitine prefers 'extended' or a^-g^- and acetylcarnitine 'folded' or g^-g^-. The important points for inhibitor design are that the C3—C4 bond has a strong preference for g^-, and the C2—C3 bond is equally populated in either the a or g^- conformation. We can lock the C3—C4 bond in this conformation by formation of a ring and not lose much in recognition. We are less certain as to whether or not to lock the conformation of the C2—C3 bond.

5. Inhibitor Design

As shown in Eq. 2, the proposed mechanism involves a direct transfer between carnitine and Coenzyme A. The reaction is specific and for the R-enantiomer of carnitine and we have proposed [14] a two-point recognition by the enzyme because the location of the cofactor gives the third point needed for chiral recognition.

Given the need for carboxylate recognition and assuming that the conformation of the bound molecule is 'folded', we can imagine how this mechanism might occur on the enzyme, Figure 6. This leads to two further topographical considerations, the locations of the Coenzyme A site and the acyl site. The carboxylate must be turned away from the oxygen on C3 to allow room for S to attack. This model further suggests a chirality for the tetrahedral intermediate; i.e., the attack must occur away from the trimethylammonium group on the Re face of the carbonyl. An additional benefit is the electrostatic catalysis [22] that results from having the developing negative charge on the carbonyl oxygen in close proximity to the trimethylammonium group.

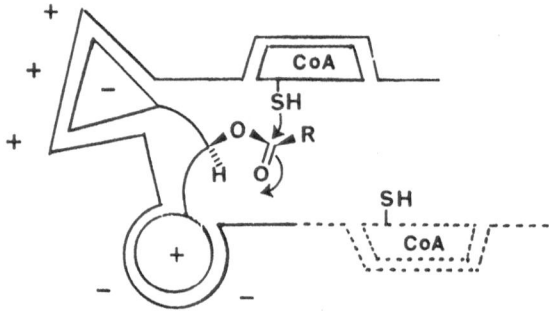

Fig. 6: Mechanism for Coenzyme A attack on acylcarnitine. The carboxylate (triangle with negative charge) is folded back to allow the thiol to approach the carbonyl. We propose the CoA attacks from this side rather than the opposite (shown as dotted lines) because of steric and electrostatic effects (see text).

We reason that the electrostatic interaction could be replaced by a covalent bond and that the S could be replaced by an H or OH, see below.

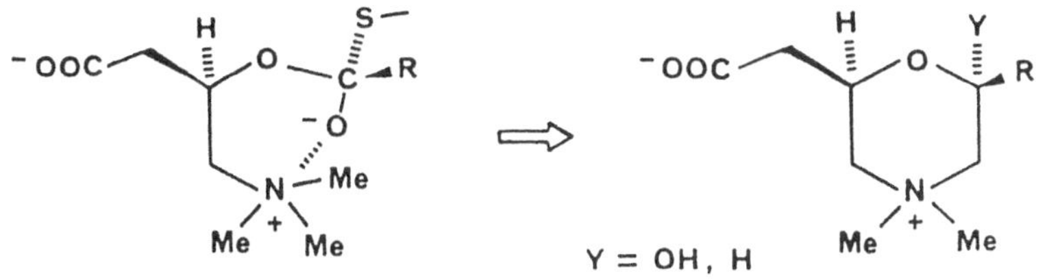

Knowing that the conformation about C3—C4 in carnitine is predominantly g^-, we felt that locking this conformation in a six-membered ring should not detract from binding. Conformational considerations in six-membered rings led us to anticipate that the carboxymethyl and the alkyl chain attached to the anomeric carbon should be *cis*-disequatorial. We approached the syntheses of morpholinium inhibitors from two directions.

6. Morpholiniums

One approach is the reaction of dimethylaminoalcohols with an ester of 4-bromo-2-butenoate, Eq. 4. Cyclization to the morpholinium ring occurs in a separate step and is stereoselective.

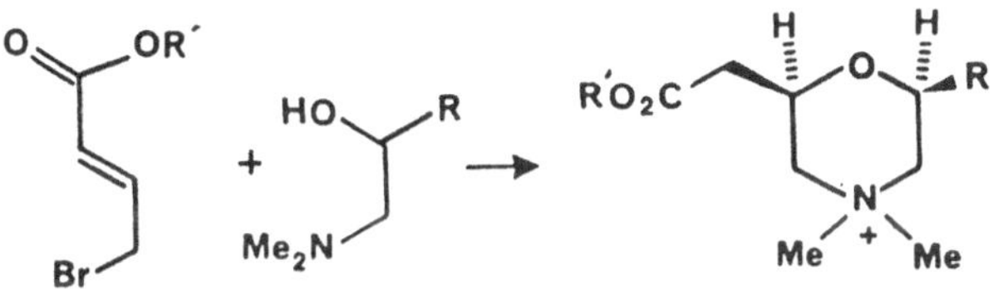

$$(4)$$

These compounds are conformationally rigid analogs of the tetrahedral intermediate proposed for acetyl transfer in CAT. CAT binds both enantiomers of carnitine equally well, but only the (R)-enantiomer is active. These analogs must have the same relative configuration as (R)-carnitine, if CAT stereoselectively binds them as it binds the tetrahedral intermediate. To verify that pigeon breast CAT recognizes only one configuration of our inhibitors, we have devised a unique approach [23], that utilizes only racemic compounds and a prochiral molecule with an achirotopal plane.

We made (*meso*)-2,6-bis(carboxymethyl)-4,4-dimethylmorpholinium, **1**, in two steps from condensation of sodium (R)-norcarnitine [24] and methyl-(E)-4-bromo-2-butenoate. Only one pair of enantiomers, presumably the *cis*-diastereomer, of **2** was present by ^{1}H NMR, indicating that the ring formed stereoselectively. Hydrolysis of **2** yielded the anticipated diequatorial *meso*-diacid, whose structure was verified by single crystal X-ray analysis. The solid-state structure of **1**, in fact, displays *meso*-symmetry, with a crystallographic mirror plane containing the O and N atoms of the ring. Hence, we call it a 'Siameso' inhibitor because of its morphological similarity to Siamese twins. Compounds **3** and **4** were prepared by similar reactions of the appropriate dimethylamino alcohol and the bromoalkenoate.

We have measured the K_is of **1–4** with pigeon breast CAT (Figure 7). Of this series **1** binds most strongly, with a K_i half that of the racemic compounds, **3** and **4**. This is because every molecule has one side with the correct configuration of (*R*)-carnitine. This two-fold improvement in binding for **1** suggests that CAT is selectively binding one configuration of these inhibitors. Compound **2** does not bind well because of the increased size of the ester or the polarity change from acid to ester.

The key features of these inhibitors are their rigidity and their similarity to the tetrahedral intermediate. Rigidity in the inhibitor reduces binding because only the enzyme can adjust, but rigidity is essential for identifying the topographical arrangement of recognition points on the enzyme, as well as the conformation of the substrate fragment of the tetrahedral intermediate. For example, the N—C—C—O torsion angle in the inhibitors is locked in the g^- conformation, which is predicted for carnitine bound to CAT [16].

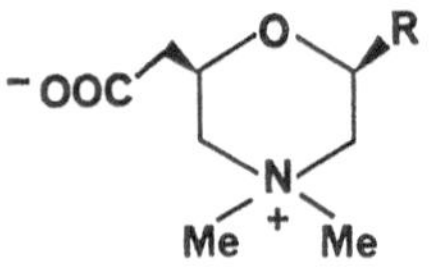

	R=	Ki (μM)
1	CH$_2$COOH	530
2	CH$_2$COOMe	8600
3	CH$_3$	1080
4	H	1000

Fig. 7. Inhibition constants for morpholinium derivatives.

7. 2-Hydroxymorpholiniums

2-Hydroxymorpholinium derivatives designed as mimics of choline have shown biological activity [25]. They are prepared by the reaction of 2-(dimethylamino)-ethanol, and a halomethyl ketone. In extending this approach to carnitine, large quantities of norcarnitine are required. We have developed a large-scale procedure for demethylating carnitine in high yield [24]. Sodium norcarnitine and the corresponding bromomethyl ketone produce the 2-hydroxymorpholinium analogs, **5**, Eq. 5.

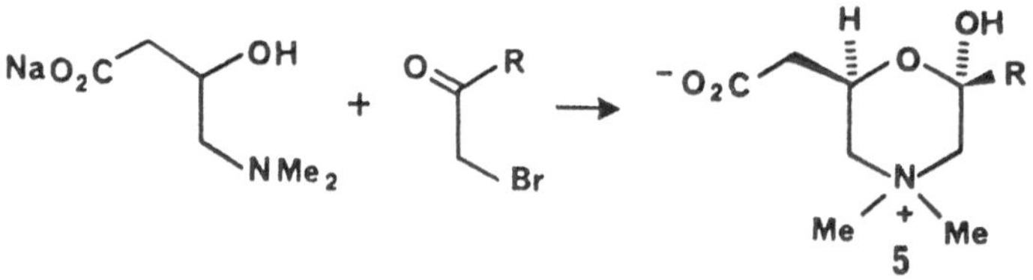

$$ (5) $$

We have prepared the carnitine analogs (hemiacylcarnitiniums) **5**, with R=CH$_3$(HAC) and (CH$_2$)$_{14}$CH$_3$(HPC) from the sodium norcarnitine and the corresponding bromomethyl ketone. The ring closure is highly favored because of the strong electron withdrawing effect of the quaternary ammonium ion. We have only seen one isomer formed and rationalize the preference for an axial OH as arising from the anomeric effect, the gauche effect, and steric effects.

The hemiketal carbon of **5** and the hemiorthothioester carbon of the proposed tetrahedral intermediate are chiral. CAT and CPT may prefer one configuration at this center, just as they prefer one configuration of the chiral center of acetylcarnitine. The preferred absolute configuration of the tetrahedral intermediate is *R*, because the most likely approach of the thiol group is on the *Re* face of the acetoxyl group. Therefore, the *2S* configuration of **5** has the same relative configuration as the proposed tetrahedral intermediate, and only (*2S,6R*)-**5** of the four possible diastereomers is expected to have activity.

Because the hydroxy is more stable in the axial position, there is complete asymmetric induction when the hemiketal is formed and only one pair of enantiomers, (*2R,6S* : *2,6R*)-**5** is produced. This stereoselectivity prevents the formation of (*2S,6S* : *2R,6R*)-**5** in which the hydroxy is equatorial. The intriguing possibility of determining the chirality of the reaction center by comparing (*2S,6R*)-**5** activity with that of (*2R,6R*)-**5** is unfortunately not possible with this inhibitor.

Racemic [26] and chiral HAC are good inhibitors of CAT. The chiral material is 3.6-fold better than the racemic, demonstrating chiral recognition of the inhibitor. This is relevant because for this enzyme the *S*-enantiomers of carnitine and acetylcarnitine bind as well as the natural substrates. Because HAC is a competitive inhibitor of both, we suggest that it occupies the same site as carnitine and acetylcarnitine, but that the enzyme is in a conformation that recognizes the chirality, presumably adopted during its catalytic activity. HAC is a good inhibitor, binding a factor of 6 better than acetyl carnitine and a factor of 2 better than carnitine. (Table I).

Table I. Binding constants for selected inhibitors and substrates of CAT and CPT.

Inhibitors (K_i, μM)	CAT	CPT
HAC (2R,6S : 2S,6R)	212[a]	—
HAC (2R,6S)	59.5[a]	—
HPC (2R,6S : 2S,6R)	—	5.1[b], 1.6[c]
Substrates (K_m, μM)		
(R)-Carnitine	120[d]	200[e]
(R)-Acetylcarnitine	350[d]	—
(R)-Palmitoylcarnitine	—	14[e]

[a] ref. [27]; [b] ref. [28], vs. (*R*)-carnitine; [c] ref. [28], vs. (*R*)-palmitoylcarnitine; [d] ref. [29]; [e] ref. [11].

Racemic-HPC, a strong inhibitor of CPT, [28] binds 9-fold better than palmitoylcarnitine. The chiral material promises to be even better. HAC competes for the carnitine site in the short-chain, but not the long-chain transferase. In summary, the success of these inhibitors attests to the design rationale that the g^- conformation

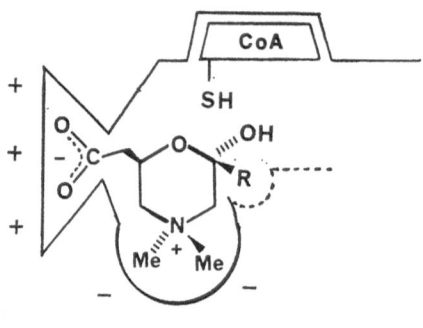

Fig. 8. Proposed fit of 2-hydroxymorpholinium in catalytic center of acyltransferase. Compare with Figure 6.

about C3—C4 can be locked. Their rigidity allows us to speculate on the relative locations of the recognition sites in each enzyme (Figure 8).

In this view of a carnitine acyltransferase, we propose that CoA is located below the inhibitor away from the quaternary ammonium. This view emphasizes the *cis*-diequatorial configuration of the alkyl (R) and carboxymethyl groups.

8. Summary and Conclusions

After structural studies on carnitine and acetylcarnitine as well as an analysis of a possible mechanism for acyl transfer, we have designed and synthesized effective inhibitors of CAT and CPT. These morpholinium derivatives provide a rigid framework from which to anchor the molecular fragments of the reaction. Consequently, we can map the topography of the active site by measuring the relative binding strength of stereoisomeric inhibitors. Our first inhibitors are very competitive, which suggests that they have the correct stereochemistry. HPC is currently the best inhibitor of purified CPT. We are working to verify that the stereochemistry of HAC and HPC are optimal by preparing other stereoisomers. The eventual goal is to add a CoA fragment to the framework in order to determine its recognition site relative to the others.

Acknowledgement

I am grateful to my collaborators W. J. Colucci, T. C. Stelly, S. P. Turnbull, Jr., and Dr. F. R. Fronczek of LSU as well as Professor L. J. Brady and Dr. P. S. Brady of University of Minnesota, whose work made this possible.

References

1. J. D. Dunitz, M. Dobler, P. Sieler, and R. P. Phizackerley: *Acta Crystallogr., Sect. B.* **B30**, 2733 (1974).
2. F. Mathieu, B. Metz, D. Moras, and R. Weiss: *J. Am. Chem. Soc.* **100**, 4412 (1978).
3. D. J. Cram and K. N. Trueblood: *Top. Curr. Chem.* **98**, 43 (1981).

4. R. A. Schultz, D. M. Dishong, and G. W. Gokel: *J. Am. Chem. Soc.* **104**, 625 (1982); G. W. Gokel, D. M. Goli, C. Minganti, and L. Echegoyen: *J. Am. Chem. Soc.* **105**, 6786 (1983).

5. R. D. Gandour, F. R. Fronczek, V. J. Gatto, C. Minganti, R. A. Schultz, B. D. White, K. A. Arnold, D. Mazzocchi, S. R. Miller, and G. W. Gokel: *J. Am. Chem. Soc.* **108**, 4078 (1986).

6. J. P. Long and F. W. Schuler: *J. Pharm. Sci.* **43**, 79 (1954).

7. D. E. Koshland, Jr.: *Proc. Natl. Acad. Sci. USA* **44** 98 (1958).

8. C. L. Hoppel: *Carnitine Palmitoyltransferase and Transport of Fatty Acids* (The Enzymes of Biological Membranes v. 2, Ed. A. Martonosi) pp. 119–143. Plenum Press (1976).

9. I. B. Fritz: *Adv. Lipid Res.* **1**, 285 (1963).

10. J. Bremer: *Physiological Rev.* **63**, 1420 (1983).

11. L. L. Bieber and S. Farrell: *Carnitine Acyltransferases* (The Enzymes v. 16, Ed. P. D. Boyer), pp. 627–644. Academic Press (1983).

12. R. Wolfenden: *Transition-State Affinity as a Basis for the Design of Enzyme Inhibitors* (Transition States of Biochemical Processes, Eds. R. D. Gandour and R. L. Schowen) pp. 555–578 (1978).

13. R. L. Schowen: *Catalytic Power and Transition-State Stabilization* (Transition States of Biochemical Processer, Eds. R. D. Gandour and R. L. Schowen) pp. 77–117 (1978).

14. R. D. Gandour, W. J. Colucci, and F. R. Fronczek: *Bioorg. Chem.* **13**, 197 (1985).

15. W. J. Murray, K. W. Reed, and E. B. Roche: *J. Theor. Biol.* **82**, 559 (1980).

16. W. J. Colucci, R. D. Gandour, and E. S. Mooberry: *J. Am. Chem. Soc.* **108**, 7141 (1986).

17. W. J. Colucci, S. J. Jungk, and R. D. Gandour: *Mag. Reson. Chem.* **23**, 335 (1985).

18. K. G. R. Pachler: *J. Chem. Soc., Perkins Trans. 2*, 1936 (1972).

19. C. A. G. Haasnoot, F. A. A. M. de Leeuw and C. Altona: *Tetrahedron* **36**, 2783 (1980).

20. M. Karplus: *J. Phys. Chem.* **30**, 11 (1959).

21. G. E. Maciel, J. W. McIver, N. S. Ostlund, and J. A. Pople: *J. Am. Chem. Soc.* **92**, 4497 (1970).

22. G. Asknes and J. E. Prue: *J. Chem. Soc.*, 103 (1959).

23. W. J. Colucci, R. D. Gandour, F. R. Fronczek, P. S. Brady, and L. J. Brady: *J. Am. Chem. Soc.* **109**, 7915 (1987).

24. W. J. Colucci, S. P. Turnbull, Jr., and R. D. Gandour: *Analyt. Biochem.* **162**, 459 (1987).

25. B. Collier and F. C. MacIntosh: *Can. J. Physiol. Pharmac.* **47**, 127 (1969).

26. R. D. Gandour, W. J. Colucci, T. C. Stelly, P. S. Brady, and L. J. Brady: *Biochem. Biophys. Res. Commun.* **138**, 735 (1986).

27. R. D. Gandour, W. J. Colucci, P. S. Brady, and L. J. Brady: unpublished.

28. R. D. Gandour, W. J. Colucci, T. C. Stelly, P. S. Brady, and L. J. Brady: *Arch. Biochem. Biophys.* **267**, 515 (1988).

29. J. F. A. Chase and P. K. Tubbs: *Biochem. J.* **99**, 32 (1966).

Journal of Inclusion Phenomena and Molecular Recognition in Chemistry 7 (1989), 53–60.
© *1989 by Kluwer Academic Publishers.*

Cyclophanes as Hosts for Aromatic and Aliphatic Guests*

KENJI KOGA** and KAZUNORI ODASHIMA‡
Faculty of Pharmaceutical Sciences, University of Tokyo, Hongo, Bunkyo-ku, Tokyo 113, Japan

(Received: 1 February 1988)

Abstract. A series of water-soluble cyclophanes, made by connecting two diarylmethane units and two bridging chains via four nitrogens, were found to provide hydrophobic cavities of definite shape and size for forming inclusion complexes with various organic compounds in aqueous solution. Some chemical modifications of these cyclophanes are described.

Key words. Cyclophane, aromatic guest, aliphatic guest, aqueous solution.

1. Introduction

Molecular recognition by host–guest complex formation is known to play a central role in biological processes, such as enzyme catalysis and inhibition, replication, immunological response, transport, drug action, etc. Since little is known about artificial reactions via a similar strategy, realization of this type of complex formation in model systems using synthetic unnatural hosts is expected to provide a novel basis for getting selectivities and efficiencies in artificial reactions.

This paper describes our studies on the design, synthesis, and properties of water-soluble cyclophanes as hosts having hydrophobic cavities of definite shape and size for forming inclusion complexes with various organic guests in aqueous solution.

2. Design, Synthesis, and Properties of CP44 (4) [1]

Organic compounds usually have hydrophobic moieties, and therefore, hosts capable of forming complexes with organic guests by hydrophobic interactions are extremely attractive. Earlier studies on cyclophanes such as **1** [2] and **2** [3] have suggested that they form inclusion complexes with hydrophobic guests in aqueous solution.

We intended to design novel macrocyclic compounds as hosts that have the following characteristics. The macrocyclic compounds should be soluble in water and have hydrophobic cavities inside to form host–guest inclusion complexes with organic compounds by hydrophobic interactions in aqueous solution. The hydrophobic cavities should be sufficiently rigid to show selectivity upon complexation. It is highly desirable that cavities of various sizes are available.

---

* This paper is dedicated to Professor D. J. Cram to celebrate his honor in receiving the 1987 Nobel Prize in Chemistry.
** Author for correspondence.
‡ Present address: Department of Chemistry, Faculty of Science, Hokkaido University, Sapporo 060, Japan.

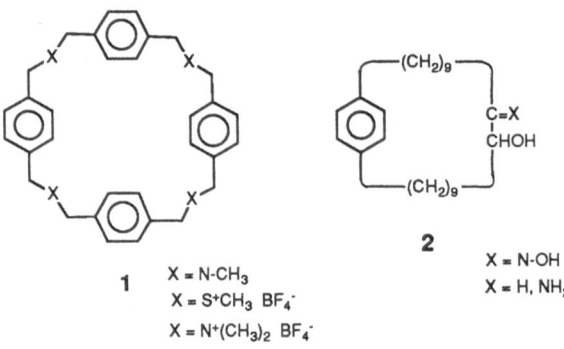

1

X = N-CH₃
X = S⁺CH₃ BF₄⁻
X = N⁺(CH₃)₂ BF₄⁻

2

X = N-OH
X = H, NH₂

Guided by CPK model studies, cyclophanes having the general structure **3** were imagined. These cyclophanes are composed of two diphenylmethane skeletons bridged by two chains. This system was chosen for the following four reasons. (i) The fixed angle of $Ar\!-\!CH_2\!-\!Ar$ is expected to make the resulting cavities reasonably rigid. At the same time, face conformation [4], in which all aromatic rings are perpendicular to the macrocyclic ring, is expected to make the resulting cavities rigid and deep. (ii) Introduction of some water-soluble functional groups somewhere in these molecules is expected to make the resulting compounds soluble in water. (iii) Substitution of two diphenylmethane skeletons and two bridging chains with other units of various length is expected to give cavities of various sizes. (iv) Chemical modifications of these compounds are considered to be possible in many ways, because these compounds are totally synthetic.

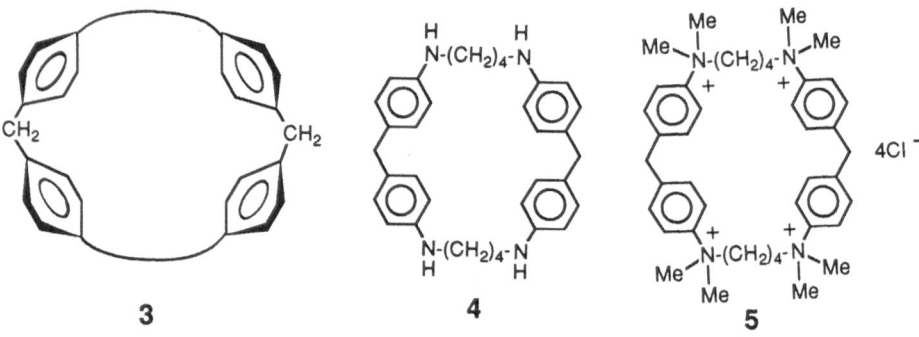

3   **4**   **5**

A cyclophane (**4**, abbreviated as CP44) was prepared starting from commercially available 4,4′-diaminodiphenylmethane, and its properties as a host were studied [5]. This compound (**4**) is soluble in water below pH 2. The corresponding permethylated tetra-ammonium tetra-chloride (**5**, QCP44) is soluble in water at any pH [6].

X-ray analyses of the crystalline complexes between these hosts and aromatic guests unequivocally demonstrated the formation of 1 : 1 inclusion complexes. For example, by shaking a mixture of a solution of **4** in aqueous hydrochloric acid and a solution of durene (**6**) in hexane vigorously, a crystalline complex was obtained. The structure of the complex (**7**) shows that the guest (**6**) is fully included at the center of the cavity of the host (protonated **4**). It is also shown that all benzene rings of the host are perpendicular to the macrocyclic ring (face conformation) as expected, and that the cavity has rectangularly shaped open ends (~ 3.5 × 7.9 Å) with a depth of 6.5 Å. The benzene ring of the guest fits well within the cavity, being nearly parallel to the inner wall, because the thickness of the aromatic ring (3.4 Å) fits well with the shorter width of the cavity open ends (~ 3.5 Å) as shown by the spacial shape of the complex (**8**). It is reasonable to assume that hydrophobic interactions play a role in this complex formation, because **6** is a nonpolar substrate, and the complex was obtained from aqueous solution.

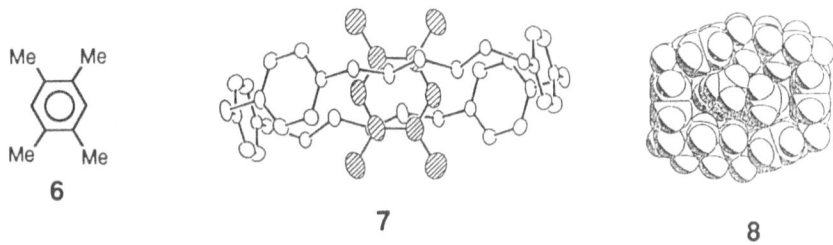

Complex formation was also observed in solution below CMC concentrations by spectroscopic methods. Figure 1 shows ^{1}H NMR spectra of (a) 2,7-dihydroxynaphthalene (**9**), (b) deutated **4**, and (c) their mixture in DCl–D$_2$O (pD 1.2). In (c), marked upfield shifts were observed for the proton signals of the guest and for the tetramethylene proton signals of the host, due to the shielding effect of the aromatic ring(s). It should be noted that each proton signal of the host and the guest shifts to a different degree, indicating that the complex is formed in a particular geometry and not in a random fashion. This feature is very important, because this phenomenon can be expected only in complexes with hosts having well-defined structures, but not with mobile systems such as micelles.

Based on the assumption that the host takes the same conformation as that in the crystalline complex with durene described above, the possible geometry of the

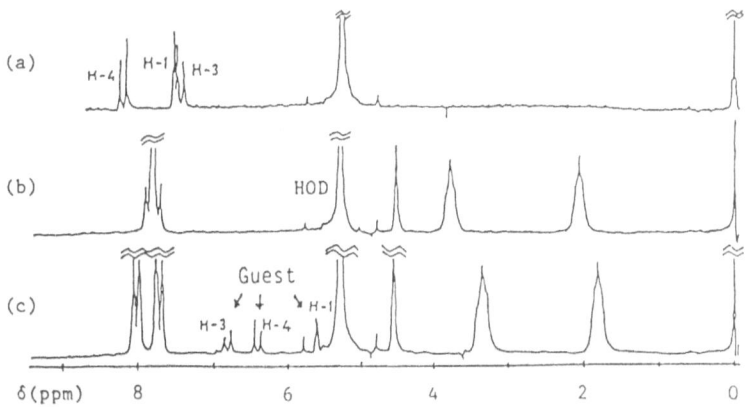

Fig. 1. ^{1}H NMR spectra of (a) **9**, (b) **4**, and (c) **9** + **4** in DCl D$_2$O (pD 1.2). [9] = 2.5 × 10^{-2} M. [4] = 5.0 × 10^{-2} M. TMS was used as an external reference.

9

10

complex in solution was determined as shown in **10** by calculation employing the method of Johnson and Bovey [7]. Since the process of this calculation is now done by a computer, examinations on the possible as well as impossible structures of the complexes in solution based on NMR data can be visualized on the three dimensional computer graphic display [8]. This method may be useful for the design of hosts specific for guests having unique structures.

3. Chemical Modifications of CP44

Chemical modifications of **4** were made in several directions to examine the relationship between the cavity structures and complexation properties.

3.1. MODIFICATIONS OF THE BRIDGING CHAINS OF CP44

Cyclophanes (**11, 12**) having different bridging chains were synthesized to change the size and the hydrophobic area of **4**, and their complexation properties with various guests (**9, 13–21**) were examined [9].

Among these hosts, **11a** (CP33) and **11f** (CP35) did not show any evidence for complex formation with all the guests examined, probably because their cavities are too small to accommodate these guests. On the other hand, hosts (**4, 11b–e, 11g–i, 12**) formed complexes with neutral and anionic aromatic guests (**9, 13–18**), but not with aliphatic guests (**19, 20**). Complex formation with the cationic aromatic guest (**21**) was found to be negligible. Some results for hosts (**4, 11h, 12**) having cavities of different size and/or hydrophobic area are shown in Table I.

Table I.   Stability constants ($K_s[\text{M}^{-1}]$) of 1:1 complexes.[a]

Guest/host	4(CP44)	11h(CP56)	12
9	2.8×10^3(11)	2.6×10^2(1.0)	4.3×10^3(17)
13	6.3×10^3(0.15)	4.3×10^4(1.0)	5.0×10^5(12)
14	2.0×10^3(2.3)	8.7×10^2(1.0)	1.4×10^4(16)
15	1.5×10^3(0.39)	3.8×10^3(1.0)	5.3×10^4(14)
16	1.9×10^4(6.4)	2.9×10^3(1.0)	3.0×10^4(10)
17	4.4×10^3(0.041)	1.1×10^5(1.0)	1.4×10^6(13)
18	1.8×10^5(5.5)	3.3×10^4(1.0)	3.2×10^5(9.6)

[a] In KCl–HCl buffer (pH 1.95) at 25°C. The values in parentheses are the relative stabilities of the complexes of **4** and **12** compared with those of **11h**.

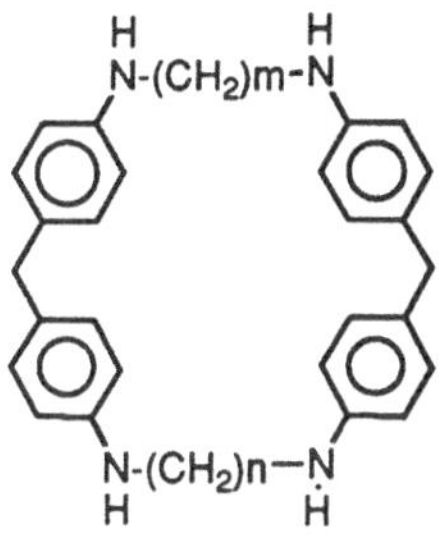

11

a: m=n=3 (CP33)
b: m=n=5 (CP55)
c: m=n=6 (CP66)
d: m=n=7 (CP77)
e: m=n=8 (CP88)
f: m=3, n=5 (CP35)
g: m=4, n=5 (CP45)
h: m=5, n=6 (CP56)
i: m=5, n=8 (CP58)

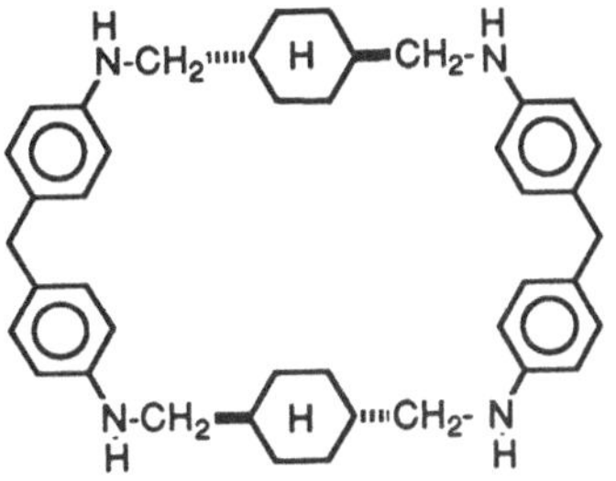

12

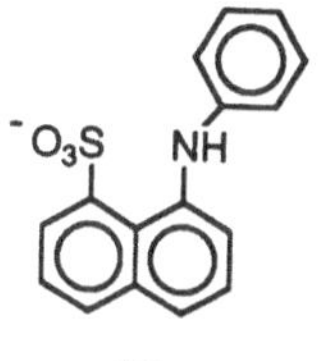

13

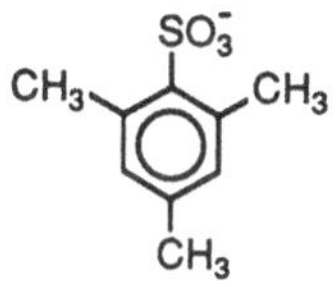

14

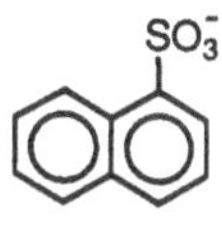

15

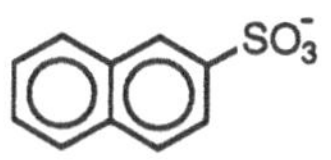

16

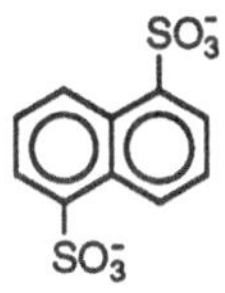

17

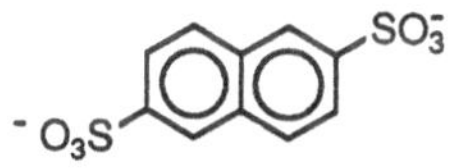

18

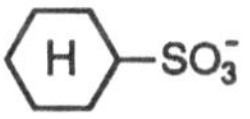

19

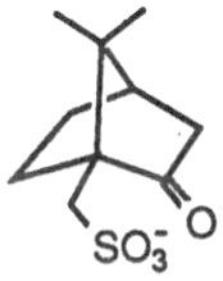

20

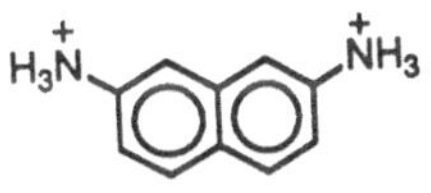

21

It is shown that the increase in hydrophobic area of the cavity greatly enhances the stability of the complex. Thus, with all the guests examined, host **12** formed more stable complexes than **11h** by a factor of 10 to 17. It is also shown that the difference in cavity size also affects the stability of the complexes. For example, **4** having a smaller cavity prefers β-substituted naphthalenes (**16**, **18**) rather than α-substituted naphthalenes (**15**, **17**), and the reverse is true for **11h** having a larger cavity.

It may be concluded that complementarity both in steric structures and electrostatic interactions between host and guest is important for strong complex formation in this type of cyclophanes.

3.2. CHIRAL MODIFICATIONS OF CP44

Optically active cyclophanes (**22**) having chiral centers at their bridging chains were synthesized starting from L-tartaric acid and their properties as hosts for chiral guests were examined [10]. As shown in Figure 2, ^{1}H NMR spectral studies have

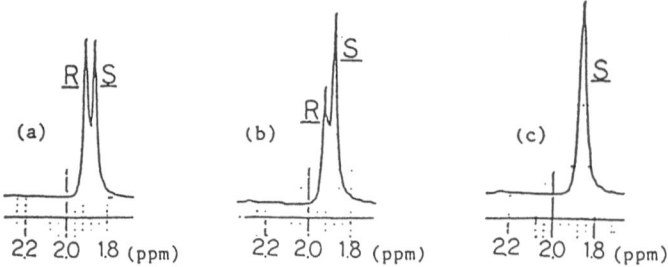

Fig. 2. ^{1}H NMR spectra (C-methyl signals) of (a) racemic **23**, (b) partially resolved **23** ($(R):(S) =$ 3 : 5), and (c) (S)-**23** in DCl–D$_2$O (pD 1.2) in the presence of **22a**. $[23] = 2.5 \times 10^{-2}$ M. $[22a] = 5.0 \times 10^{-2}$ M. TMS was used as an external reference.

shown that diastereomeric host–guest complexes are formed between **22a** and (R)- and (S)-atrolactic acid (**23**). It is also shown that the reduction of 1-naphthylgly-oxylic acid (**24**) by complex formation with **22a** followed by treatment with sodium borohydride in acidic water afforded (R)-1-naphthylglycolic acid ((R)-**25**) in 9.7% ee.

Although not very efficient at present, these results clearly show that chiral cyclophanes provide chiral cavities which can recognize and induce chirality in the bound guests.

3.3. BIS-CYCLOPHANES

Bis-cyclophanes (**26**) having two independent binding sites, capable of forming complexes at two sites, were synthesized by connecting two cyclophane units [11].

It is shown that these hosts form complexes with aromatic guests such as **27** at two sites simultaneously. In cases where molecules having two aromatic groups such as **28** are used as guests, examples of cooperative binding at two sites were found, when the distances between the two cavities of the host and the two aromatic groups of the guests are complementary. These hosts are of particular interest in relation to the methods for assembling and recognizing guests by multiple complexation.

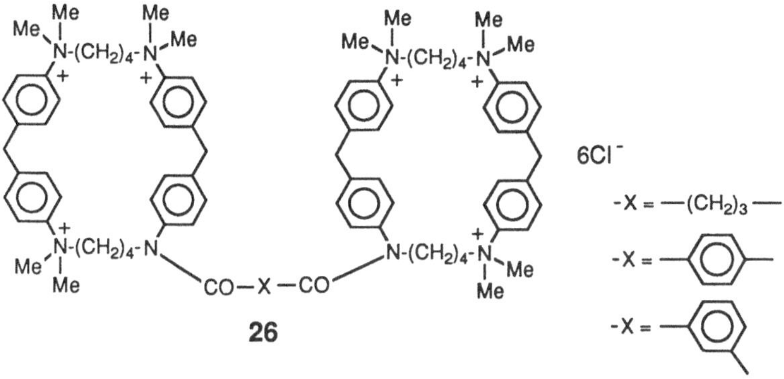

26

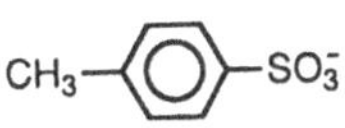

$6Cl^-$

$-X = -(CH_2)_3-$

$-X =$ (aromatic ring)

$-X =$ (aromatic ring)

CH_3-(ring)$-SO_3^-$

27

$^-O_3S-$(ring)$-O-(CH_2CH_2O)_{\overline{n}}$(ring)$-SO_3^-$

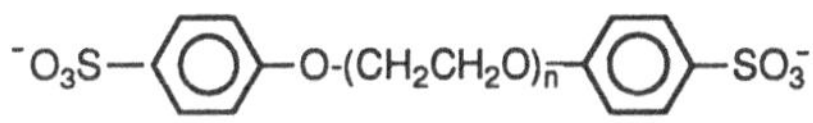

28 a : n = 1
        b : n = 5

3.4. HOSTS FOR ALIPHATIC GUESTS

Cyclophanes such as **4** and **5**, composed of two diphenylmethane skeletons and two bridging chains are thus shown to work as hosts selectively for aromatic guests, but not for aliphatic guests.

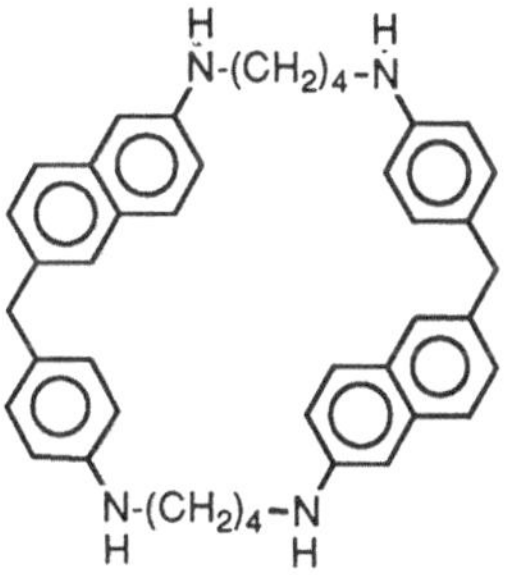

29

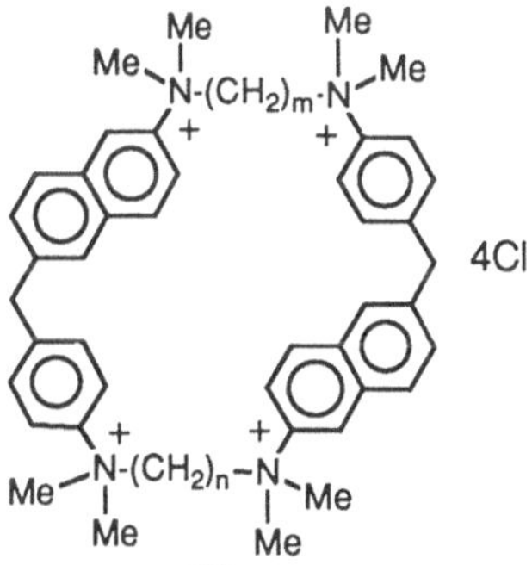

30

$4Cl^-$

$CH_3-(CH_2)_5-SO_3^-$

31

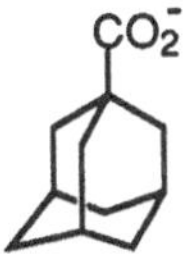

32

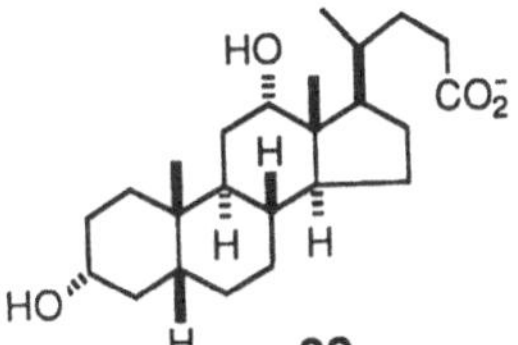

33

To get hosts capable of binding aliphatic guests, novel cyclophanes (**29, 30**) having naphthylphenylmethane skeletons instead of diphenylmethane skeletons were designed based on CPK model studies [12]. In their most expanded conformations, these hosts are expected to provide wider cavities of about 5.4 Å in their shorter width of their open ends.

Although these hosts do not show any evidence for complex formation with smaller aliphatic guests such as **19** and **31**, they form reasonably stable complexes (K_s values of $10^2 \sim 10^4 (M^{-1})$) with more bulky aliphatic guests such as **20, 32**, and **33**. It is also shown that these hosts form complexes with aromatic guests.

4. Conclusion

Water-soluble cyclophanes made by connecting two diarylmethane skeletons and two bridging chains via four nitrogens are shown to provide cavities of definite shape and size and constitute a group of artificial hosts that form 1 : 1 inclusion complexes with various organic guests in water. For complexation to occur in particular geometries with remarkable selectivities, the importance of the fit in steric structure and charge between the host and the guest has been recognized.

Since these hosts are totally synthetic, design for selective and efficient artificial systems via host–guest complex formation is open as one of the most challenging and exciting fields in synthetic organic chemistry.

References

1. K. Odashima and K. Koga: in 'Cyclophanes', Vol. 2, ed. by P. M. Keehn and S. M. Rosenfeld, Academic Press, New York, 1983, Chapter 11.
2. I. Tabushi and K. Yamamura: *Top. Curr. Chem.* **113**, 145 (1983).
3. Y. Murakami: *Top. Curr. Chem.* **115**, 107 (1983).
4. (a) D. J. Cram and M. F. Antar: *J. Am. Chem. Soc.* **80**, 3103 (1958); (b) I. Tabushi, H. Yamada, and Y. Kuroda: *J. Org. Chem.* **40**, 1946 (1975).
5. (a) K. Odashima, A. Itai, Y. Iitaka, and K. Koga: *J. Am. Chem. Soc.* **102**, 2504 (1980); (b) K. Odashima, A. Itai, Y. Iitaka, and K. Koga: *J. Org. Chem.* **50**, 4478 (1985).
6. (a) A. Miwa, K. Odashima, and K. Koga: unpublished data; (b) J. Winkler, E. Coutouli-Argyropoulou, R. Leppkers, and R. Breslow: *J. Am. Chem. Soc.* **105**, 7198 (1983).
7. K. Odashima, A. Itai, Y. Iitaka, Y. Arata, and K. Koga: *Tetrahedron Lett.* **21**, 4347 (1980).
8. A. Itai, M. Sakamoto, Y. Iitaka, K. Odashima, and K. Koga: unpublished data.
9. (a) T. Soga, K. Odashima, and K. Koga: *Tetrahedron Lett.* **21**, 4351 (1980); (b) K. Odashima, T. Soga, and K. Koga: *Ibid.* **22**, 5311 (1981).
10. (a) I. Takahashi, K. Odashima, and K. Koga: *Tetrahedron Lett.* **25**, 973 (1984); (b) I. Takahashi, K. Odashima, and K. Koga: *Chem. Pharm. Bull.* **33**, 3571 (1985).
11. C. F. Lai, K. Odashima, and K. Koga: *Tetrahedron Lett.* **26**, 5179 (1985).
12. (a) H. Kawakami, O. Yoshino, K. Odashima, and K. Koga: *Chem. Pharm. Bull.* **33**, 5610 (1985); (b) H. Kawakami, K. Odashima, and K. Koga: unpublished data.

Journal of Inclusion Phenomena and Molecular Recognition in Chemistry 7 (1989), 61–72.   61
© 1989 *by Kluwer Academic Publishers.*

Topics in Calixarene Chemistry

C. DAVID GUTSCHE*, IFTIKHAR ALAM, MUZAFFAR IQBAL,
THOMAS MANGIAFICO, KYE CHUN NAM, JANET ROGERS, and KEAT AUN SEE
Department of Chemistry, Washington University, St. Louis, MO 63130, U.S.A.

(Received: 1 February 1988)

Abstract. Several facets of calixarene chemistry have been investigated including the mechanism of their formation by the base-induced condensation of phenols and formaldehyde, procedures for introducing functional groups onto the 'upper rim' and 'lower rim' of calixarenes, the conformational behavior of calixarene oxyanions, the formation of complexes in aqueous and nonaqueous systems, and the application of calixarenes as catalysts.

Key words. Calixarenes, calixarene oxyanions, conformation, complex formation, catalysis.

1. Mechanism of Formation of Calixarenes

Calixarenes, represented by the general structure **1**, comprise a family of macrocyclic compounds that can be prepared by the base-induced reaction of certain *p*-substituted phenols with formaldehyde. When hydroxymethylated compounds of the general structure **2** with *n* ranging from 1–4 and $R = H$ or CH_2OH are used

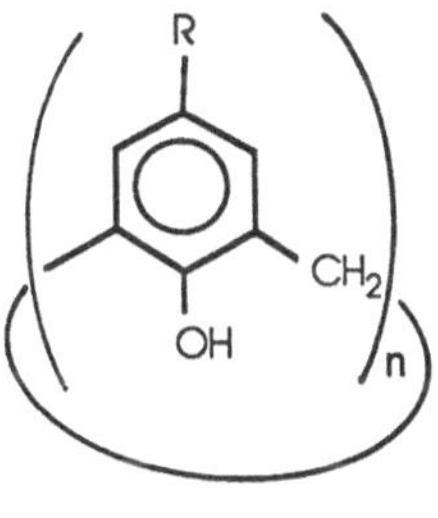

1 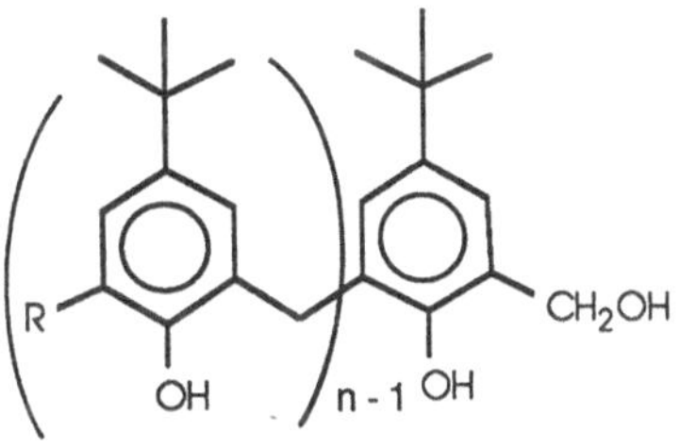   **2**

instead of phenols the mixture of calixarenes produced is independent of the starting material and dependent mainly on the reaction conditions, suggesting that a mobile equilibrium exists between various oligomeric precursors to the calixarenes. In a reaction carried out under mild conditions and monitored at various stages it was determined that the initially formed cyclic product is the cyclic octamer and that only under more strenuous conditions is it converted to the cyclic tetramer. This sequence of events is attributed to the existence of the linear oligomer **2** ($n = 4, R = CH_2OH$) in a 'zig-zag' conformation [2] which must undergo a

---

* Author for correspondence.

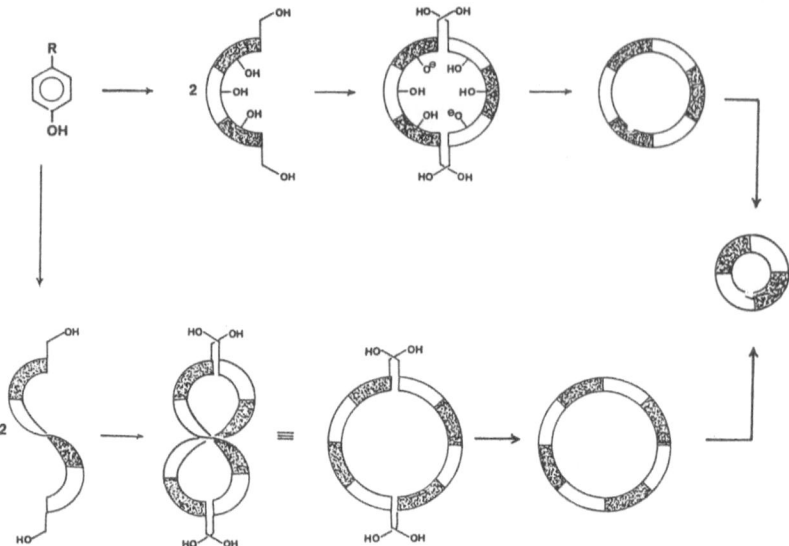

Fig. 1.   Pathways of the one-step calixarene formation.

conformational change in order to cyclize. However, it can associate via intermolec-ular hydrogen bonding to a 'hemicalix[8]arene' in which a circular array of hydrogen bonds is present in essentially the same orientation as in the cyclic octamer [3] thus predisposing this intermediate to undergo conversion to the cyclic tetramer, as illustrated in Figure 1. In similar fashion, cyclic hexamer formation is attributed to the association of a pair of linear trimers to form a 'hemi-calix[6]arene'. Among the unanswered questions in the one-step formation of calixarenes is (a) why the reaction is so extremely sensitive to the identity of the *p*-substituent, only *tert*-butylphenol or closely related phenols providing useful yields of pure calixarene and (b) why the odd-numbered calixarenes are formed in so much smaller an amount than the even-numbered calixarenes.

2. 'Upper-rim' Functionalization Procedures

Functionalized calixarenes are of interest as potential catalysts, and considerable effort has been directed to the attachment of various entities onto both the 'upper' and 'lower' rims of the cyclic oligomers. 'Upper rim' functionalization has been achieved by aluminum chloride-catalyzed debutylation followed by functional group introduction via the electrophilic substitution route [4], the *p*-Claisen rear-rangement route [5], and the *p*-quinonemethide route. The last of these methods, illustrated in Figure 2, involves the condensation of a debutylated calixarene (3) with formaldehyde and any of a wide variety of dialkylamines to give Mannich bases (4). Quaternization of 4 (e.g. with MeI) to give 5 followed by treatment with two equivalents of a nucleophile (the first equivalent acting as a base) provides good yields of products of structure 6 where Nu can be a variety of functions including OR, CN, $CH(CO_2Et)_2$, $CH(NO_2)(CO_2Et)$, SR, and N_3. Weak nucleophiles (e.g. imidazole or 2-nitrophenoxide) fail to react, and very strong nucleophiles (e.g.

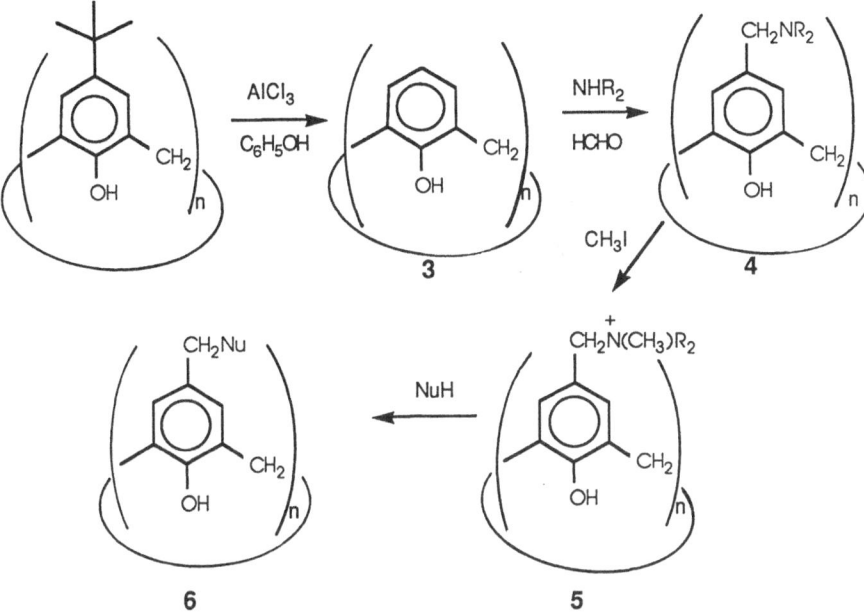

Fig. 2.   *p*-Quinonemethide route to functionalized calixarenes.

acetylides, hydride) give mixtures in which some of the reaction involves displacement on nitrogen.

3. 'Lower-rim' Functionalization of Calixarenes

The OH groups at the 'lower-rim' of the calixarenes provide a ready handle for the introduction of substituents, and numerous esters and ethers have been made in this laboratory and elsewhere. The aroylation of calix[4]arenes has now been studied in some detail using the aroyl halides shown in Table I. The conformations of the products have been determined by ^{1}H NMR measurements. The product mixtures

Table I.   Product composition from the aroylation of *p*-allylcalix[4]arene and *p*-*tert*-butylcalix[4]arene with *p*-substituted ArCOCl.

p-R group of of ArCOCl	σ_{para}	Product	
		p-allylcalix[4]arene	*p*-*tert*-butylcalix[4]arene
OCH$_3$	−0.27	1,3-alt	cone (90%); 1,3-alt (5%)
C(CH$_3$)$_3$	−0.20	1,3-alt	cone (80%); 1,3-alt (20%)
CH$_3$	−0.17	1,3-alt	cone (90%); 1,3-alt (5%)
H	0.0	1,3-alt	cone (98%); 1,3-alt (trace)
Br	0.23	1,3-alt	cone
CN	0.66	cone	cone
NO$_2$	0.76	cone	cone

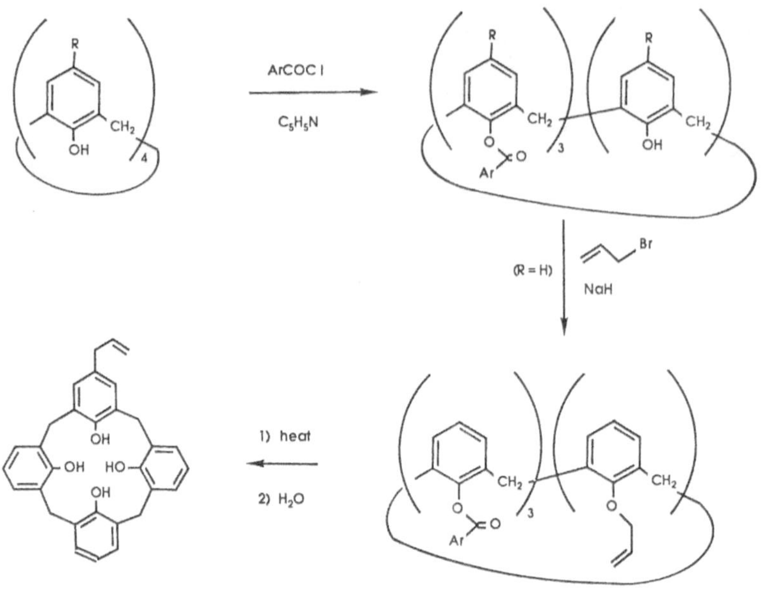

Fig. 3.   Synthesis of *p*-monoallylcalix[4]arene via the tribenzoate.

indicated in Table I are interpreted as the result of a competition between the rate of conformational interconversion and the rate of derivatization, the slower the former and the more rapid the latter the more likely will be the formation of the cone conformer as the major product.

Solvent effects are shown to play a part in the reaction. For example, the use of pyridine in place of THF provides a means for obtaining triaroylates from calix[4]arenes and makes available a starting material for the preparation of 'upper-rim' mono-functionalized calix[4]arenes [6], as shown in Figure 3.

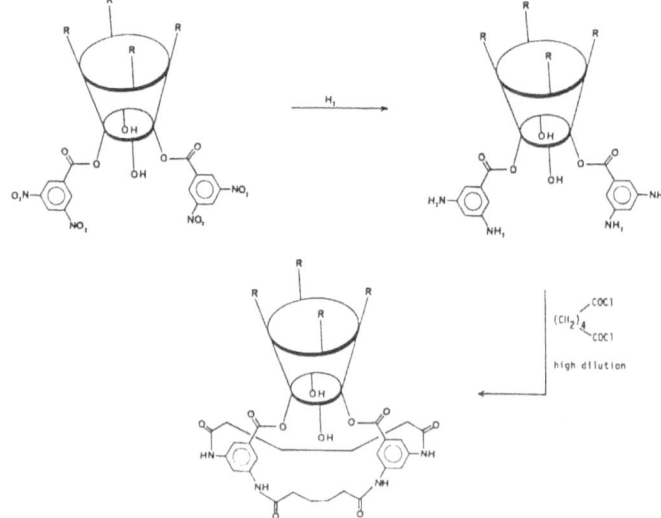

Fig. 4.   Synthesis of a 'double cavity' calixarene.

Steric effects are shown to affect the outcome of the aroylation reaction as illustrated, for example, by the reaction with 3,5-dinitrobenzoyl chloride which affords an excellent yield of a di-aroylate and provides a starting material for the synthesis of a 'double cavity' calixarene, as illustrated in Figure 4.

4. Conformational Behavior of Calixarene Oxyanions

Treatment of a DMSO solution of a calix[4]arene with incremental amounts of n-butyllithium generates solutions that have been analyzed by means of ^{1}H NMR and ^{7}Li NMR with the results shown in Figure 5. In both cases spectral changes are observed until four equivalents of base have been added, at which point the spectra remain essentially invariant upon addition of more base. The conformational identity of the various species can be established by inspection of the δ 3–5 ppm region of the ^{1}H NMR spectra which arises from the CH$_2$ protons, a pair of doublets signifying a cone conformation, two sets of pairs of doublets signifying a partial cone conformation, and a singlet signifying a 1,3-alternate conformation. Inspection of Figure 6 reveals that the mono- and tetra-anions are cone conformers and the tri-anion is a partial cone conformer. The spectrum of the system generated from two equivalents of base, however, is best interpreted as a superposition of the spectra of the mono- and tri-anions, suggesting that a dianion is unstable and disproportionates to equal amounts of these anions. These conclusions are corroborated by the ^{7}Li NMR spectra which show single resonances for the mono- and tetra-anions, a three line spectrum for the tri-anion, and a spectrum for the system containing two equivalents of base that is the superposition of the spectra of the mono- and tri-anions.

Temperature-dependent NMR measurements of solutions containing calix-[4]arenes provide a means for measuring rates of conformational interconversion;

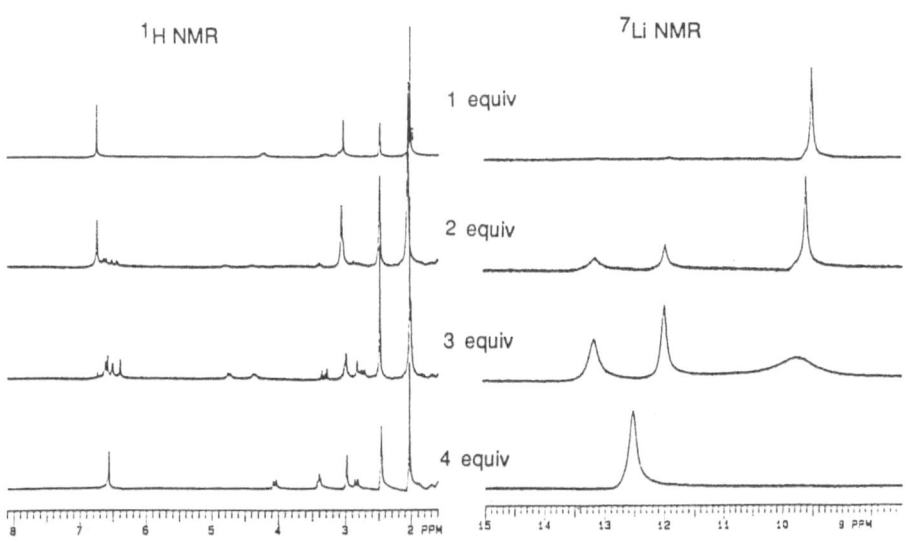

Fig. 5. ^{1}H NMR and ^{7}Li NMR spectra of lithium oxyanions of calix[4]arenes.

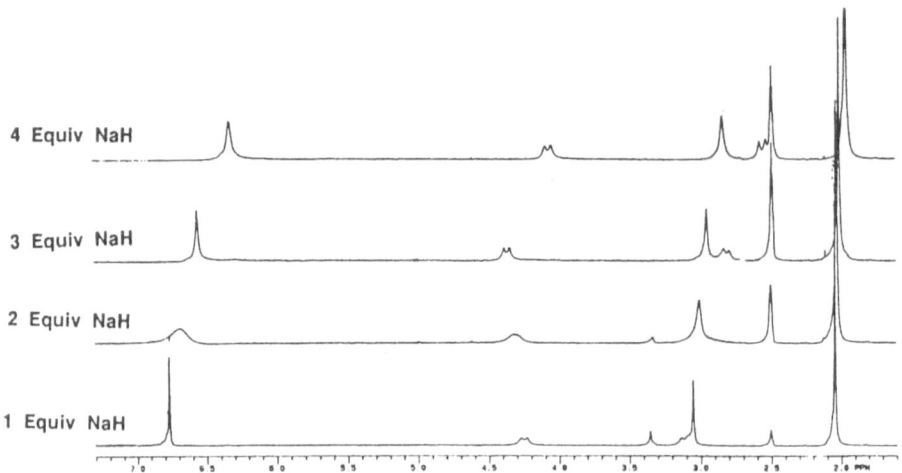

Fig. 6.   ¹H NMR spectra of the sodium oxyanions of calix[4]arenes.

e.g. in the case of the cone conformation a pair of doublets (from CH_2 protons) indicates an interconversion that is slow on the NMR time scale, a singlet indicates an interconversion that is fast, and a broad signal characterizes the coalescence temperature T_c. The coalescence temperature for the mono-anion is *ca* 90°C, while the tri- and tetra-anions show no hint of coalescence at 140°C. To test the premise that a lithium cation is complexed by a calix[4]arene in crown ether-like fashion, a similar study was carried out with sodium as the cation, with the results shown in Figure 6. As in the lithium system, the mono- and tetra-anions exist in the cone conformation. In contrast, however, the tri-anion appears to exist in the cone rather than the partial cone conformation, and the system generated from two equivalents of base does not appear to be a mixture of the mono- and tri-anions. A temperature-dependent study of the sodium system showed that it is more conformationally mobile than the lithium system. For example, T_c for the tetra-anion is only 80°C. To further explore the relation between conformational mobility of the oxyanions and the size of the cation, the potassium tetra-anion was generated; its T_c was determined to be even lower, at 20°C.

These studies have been extended to the calix[6]arenes, which also have been observed to show changes in the ¹H NMR spectra as incremental amounts of strong base, up to six equivalents, are added to a DMSO solution. With the calix[6]arene systems the spectra are considerably more complex than those of the calix[4]arenes, and conformational assignments are much more difficult to make. However, a comparison of the temperature-dependent behavior of the lithium and sodium oxyanions of the calix[6]arenes indicates that the sodium system is less conformationally mobile and suggests that here, also, a crown ether-like complexation of the metal ion may be occurring. The annulus of the calix[6]arene system is complementary in size to the sodium cation, while that of the calix[4]arene is complementary in size to the lithium cation.

5. Aqueous Solution Complexation of Organic Molecules

The p-quinonemethide method of functionalization has made carboxyl and amino-containing calixarenes available in all of the ring sizes 4–8, thus allowing the complexation behavior of these compounds in dilute aqueous base and acid to be measured. The guest molecules used were naphthalene, anthracene, pyrene, fluoranthene, and perylene (left-hand column in Figure 7), and the association constants for complex formation of these hydrocarbons with the various calixarenes are shown in Figure 7. The similarity in K_{assoc} values of the carboxy and amino calixarenes for a given hydrocarbon and a given ring size suggests that the site of complexation is not proximate to these functional groups but involves the hydroxyl end of the system (referred to in Section 3 as the 'lower rim'). There is, in fact, a rough correlation between the size of the guest aromatic hydrocarbon and the size of the opening of the 'lower rim'.

6. Non-aqueous Complexation of Organic Molecules

The interaction of calixarenes with amines in acetonitrile solution has been studied in some detail [7] and it has been postulated that a two-step sequence takes place in which a proton is transferred from the calixarene to the amine, producing a calixarene anion and an ammonium ion, followed by ion-pairing between these species. Toward most other types of organic molecules in organic solution, however, the simple calixarenes or calixarene derivatives have shown little tendency to form

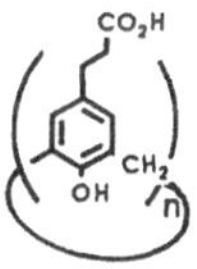

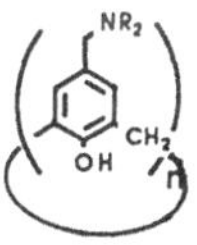

	n = 5	n = 6	n = 7	n = 8	n = 5	n = 6	n = 7	n = 8
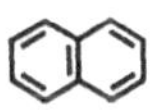 naphthalene	3.7×10^3	3.7×10^3	3.9×10^3	6.1×10^2	3.3×10^3	4.5×10^3	3×10^3	1.1×10^3
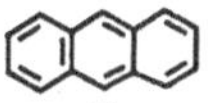 anthracene	9.1×10^3	1.3×10^4	1.1×10^4	$<10^2$	9×10^3	1.6×10^4	8.3×10^3	$<10^2$
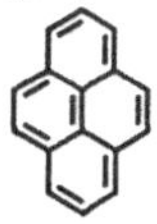 pyrene	$<10^2$	$<10^2$	1.1×10^4	4.4×10^4	$<10^2$	$<10^2$	9×10^3	3.6×10^4
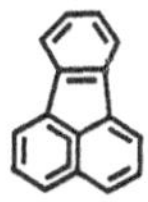 fluoranthene	4×10^3		3.6×10^3	1.4×10^4	2×10^3	4×10^3	3×10^3	1.5×10^4
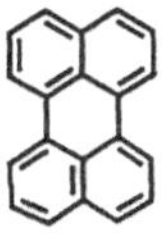 perylene	$<10^2$	$<10^2$	9×10^3	8.4×10^3	$<10^2$	$<10^2$	1×10^4	1×10^4

Fig. 7. Complexation of aromatic hydrocarbons in aqueous solution by carboxy and amino calixarenes.

strong complexes. For example, the tetra-(p-nitrobenzoyl) ester of calix[6]arene was considered to be a likely candidate, because CPK models indicated it to contain a long trench into which an aromatic hydrocarbon might fit. This compound, prepared by treating a calix[4]arene with an excess of NaH and p-nitrobenzoyl chloride, was shown by ^{1}H NMR studies to possess the conformation pictured in Figure 8. Solutions of the tetra-ester in CHCl$_3$ containing naphthalene or anthracene, however, showed no change in the resonance lines of the hydrocarbons, and a closer investigation of the ^{1}H NMR characteristics of the tetra-ester, using transient NOE techniques, reveals that it probably exists in a form in which the aromatic moieties is tipped inward to fill the crevice created by the other aromatic moieties, thus excluding external guest molecules.

In contrast, 'double cavity' calixarenes, synthesized as described in Section 3, show a significant tendency to form complexes in organic solvents. In particular, they appear to interact with acidic compounds, as illustrated by the data in Figure 9 for carboxylic acids and in Figure 10 for phenols, where the magnitudes of the changes in chemical shifts of the resonances of the guest molecule are taken as a measure of the degree of complexation. Even molecules as weakly acidic as acetonitrile show significant ^{1}H NMR shifts in the presence of the 'double cavity' calixarene, as illustrated by the data in Figure 11. Moderate complexation (shifts *ca* 0.05 ppm) was observed with ethanol, n-butanol, phenol, bromophenol, nitrobenzene, and CH$_2$Cl$_2$. Weak or no complexation (shifts less than 0.02 ppm) was observed with aniline, benzene, acetone, toluene, bromobenzene, adiponitrile, quinuclidine, collidine, ethyl cyanoacetate, and ethyl nitroacetate. That the apparent degree of complexation is not directly proportional to the acid strength of the guest is shown by a variety of comparisons. For example, isobutyric acid shows less

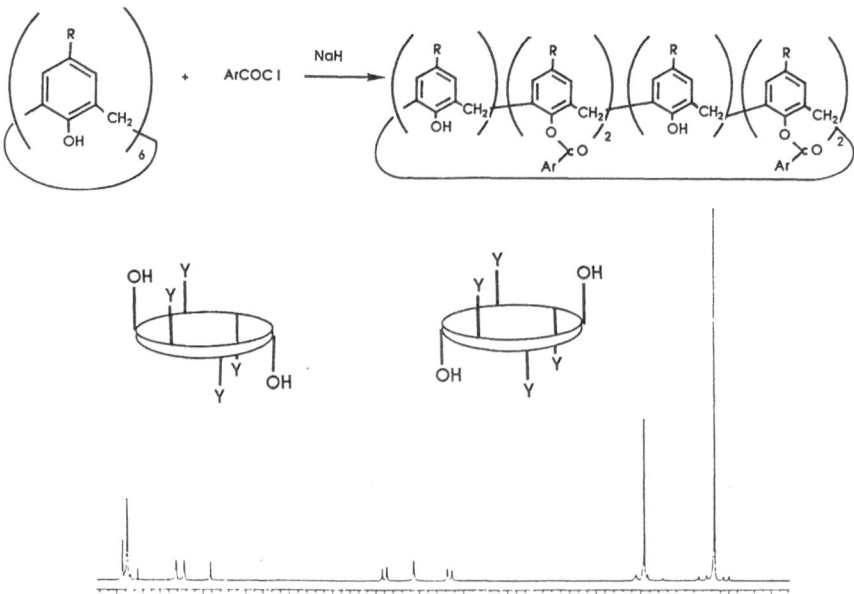

Fig. 8. Conformation of tetra-(p-nitrobenzoyloxy)-calix[6]arenes.

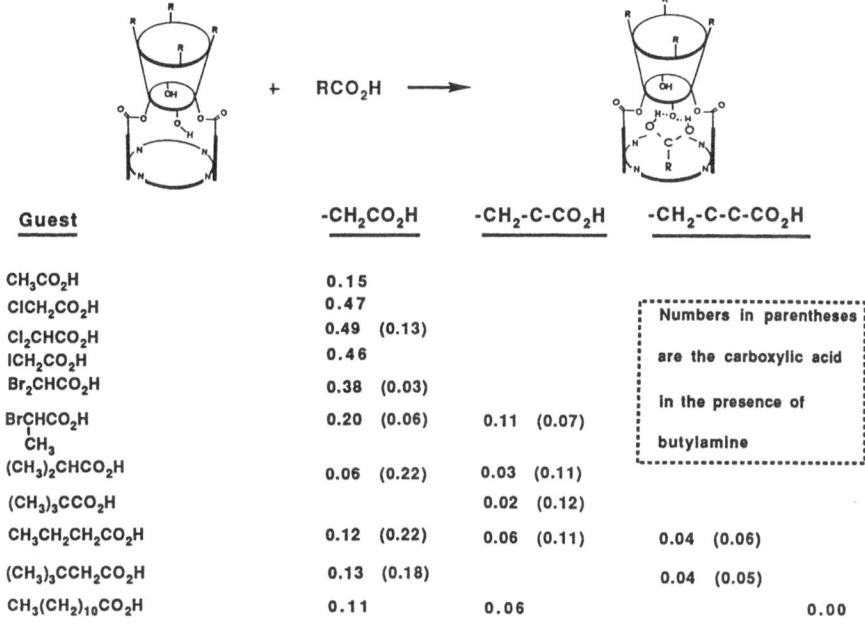

Guest	$-CH_2CO_2H$	$-CH_2-C-CO_2H$	$-CH_2-C-C-CO_2H$
CH_3CO_2H	0.15		
$ClCH_2CO_2H$	0.47		
Cl_2CHCO_2H	0.49 (0.13)		
ICH_2CO_2H	0.46		
Br_2CHCO_2H	0.38 (0.03)		
$BrCHCO_2H$ CH_3	0.20 (0.06)	0.11 (0.07)	
$(CH_3)_2CHCO_2H$	0.06 (0.22)	0.03 (0.11)	
$(CH_3)_3CCO_2H$		0.02 (0.12)	
$CH_3CH_2CH_2CO_2H$	0.12 (0.22)	0.06 (0.11)	0.04 (0.06)
$(CH_3)_3CCH_2CO_2H$	0.13 (0.18)		0.04 (0.05)
$CH_3(CH_2)_{10}CO_2H$	0.11	0.06	0.00

Numbers in parentheses are the carboxylic acid in the presence of butylamine

Fig. 9. Complexation of a 'double cavity' calixarene with carboxylic acids.

Guest	H_{ortho}	H_{meta}	H_{para}
Phenol	0.04	0.05	0.06
o-Nitrophenol	0.01 (0.02)	002 (0.02)	0.02 (0.02)
m-Nitrophenol	0.02 (0.09); 0.03 (0.12)	0.13 (0.08)	0.25 (0.13)
p-Nitrophenol	0.04 (0.14)	0.10 (0.08)	
2,4-Dinitrophenol	0.00	0.01; 0.03	
2,5-Dinitrophenol	0.01	0.02	0.00
Picric Acid		0.04	

Numbers in parentheses are the phenol in the presence of butylamine

Fig. 10. Complexation of a 'double cavity' calixarene with phenols.

interaction than acetic acid, and *m*-nitrophenol and *p*-nitrophenol show a stronger interaction than *o*-nitrophenol, 2,4-dinitrophenol, or picric acid. To explore this phenomenon in greater detail ^{1}H NMR relaxation values (T_1) were measured for the 'double cavity' calixarene in the presence of several carboxylic acids as well as *m*-nitrophenol, with the results shown in Figure 12. These values were compared with those of the guest molecule both in the absence of the calixarene and in the presence of butylamine, to which a proton transfer occurs. In the case of butyric acid the T_1 values are significantly lower in the presence of the calixarene than in the presence of butylamine, whereas this is not true with pivalic acid,

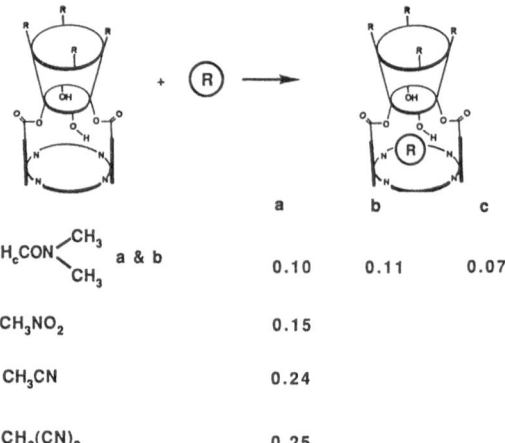

	a	b	c
H$_c$CON(CH$_3$)(CH$_3$) a & b	0.10	0.11	0.07
CH$_3$NO$_2$	0.15		
CH$_3$CN	0.24		
CH$_2$(CN)$_2$	0.25		

Fig. 11. Complexation of a 'double cavity' calixarene with small molecules.

indicating that butyric acid forms a complex while pivalic acid does not. The ^{1}H NMR chemical shift data indicating m-nitrophenol forms a complex are corroborated by the T_1 measurements, which show values that are considerably lower than those of the uncomplexed guest and also lower than those of the guest in the presence of butylamine.

Guest		T_1 (neat)	T_1 (with calixarene)	T_1 (with butylamine)
CH$_3$CH$_2$CH$_2$CO$_2$H	H$_a$	2.71	0.62	1.32
	H$_b$	3.03	0.61	1.50
	H$_c$	2.73	0.80	1.85
(CH$_3$)CCO$_2$H		2.50	1.81	1.45
CH$_3$(CH$_2$)$_5$CH$_2$CH$_2$CO$_2$H	H$_a$	1.26	0.47	0.59
	H$_b$	1.29	0.49	0.87
	H$_{CH_3}$	2.42	1.23	2.00
CH$_3$(CH$_2$)$_8$CH$_2$CH$_2$CO$_2$H	H$_a$	1.19	0.58	0.66
	H$_b$	1.32	0.50	1.05
	H$_c$	2.57	1.87	2.26
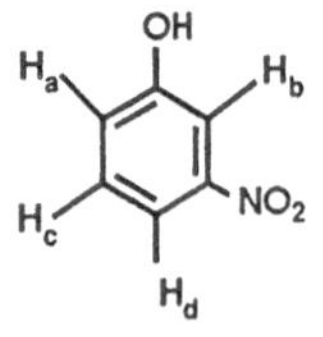 (phenol with OH, H$_a$, H$_b$, H$_c$, H$_d$, NO$_2$)	H$_a$	4.11	1.48	2.88
	H$_b$	5.00	1.60	2.76
	H$_c$	4.27	1.67	3.37
	H$_d$	4.86	3.23	4.38

Fig. 12. T_1 values for guest molecules, free and complexed with a 'double cavity' calixarene.

Since the most basic group in the calixarene is the amide moiety, the pK_{BH} of which would be expected to be quite low, the interaction with acids is unlikely to involve complete proton transfer. Hydrogen bond formation, combined with van der Waals interactions between host and guest, are probably responsible for the association. The initial concept of the 'double cavity' calixarene considered it to be a molecule in which the 'lower cavity' is established by the macrocyclic tetraamide ring, complex formation involving the insertion of the guest molecule through this annulus so as to bring its polar end proximate to the phenolic groups, as suggested in Fig. 9. The ^1H NMR spectrum of the 'double cavity' calixarene, however, shows a pair of resonances from N—H, and an NOE experiment indicates that one (or two?) of the amides is proximate to an OH group. A CPK model conformationally adjusted to accommodate these data shows a cavity not in the bottom of the molecule but at the side of the molecule. It is hoped that X-ray crystallography of the 'double cavity' calixarene will allow an accurate assignment of the detailed architecture of this molecule.

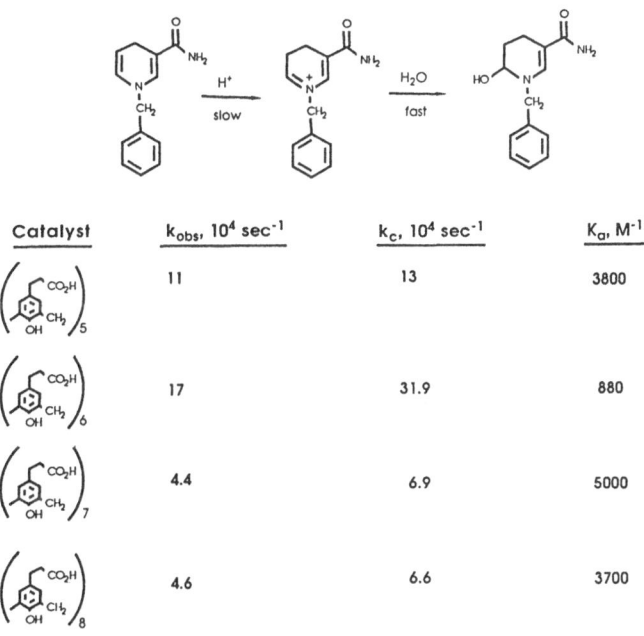

Catalyst	k_{obs}, 10^4 sec^{-1}	k_c, 10^4 sec^{-1}	K_a, M^{-1}
(CO₂H / CH₂ / OH)₅	11	13	3800
(CO₂H / CH₂ / OH)₆	17	31.9	880
(CO₂H / CH₂ / OH)₇	4.4	6.9	5000
(CO₂H / CH₂ / OH)₈	4.6	6.6	3700

Fig. 13. Catalysis of the hydration of a dihydropyridine.

7. Catalysis of the Hydration of a Dihydropyridine

Shinkai and coworkers [8] have prepared the water soluble hexasulfonate of calix[6]arene and have shown that it enhances the rate of the acid-catalyzed addition of water to N-benzyldihydronicotinamide. With the water soluble carboxy calixarenes now available by the procedure described in Section 2 it is of interest to compare these with the sulfonate and also to ascertain the effect of ring size on the catalysis. The observed rates of hydration, the values of the catalytic constant k_c, and the values of the association constant K_a for carboxycalix[n]arenes with $n = 5$–8 are shown in Figure 13. The greatest catalysis is observed for the cyclic hexamer, the rates dropping off as the ring is made smaller or larger than six aromatic moieties. Carboxycalix[4]arene shows no catalysis whatsoever. Although carboxycalix[6]arene shows a significant level of catalysis it is less effective than the sulfonatocalix[6]arene, for which Shinkai obtained a catalytic constant of 131 and an association constant of 564. The difference can probably be attributed to the greater acid strength of the arenesulfonate groups, in keeping with Shinkai's rationalization of the catalysis as a complexation-polarization phenomenon.

References

1. B. Dhawan, S. I. Chen, and C. D. Gutsche: *Makromol. Chem.* **188**, 921 (1987).
2. V. Böhmer, R. Funk, and W. Vogt: *Makromol. Chem.* **185**, 2195 (1984).
3. C. D. Gutsche, A. E. Gutsche, and A. I. Karaulov: *J. Incl. Phenom.* **3**, 447 (1985).
4. C. D. Gutsche and P. F. Pagoria: *J. Org. Chem.* **50**, 5795 (1985).
5. C. D. Gutsche, J. A. Levine, and P. K. Sujeeth: *J. Org. Chem.* **50**, 5802 (1985).
6. C. D. Gutsche and L.-g Lin: *Tetrahedron* **42**, 1633 (1986).
7. C. D. Gutsche, M. Iqbal, and I. Alam: *J. Am. Chem. Soc.* **109**, 4314 (1987).
8. S. Shinkai, S. Mori, H. Koreishi, T. Tsubaki, and O. Manabe: *J. Am. Chem. Soc.* **108**, 2409 (1986).

Journal of Inclusion Phenomena and Molecular Recognition in Chemistry 7 (1989), 73–81.

Studies Directed Toward the Fabrication of a Synthetic Cation-conducting Channel Based on Lariat Ethers: the Feeble Forces Concept for Self-assembly

GEORGE W. GOKEL*, LUIS ECHEGOYEN, MINSOOK KIM,
JEANETTE C. HERNANDEZ, and MAYRA DE JESUS
Department of Chemistry, University of Miami, Coral Gables, FL 33124, U.S.A.

(Received: 1 February 1988)

Abstract. The work described here derives from our observation that while Nature certainly utilizes covalent interactions and other forces of great strength in the construction of biomolecules, many 'feeble forces' are involved as well. We have used feeble forces in model studies directed eventually to the synthesis of a self-assembling, cation-conducting channel. Specifically, we have prepared three models for portions of such a cation-conducting channel. The first is based on the steroidal lariat ethers that clearly self-assemble. This process involves *inter alia* entropy as a feeble force of cumulative importance. Second, we have prepared three *tris*(macrocyclic) systems that are simplified versions of the proposed channel-former. Third, we have demonstrated the strength of carefully-conceived hydrogen bonding interactions by constructing a molecular box based on the interaction between adenine and thymine.

Key words. Lariat ether, self-assembly, channel, cation-conduction.

1. Introduction

During the past decade, numerous studies have been undertaken to develop synthetic ionophores that might permit cations or molecules to pass through a lipid bilayer [1]. Naturally-occurring gramicidin A is known to form transmembrane channels [2] and efforts to prepare a cation-conducting channel have been reported as well [3]. In our work, we have studied the selectivity of numerous crown ethers, lariat ethers [4, 5], and multi-armed versions of the latter [6]. Much has been learned about flexible ionophores and we have now attempted to utilize the concepts of flexibility and self-assembly to permit construction of a cation- or molecule-conducting channel.

Although Nature utilizes covalent bonds and rigid structural arrangements to establish steric relationships, weak or 'feeble' forces such as conformational change, hydrogen bonding, the formation of salt bridges, entropic forces, and even chirality play a major role in determining the three-dimensional structures of natural systems. In our approach to the self-assembly problem, we have utilized two of these feeble forces. By feeble forces, we mean interactions that individually contribute a small amount of energy, but in concert can lead to well-defined structural arrangements. Specifically, we have utilized entropic driving forces, to prepare novel cholesteryl lariat ether-based bilayers, and hydrogen bonding, to induce the formation of a molecular box. These are preliminary steps required for development of the cation channel molecule.

* Author for correspondence.

2. Experimental Methods

Melting points were determined on a Thomas-Hoover capillary device and are corrected. Infrared spectra were recorded as neat films on NaCl plates and are calibrated against the 1601 cm^{-1} band of polystyrene. Proton NMR spectra were recorded in CDCl$_3$ or DMSO-d_6 using internal Me$_4$Si as standard. Combustion analyses were performed by Atlantic Microlab, Atlanta, GA and molecular weights were determined using a Wescor 5100C vapor pressure osmometer. Vapor pressure osmometer readings were evaluated using calibration curves described in the text.

2.1. SYNTHESES OF tris(MACROCYCLIC) COMPOUNDS

To a stirred solution of N,N'-bis(6-chlorohexyl)-4,13-diaza-18-crown-6 (1 mmole) in MeCN (15 mL) was added 2 mmole of aza-3n-crown-n in which n = 4, 5, or 6 in 5 mL of MeCN. The solution was heated at reflux for 3 days. Workup involved extraction and chromatography over alumina (2-propanol : hexane mixtures as eluents).

2.2. N,N'-bis[6-(N-AZA-12-CROWN-4)HEXYL]-DIAZA-18-CROWN-6

The compound was obtained as described above (0.45 g, 58%). Anal. calcd. for C$_{40}$H$_{80}$N$_4$O$_{10}$: C, 61.81; H, 10.40%. Found: C, 60.58; H, 10.42%.

2.3. N,N'-bis[6-(N-AZA-15-CROWN-5)HEXYL]-DIAZA-18-CROWN-6

The compound was obtained as described above (0.53 g, 61%). IR: 3390, 2870, 1450, 1350, 1240, 1105 cm^{-1}. Anal. calcd. for C$_{44}$H$_{88}$N$_4$O$_{12}$ · H$_2$O: C, 59.82; H, 10.29%. Found: C, 59.88; H, 10.28%.

2.4. N,N'-bis[6-(N-AZA-18-CROWN-6)HEXYL]-DIAZA-18-CROWN-6

The compound was obtained as described above (0.64 g, 67%). Anal. calcd for C$_{48}$H$_{96}$N$_4$O$_{14}$: C, 60.46; H, 10.17%. Found: C, 60.21; H, 10.23%. Osmometric molecular weight (3 trials, 1, 2-dichloroethane), theory: 953. Found 988 (3.7% error).

3. Results and Discussion

3.1. SCHEMATIC FOR THE SYNTHESIS OF A CATION CHANNEL

The vast majority of cation transport model studies have been conducted using crown ethers as ionophores and bulk liquid membranes as bilayer models. Much information has been obtained about carrier-mediated transport [1, 7] but the mechanism for transport of molecules through biological membranes is predominantly of the channel type [8]. The synthesis of a cation-conducting channel is a daunting task. We felt that a suitable compound would require at least three basic features. First, it must be capable of insertion into a lipid bilayer. Second, it must span the membrane. Third it must have some residue integral to it that would exhibit a cation or molecule affinity. An additional consideration is that it must have polar groups positioned at distances appropriate for the relay of a cation through it, should the surface polar groups be spaced too far apart for a direct jump.

Based on our experience with macrocyclic polyether compounds, we reasoned that three parallel crown ether rings might form the basis of such a channel. As envisioned, two of the rings would be on the external and internal surfaces of the bilayer and a third would lie halfway between the other two. Since membrane thickness varies considerably, we based our first design on the well-characterized rat liver plasma membrane. It contains more than 50% phosphatidyl choline and nearly 20% cholesterol [9]. Phosphatidyl choline (PC) is derived from palmitic and oleic acids, each of which contains 16 carbons. The extended length of PC is thus about 26 atoms including the trimethylammonium and phosphate residues. A straight span of 26 carbon atoms would be about 40 Å long so the dimer of such a system falls well within the normal 60–100 Å membrane thickness. It should not be overlooked, however, that the crystal structure of coiled gramicidin A shows a surface-to-surface distance of only 32.5 Å [2]. The basic, schematic outline of a cation-conducting channel might thus be as shown in the figure below.

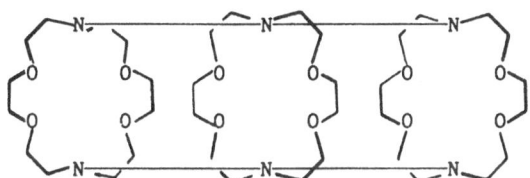

3.2. STEROIDAL LARIAT ETHERS

As noted above, cholesterol comprises nearly 20% of rat liver plasma membrane [9]. Its presence controls membrane fluidity and cholesterol is obviously quite lipophilic. It seemed reasonable to consider incorporating cholesterol directly into the spacer units, shown above by a simple, straight line. Since cholesterol exists in many membranes in free form, it must be a highly organizing molecule, capable of significantly stabilizing a self-assembling structure.

Entropy is the major force for membrane formation. If enough elements of a membrane itself were incorporated in a model cation-channel-former, the normally feeble entropic force might assist the assembly and/or insertion process. If so, then the compound illustrated schematically could be reduced by two covalent linkages to that shown below. It, in turn, could exist in a fully extended conformation as shown.

In principle, the equilibrium between the *tris*(macrocylic) compound shown overleaf and the folded form shown above it, would be driven *inter alia* by entropy. The credibility of such a proposition would be enhanced if a cholesteryl lariat ether, a compound having the macroring and spacer portions of the system shown, could be prepared. Such a structure has been synthesized [10] and its self-assembly into niosomes, in the absence of cations, and vesicles, in their presence, has now been demonstrated [11]. The self-assembling, steroidal lariat ethers, are based upon both carbon-pivot [5] and nitrogen-pivot [6] systems. The most versatile structure has thus far proved to be the steroidal ester of an N-glycyl-azacrown. The structure of a typical steroidal lariat ether compound is shown below.

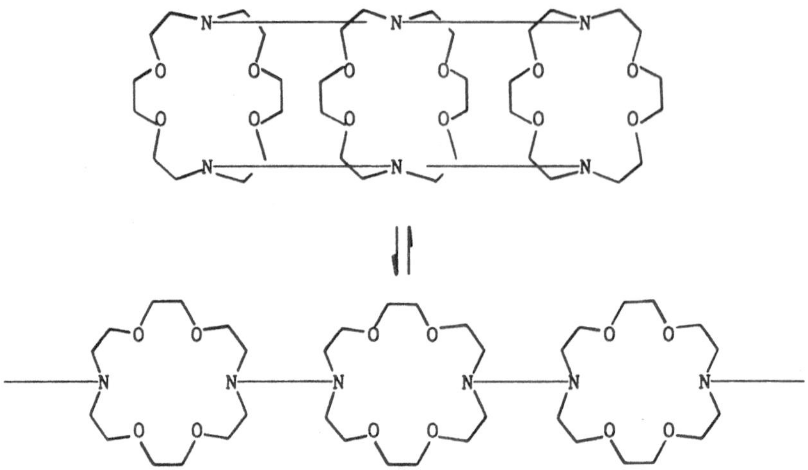

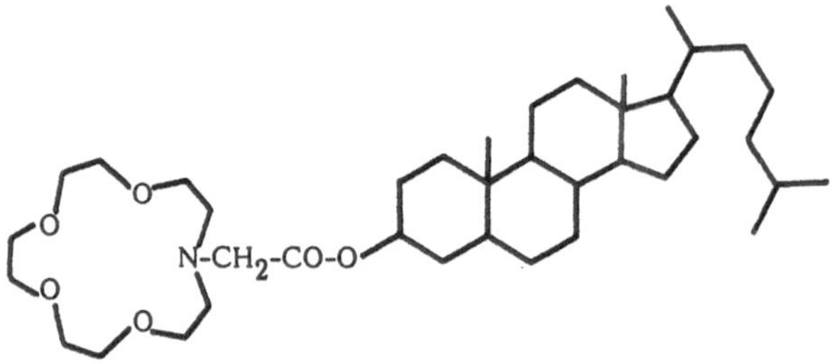

The self-assembly of steroidal lariat ethers bodes well for the possible self-assembly of a cation-conducting channel. Clearly, the entropic driving force, although feeble compared to a covalent bond, manifests itself with considerable vigor in this system. It is also interesting to note that ESR studies of these membrane systems utilizing a steroidal nitroxide spin label indicates that the resulting bilayers are very rigid (non-fluid) structures [12].

3.3. Tris(MACROCYCLIC) COMPOUNDS

The tris(macrocylic) compounds required for the present studies pose some significant difficulties. As models, somewhat simpler systems were prepared in two steps from diaza-18-crown-6. 1-Bromo-6-chlorohexane, 4,13-diaza-18-crown-6, and sodium carbonate were stirred at reflux in acetonitrile solution for 48 hours. N,N'-bis(6-chlorohexyl)-4,13-diaza-18-crown-6 was obtained as a colorless oil in

70% yield after chromatography. This dichloride could then be converted into compounds that we can illustrate schematically as O–O–O or o–O–o to indicate that the same or differently-sized macrocycles may be attached to each other. The former structure could represent three 18-membered rings while the latter could represent an array of three rings as follows: 12–18–12. A schematic representation of the channel-forming precursor containing three 18-membered rings might be –O–O–O–. The general synthetic sequence is straightforward and an example is shown below.

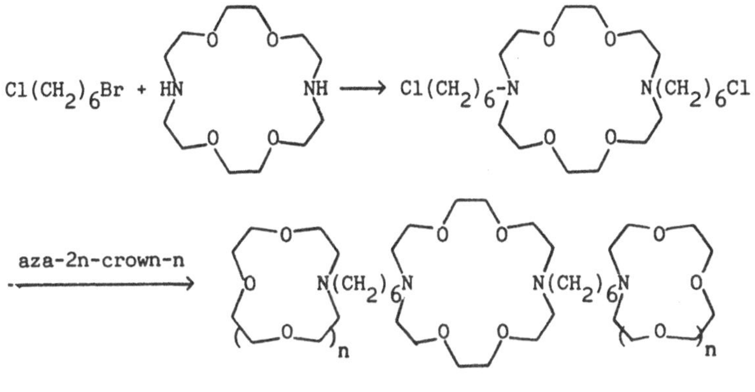

Of course, the molecules shown above lack two crucial features required for the envisioned channel formation in membranes. First, the interring spacers are far too short but represent, in most respects, the required spacers in a synthetically accessible form. Second, there is no second chain on the terminal (first and third) macrocycles. Nevertheless, the essential features of the desired system have been brought to hand and this shows, at least in general, synthetic feasibility.

3.4. HYDROGEN BONDING IN MODEL SELF-ASSEMBLY SYSTEMS

Hydrogen bonding is a well-known and much studied phenomenon [13]. In general, hydrogen bonds exhibit strengths from 1–6 kcal/mole although higher and lower values are known. In addition, they generally exhibit a predictable directionality. Although not specifically incorporated in the self-assembling, cation-conducting channel model illustrated schematically above, hydrogen bonding is a feeble force whose cumulative effect on the assembly process could be considerable. Our model to assess cumulative hydrogen bonding strength and to incorporate potential molecular recognition was based on the well-established base-pairing of purines and pyrimidines. In the matrix of DNA, two hydrogen bonds hold together adenine and thymine while cytosine and guanine base-pair use three such interactions. While the latter is more promising from the energetic point of view, synthetic access to model systems involving these two residues is less propitious.

We envisioned the formation of a molecular box that comprises essentially half (although not to scale) of the channel molecule shown above. We anticipated that a diaza-18-crown-6 macroring would be substituted by two sidearms, each termi-nated in adenine or thymine. Leonard and coworkers [14] had reported the syntheses of adenine- and thymine-terminated propyl chains so these were chosen as bibracchial lariat ether (BiBLE) [6] sidearms. Although the preparation of a single macrocycle having one adenine- and one thymine-terminated chain was the most intellectually pleasing prospect, its synthesis appeared beyond our grasp. Instead, two BiBLE subunits were prepared, each having two identical propyl sidearms terminated either in adenine or thymine. These may be represented schematically as A–O–A and T–O–T in which A and T stand for adenine and thymine respectively and O is 4,13-diaza-18-crown-6.

Once these compounds were brought to hand [15] their affinity for each other in aqueous solution was assessed using vapor pressure osmometry (VPO, see experimental section). A calibration curve was obtained using (65, 114, 155, 203, and 265 mmolar, 5 trials each) 18-crown-6 in water. The calibration curve was a straight line: $Y = 0.785X + 59.43$. Based on this, the apparent molecular weight of the particles in solution was assessed. When A–O–A and T–O–T (each approximately 100 mmolar) were dissolved in pure water, an apparent molecular weight of 771.2 was observed. The molecular weights of the individual species are: A–O–A, 612.84; T–O–T, 594. In the absence of any association, a simple average of 603 should be observed. The apparent molecular weight of 771.2 suggests that there is about 25% association. Thus the feeble hydrogen bonding forces compete quite effectively with water, a solvent of considerable organization and hydrogen bonding strength itself. The envisioned complex is illustrated below.

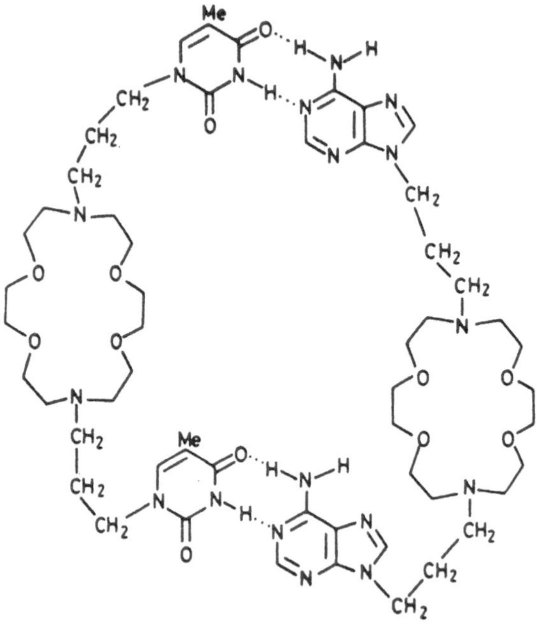

For comparison, we examined the behavior of N,N'-bis-(2-hydroxyethyl)-diaza-18-crown-6. This differs from the 'DNA-box' precursors in the chain length (two, rather than three, carbons) and in the terminal hydroxyl group. Even so, the systems are closely related and the hydroxyl function is known to be both a hydrogen-bond receptor and donor. Thus, its similarity to A–O–A or T–O–T is clear. In principle, a dimer of the type shown below could form.

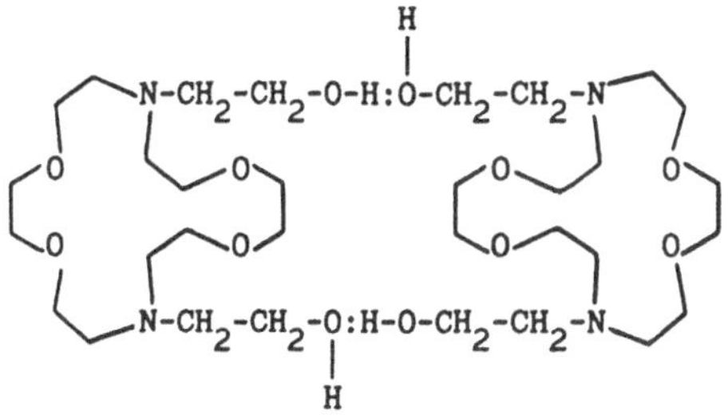

Studies similar to those described for the DNA-box (above) were undertaken using *N,N'-bis*-(2-hydroxyethyl)-diaza-18-crown-6 [16]. Thus, 12.5 mg of *N,N'-bis*-(2-hydroxyethyl)-diaza-18-crown-6 was dissolved in 0.6249 g of H_2O. Again using the 18-crown-6 calibration curve, an apparent molecular weight was determined. *N,N'-bis*-(2-hydroxyethyl)-diaza-18-crown-6 has a molecular weight of 350.52. The observed molecular weight of 346.38 differs from the expected weight for a completely monomeric system by less than 2%. Thus, the self-association of *N,N'-bis*-(2-hydroxyethyl)-diaza-18-crown-6 is $\leqslant 2\%$ and well within experimental error for a monomeric species. Compared to this, the association of A–O–A with T–O–T is substantial.

It occurred to us that the DNA-box is, in a sense, an induced-fit receptor. The affinity of primary ammonium salts for 18-membered crown ethers is well known [17] and the enthalpy of this weak interaction has been assessed at 2–3 kcal/mole in water [18]. Studies of CPK molecular models suggested that the macroring-to-macroring span across the DNA box could effectively accommodate 1,12-dodecane-diammonium salt. The ideal interaction would increase the number of hydrogen bonds from 4 to 10. Addition of 1,12-dodecanediammonium dichloride to a solution containing equimolar A–O–A and T–O–T, affords an apparent molecular weight of 1123 while the ternary complex should have a molecular weight of 1498. The meaning of these numbers is unclear. One might assume that the simple ratio $1123/1498 = 75\%$ suggests extensive association and, indeed, it may. Unfortunately, the problem is far too complicated to quote such a percentage based on current data.

The greatest difficulty is that any salt in the structure $ClH_3N(CH_2)_nNH_3Cl$ dissociates into three particles in water. Thus the apparent molecular weight of these species should be one-third of the expected value. Measurements of $ClH_3N(CH_2)_3NH_3Cl$, $ClH_3N(CH_2)_6NH_3Cl$, $ClH_3N(CH_2)_{10}NH_3Cl$, and $ClH_3N(CH_2)_{12}NH_3Cl$ all gave correct molecular weights when a calibration curve based on $ClH_3N(CH_2)_3NH_3Cl$ rather than 18-crown-6 was used. Thus correct (by a factor of 2.0–2.5 rather than the expected factor of 3) molecular weights were obtained for diammonium salts using the crown calibration curve and *vice versa*. In our original report of DNA box formation, we also noted that the system was, so to say, complex [15]. There are thus two essential difficulties in evaluating complex formation between ammonium salts and neutral species. First, complexes having either arrangements or stoichiometries other than the expected one may form. Second, the analytical method must be fully compatible with and respond similarly to all chemical species present. These problems do not affect the results for association between A–O–A and T–O–T in the absence of $ClH_3N(CH_2)_n NH_3Cl$. In the presence of the latter, conclusions must currently remain tentative.

In order to obtain some additional calibration on the association problem, we examined apparent molecular weights for mixtures of 4,10,16-triaza-18-crown-6 and $CH_3NH_3Cl^-$. Our assumption was that the ammonium salt should bind strongly to the triaza-macrocycle even in water because of the excellent threefold N—H—N complementarity. Equimolar amounts of triaza-18-crown-6 and $CH_3NH_3^+Cl^-$ in water should give an apparent molecular weight (by VPO) of 328.8 D if fully associated. Using the 18-crown-6 calibration curve to evaluate the VPO readings, the apparent molecular weight was 119 ± 5 D. When the independently determined calibration curves for triaza-18-crown-6 or methylammonium chloride were used, the apparent molecular weights were, respectively, 216 ± 10 D and 216 ± 10 D. The difference of the above results from those obtained using the 18-crown-6 calibration curve is less troubling than the similarity of the latter two values. We have thus obtained clear evidence of association driven, at least in part, by hydrogen bonding. A quantitative evaluation of the association remains elusive.

3.5. SUMMARY

The basis of the effort described here is our observation that while Nature certainly uses covalent interactions and other forces of great strength, many feeble forces are involved as well. We have attempted to utilize these feeble forces in model studies directed eventually to the synthesis of a self-assembling, cation-conducting channel. In particular, we have prepared three types of models for such a cation-conducting channel. The first is the steroidal lariat ethers that clearly self-assemble. This self-assembly involves entropy as a feeble force of cumulative importance. Second, we have prepared three *tris*(macrocyclic) systems that are simplified versions of the desired channel-former. Third, we have demonstrated the strength of carefully-conceived hydrogen bonding interactions by constructing a molecular box based on hydrogen bonds formed between adenine and thymine.

Acknowledgement

We warmly thank the National Institutes of Health for a grant (GM 36262) that supported most of the work reported herein.

References

1. J. D. Lamb, R. M. Izatt, and J. J. Christensen: *Progr. Macrocyclic Chem.* **1**, 1 (1979).
2. (a) R. Sarges and B. Witkop: *J. Am. Chem. Soc.* **86**, 1862 (1964); (b) S. B. Hladky and D. A. Haydon: *Nature* **225**, 451 (1970); (c) D. W. Urry: *Proc. Nat. Acad. Sci. USA* **68**, 672 (1971); (d) D. W. Urry, M. C. Goodall, J. D. Glickson, and D. F. Mayers: *Proc. Nat. Acad. USA* **68**, 1907 (1971); (e) E. Bamberg and P. Laeuger: *J. Membrane Biol.* **35**, 351 (1977); (f) R. E. Koeppe II, K. O. Hodgson, and L. Stryer: *J. Mol. Biol.* **121**, 41 (1978); (g) R. E. Koeppe II, J. M. Berg, K. O. Hodgson, and L. Stryer: *Nature* **279**, 723 (1979).

3. J.-M. Lehn: *Science* **227**, 849 (1985).
4. (a) G. W. Gokel, D. M. Dishong, and C. J. Diamond: *J. Chem. Soc. Chem. Commun.* 1053 (1980); (b) D. M. Dishong, C. J. Diamond, M. I. Cinoman, and G. W. Gokel: *J. Am. Chem. Soc.* **105**, 586 (1983).
5. (a) R. A. Schultz, D. M. Dishong, and G. W. Gokel: *Tetrahedron Lett.* 2623 (1981); (b) R. A. Schultz, D. M. Dishong, and G. W. Gokel: *J. Am. Chem. Soc.* **104**, 625 (1982); (c) R. A. Schultz, E. Schlegel, D. M. Dishong, and G. W. Gokel: *J. Chem. Soc. Chem. Commun.* 242 (1982); (d) B. D. White, D. M. Dishong, C. Minganti, K. A. Arnold, D. M. Goli and G. W. Gokel: *Tetrahedron Letters* 151 (1985); (e) R. A. Schultz, B. D. White, D. M. Dishong, K. A. Arnold, and G. W. Gokel: *J. Am. Chem. Soc.* **107**, 6659 (1985).
6. (a) V. J. Gatto and G. W. Gokel: *J. Am. Chem. Soc.* **106**, 8240 (1986); (b) V. J. Gatto, K. A. Arnold, A. M. Viscariello, S. R. Miller, and G. W. Gokel: *J. Org. Chem.* **51**, 5373 (1986); (c) V. J. Gatto, K. A. Arnold, A. M. Viscariello, S. R. Miller, and G. W. Gokel: *Tetrahedron Lett.* 327 (1986); (d) D. A. Gustowski, V. J. Gatto, J. Mallen, L. Echegoyen, and G. W. Gokel: *J. Org. Chem.* **52**, 5172 (1987); (e) B. D. White, K. A. Arnold, and G. W. Gokel: *Tetrahedron Lett.* 1749 (1987); (f) B. D. White, F. R. Fronczek, R. D. Gandour, and G. W. Gokel: *Tetrahedron Lett.* 1753 (1987).
7. (a) T. M. Fyles: *J. Chem. Soc., Faraday Trans. 1* **82**, 617 (1986). (b) T. M. Fyles: *Can. J. Chem.* **65**, 884 (1987).
8. (a) L. Stryer: *Biochemistry*, Second edition, Freeman, San Francisco, 1981, p. 861; (b) I. C. West: *The Biochemistry of Membrane Transport*, Chapman and Hall, London, 1983; (c) W. D. Stein: *Transport and Diffusion across Cell Membranes*, Academic Press, New York, 1986.
9. E. Sim: *Membrane Biochemistry*, Chapman and Hall, London, 1982, p. 26.
10. G. W. Gokel, J. C. Hernandez, A. M. Viscariello, K. A. Arnold, C. F. Campana, L. Echegoyen, F. R. Fronczek, R. D. Gandour, C. R. Morgan, J. E. Trafton, C. Minganti, D. Eiband, R. A. Schultz, and M. Tamminen: *J. Org. Chem.* **52**, 2963 (1987).
11. (a) L. E. Echegoyen, J. C. Hernandez, A. Kaifer, G. W. Gokel, and L. Echegoyen: *J. Chem. Soc., Chem. Commun.* 836 (1988); (b) L. E. Echegoyen, L. Portugal, S. R. Miller, J. C. Hernandez, L. Echegoyen, and G. W. Gokel: *J. Chem. Soc., Chem. Commun.*, submitted.
12. L. E. Echegoyen: unpublished results.
13. (a) P. Schuster and G. Zundel: C. Sandorfy; (Eds): *The Hydrogen Bond, Part II. Structure and Spectroscopy*, North Holland Publishing Compnay, Amsterdam, 1976; (b) J. N. Israelachvilli: *Intermolecular and Surface Forces*, Academic Press, London, 1985.
14. (a) N. J. Leonard, T. G. Scott, and P. C. Huang: *J. Am. Chem. Soc.* **89**, 7137 (1967); (b) S. T. Browne, J. Eisinger, and N. J. Leonard: *J. Am. Chem. Soc.* **90**, 7302 (1968).
15. M. Kim and G. W. Gokel: *J. Chem. Soc., Chem. Commun.* 1686 (1987).
16. (a) K. A. Arnold, L. Echegoyen, and G. W. Gokel: *J. Am. Chem. Soc.* **109**, 3713 (1987): (b) K. A. Arnold, L. Echegoyen, F. R. Fronczek, R. D. Gandour, V. J. Gatto, B. D. White, and G. W. Gokel: *J. Am. Chem. Soc.* **109**, 37176 (1987).
17. C. J. Pedersen: *J. Am. Chem. Soc.* **89**, 7017 (1967).
18. R. M. Izatt, R. E. Terry, B. L. Haymore, L. D. Hansen, N. K. Dalley, A. G. Avondet, and J. J. Christensen: *J. Am. Chem. Soc.* **98**, 7620 (1976).

Journal of Inclusion Phenomena and Molecular Recognition in Chemistry 7 (1989), 83–90.

Cation Selective Complexation-Coloration with Chromophoric Crowns

SOICHI MISUMI and TAKAHIRO KĀNEDA
The Institute of Scientific and Industrial Research, Osaka University, Ibaraki, Osaka 567, Japan

(Received: 1 February 1988)

Abstract. Amine selective complexation-coloration of some azophenol-dyed crowns and X-ray structures of two secondary amine complexes are reported as well as lithium ion specific coloration with an azophenol spherand.

Key words. Azophenol crown ether, cation, lithium.

1. Introduction

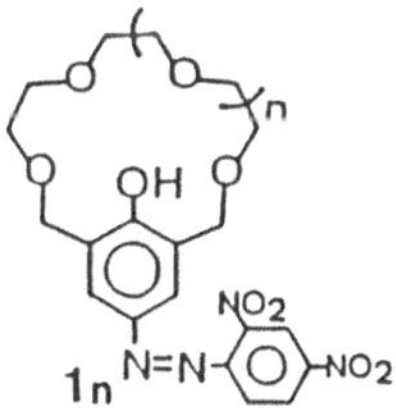

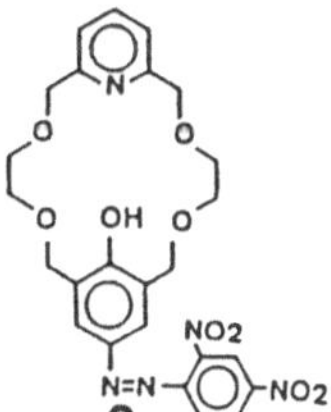

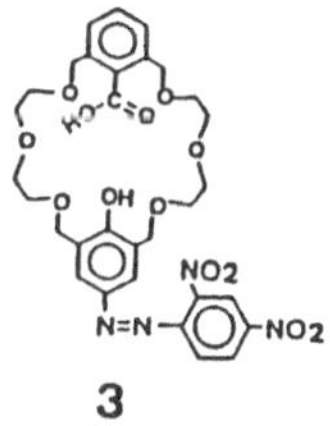

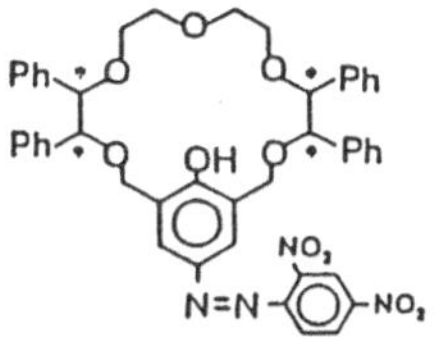

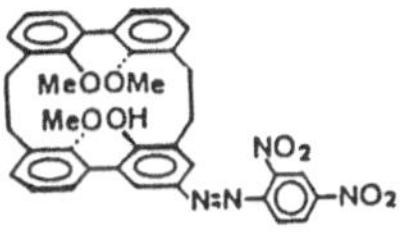

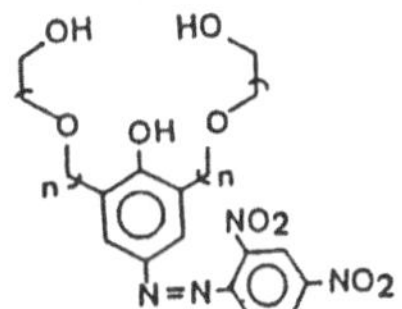

The design and synthesis of a series of ligands which can selectively bind a given metal or organic ion and undergo concurrently a color change are   subjects in host-guest chemistry which have been highlighted recently [1]. Previously we reported lithium ion selective coloration with azophenol-dyed crowns [2], fluorescent emission with benzothiazolylphenol crowns [3], and their application to lithium ion analysis. We attempted the structural recognition of protonated amines by the use of azophenol crowns 1–4, spherand 5, and podand 6, which have a phenolate anion in the center. The colored complexes, 7 and 9, of these crowns with metal or ammonium ion are differentiated from simple ion-dipole type complexes by an additional binding force, that is, coulombic interaction between phenolate anion and the guest cation. Now we report amine-selective and enantiomer-selective complexation-coloration with azophenol crowns as well as lithium ion specific coloration with a spherand analog.

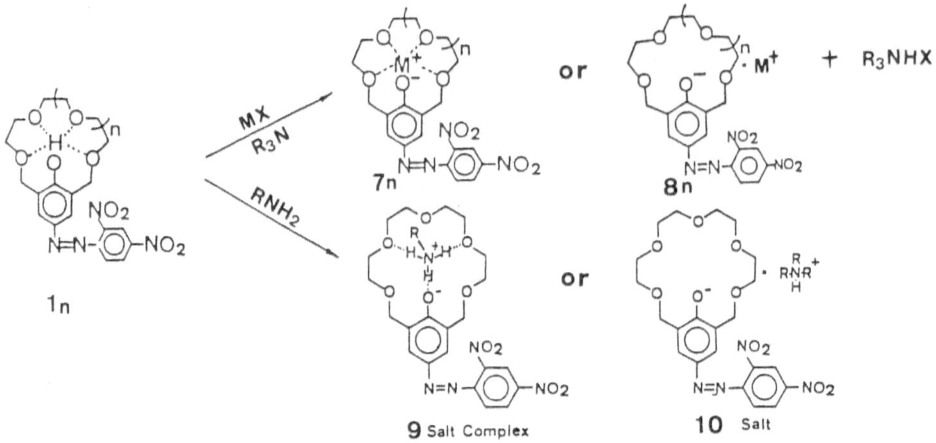

2. Discussion

(2.1) With the intention of achieving much higher selectivity for lithium ion complexation, a spherand type host **5** containing a dinitrophenylazophenol moiety was designed and synthesized by the use of a key step, photodeselenation which we developed previously (Scheme 1) [5].

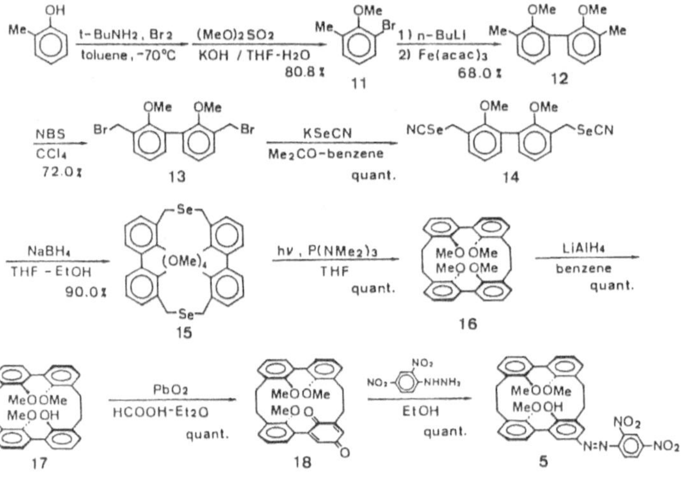

Scheme 1.

In chloroform, no phenolate anion of **5** was detected even in the presence of excess piperidine as a base. When crystalline lithium salts were added to this solution, a dramatic color change from yellow to violet took place rapidly, except for nitrate, fluoride, and sulfate. This phenomenon indicates evidently the formation of the lithium phenolate. On the other hand, no tendency of the interaction was observed with any of the other 58 metal salts listed in Fig. 1.

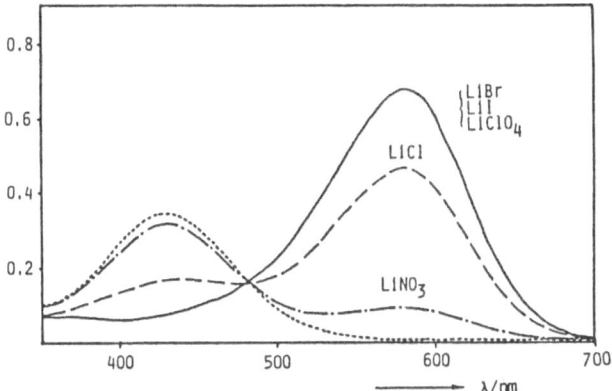

Fig. 1. Visible spectra of spherand-salt-piperidine systems in $CHCl_3$. - - - - : Salt-free, LiF, Li_2SO_4, $BeCl_2$, MX (M = Na, K, Rb, Cs; X = F, Cl, Br, I, ClO_4, NO_3, SO_4), MX_2 (M = Mg, Ca, Sr, Ba; X = Cl, Br, ClO_4), MX_2 (M = Mn, Co, Ni, Zn, Cd, Sn, Hg; X = Cl or Br), MCl_3 (M = Al, Cr, Fe, Sb, Ce, Bi), $Pb(OAc)_2$, $AgClO_4$.

In the hydrophobic solvent, the lithium ion of the colored species must be accommodated in the hydrophilic cavity of the host anion on the basis of the following facts: (i) the cavity is expected to be small but able to accommodate the guest, (ii) the coloration takes place only when the specimen is in contact with Li salts, (iii) the monodemethylation of spherand 16 to 17 occurs exclusively in aprotic solvent such as benzene even in the presence of excess $LiAlH_4$, (iv) no cation exchange between Li^+ and Ca^{2+} or Ba^{2+} was observed.

The observed perfect lithium selectivity under the given conditions is described in terms of binding the guest ion by coulombic attraction and ion–dipole interactions and rejecting larger or multivalent cations by the steric effect caused by the narrow entrances to the small cavity in the spherand 5.

(2.2) Another type of complexation with crown ethers is concerned with a tripod arrangement by three ether oxygens of the crown ether. We recently observed such a perching type complexation of protonated amines by the use of the above-stated azophenol crowns.

The azophenol crown 1 does not dissociate in chloroform, but shows the formation of the phenolate-ammonium complex in the presence of amines in their absorption spectra. Thus the formation of the salt complex 9 with amine is strongly dependent upon the structure of the amines but not their basicity. For example, absorption maxima of 1 ($n = 1$–3)-amine complexes appear in the shorter wavelength region compared to those of acyclic reference azophenols like A and B in Figure 2. The figure also shows that acyclic hosts reveal nearly constant values of the maxima regardless of the amine species. The complexes of dimethylamine show, in general, blue shifts of their maxima in any solvent. On the other hand, 1_1 shows larger absorbances of the salt complexes with relatively less bulky amines compared to those of bulky amines. The 1_1-piperidine (1:1) complex was purified and determined by X-ray analysis to be a typical perching type structure where the whole chromophore is coplanar and nearly perpendicular to the crown ring (Figure 3) [5].

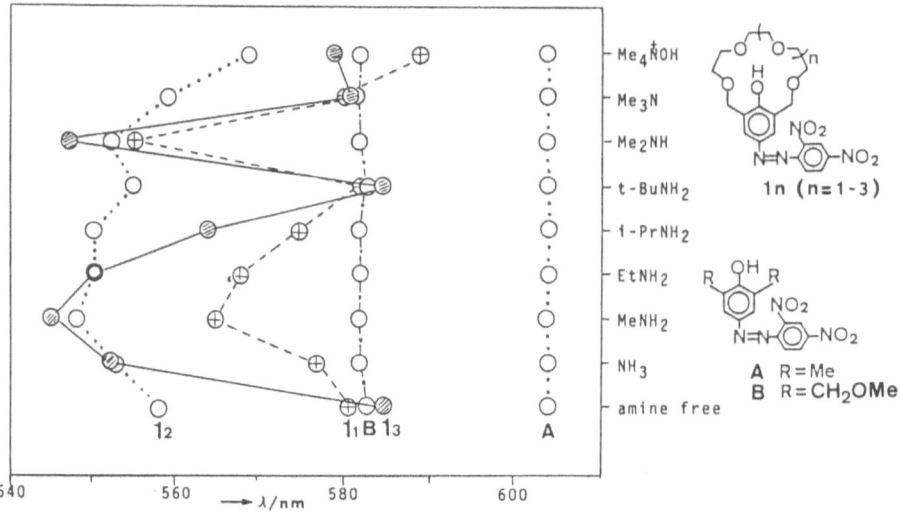

Fig. 2. Absorption maxima of azophenol dye–amine systems in EtOH.

Next, the complexation of azophenol crown **3** containing a benzoic acid moiety
was studied. The dye **3** is also yellow (λ_{max} 400 nm) in chloroform and shows a color
change to blue by the addition of monoamines but a change to pink by diamines,
indicating a remarkably different color based on the two ionic binding sites,
carboxylate and phenolate anions, in the crown cavity.

Such a coloration (λ_{max} 535–800 nm) of dye crown **3**–amine complexes is related
to the structure of the complex and not strongly to the relative basicity of the

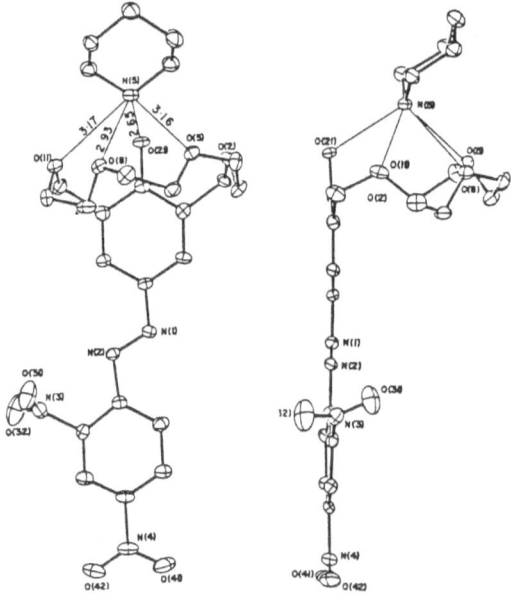

Fig. 3. ORTEP views of $\mathbf{1_1}$–piperidine (1 : 1) complex.

amines. From the study of the relationship between the positions of absorption maxima and the concentrations of amines, it was observed that the dye **3** forms a 1 : 1 complex with some equivalents of diamine (0.5–500 eq. in the case of piperazine) and a 1 : 2 complex with a large excess of diamine (more than 500 eq. for piperazine) (Figure 4). Especially, piperazine forms a 1 : 1 complex in quantitative yield.

The molecular structure of the complex was determined by X-ray crystallography. As shown in Figure 5, the diprotonated piperazine is sandwiched between phenolate and benzoate planes in the chair form, and is bound by markedly short hydrogen bonds. This is the first example of the X-ray analysis of a crown ether-sec. amine complex, to our knowledge, and the figure indicates an intercharge hydrogen bond $(O^- \cdots H-N^+)$ to be useful to bind the guest ion as well as ion–dipole interaction and hydrogen bonding.

(2.3) As an application of amine selective complexation, it was observed that the substitution pattern of amines, i.e. primary, secondary, and tertiary, may be discriminated by combined use of two azophenol hosts, pyridine-O_4-crown **2** and O_4-podand **6** $(n = 1)$.

Dye crown **2** reacts with all kinds of amines in acetonitrile to form ammonium phenolates. The absorption maxima of **2**-prim. amine systems appear in the region of 574–586 nm, which is definitely distinct from those of sec. and tert. amine systems (λ_{max} 602–606 nm except for dimethylamine, 592 nm) as shown in Figure 6.

On the other hand, azophenol O_4-podand **6** $(n = 1)$ shows the complexation-coloration with prim. and sec. amines, included sterically bulky amines such as t-BuNH$_2$, $(i$-Pr)$_2$NH, and 2,2,6,6-tetramethylpiperidine. Consequently, the combined use of two types of azophenol hosts, **2** and **6**, is highly useful to discriminate the substitution pattern of amines by means of coloration. The two end hydroxyl groups in the dyed podands are required for such discrimination of the amine pattern because no complexation-coloration is observed by addition of various amines when the hydroxyl groups are substituted with methoxyl groups.

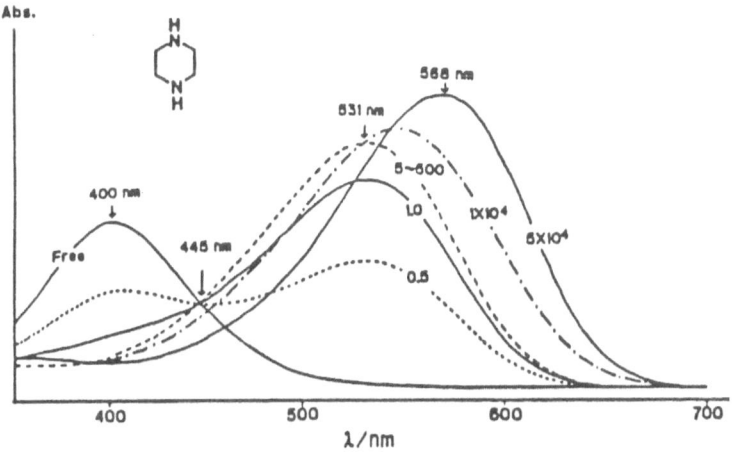

Fig. 4.   Visible spectra of dye **3**-piperazine system in CHCl$_3$.

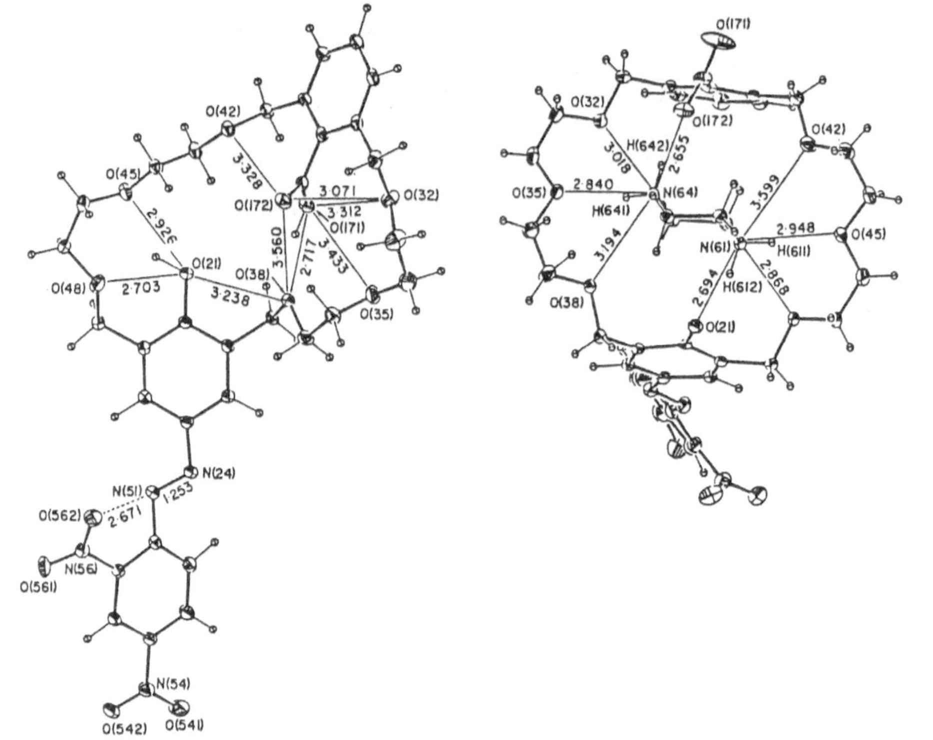

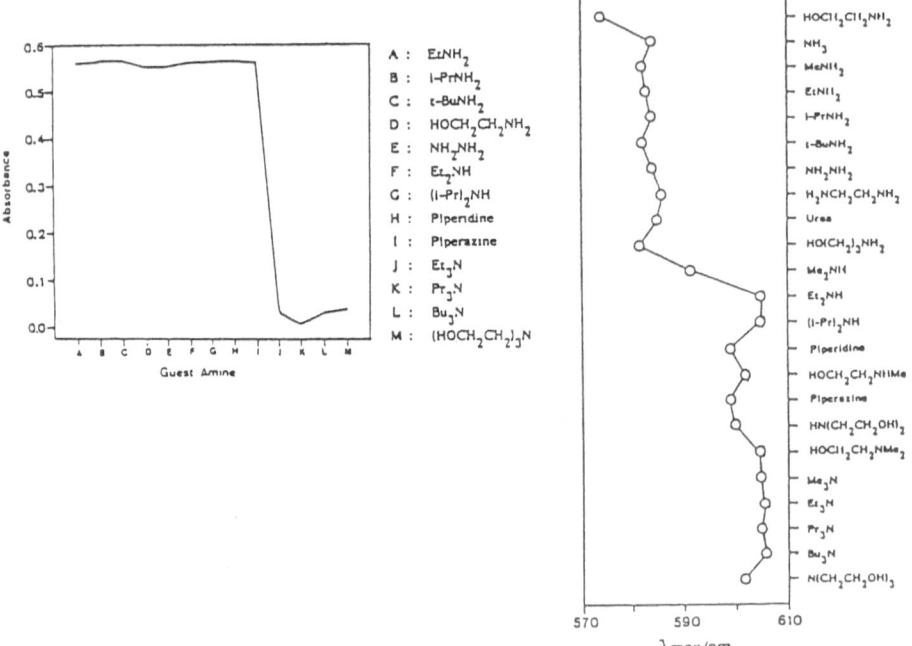

Fig. 5.   ORTEP views of dye **3** (left) and **3**-piperazine (1 : 1) complex (right).

A : EtNH₂
B : I-PrNH₂
C : t-BuNH₂
D : HOCH₂CH₂NH₂
E : NH₂NH₂
F : Et₂NH
G : (I-Pr)₂NH
H : Piperidine
I : Piperazine
J : Et₃N
K : Pr₃N
L : Bu₃N
M : (HOCH₂CH₂)₃N

Fig. 6.   Absorbances at absorption maxima of **6₁**-amine complexes in CHCl₃ (left) and absorption maxima of **2**-amine complexes in CH₃CN (right).

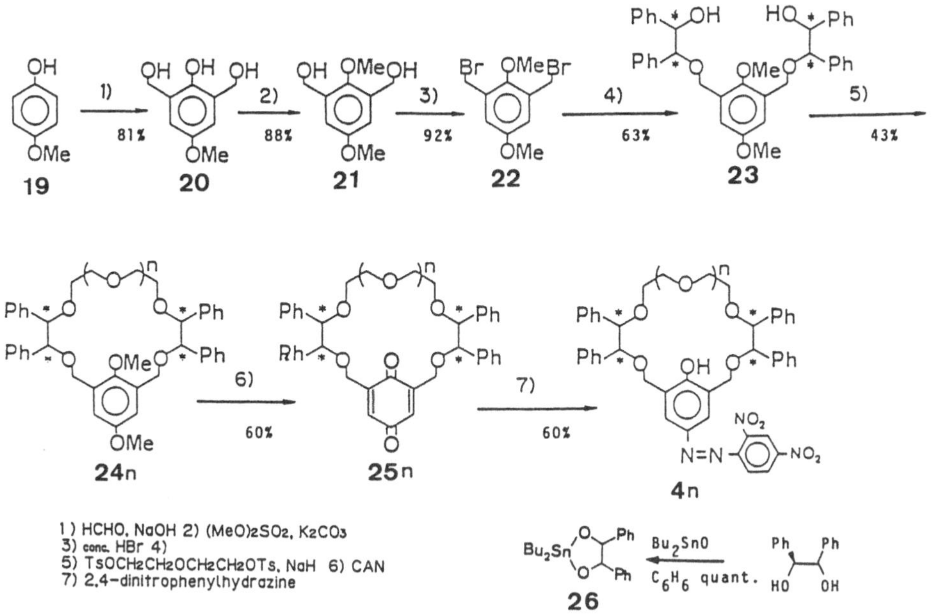

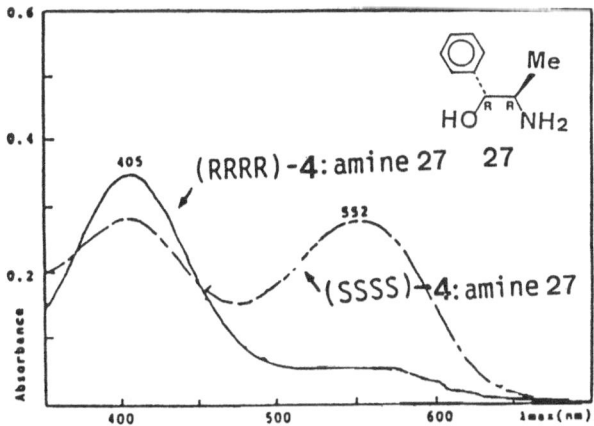

Scheme 2.

(2.4) Enantiomer selective coloration of optically active amines, our important project, was realized by chiral azophenol crown **4** incorporated with two units of optically active hydrobenzoin. The synthetic route is shown in Scheme 2. Reaction of 2,6-bis(bromomethyl)-1,4-dimethoxybenzene **22**, which is derived from hydroquinone monomethylether **19** by a three-step procedure, with the dibutyltin derivative **26** of optically active dihydrobenzoin gives optically active podand **23** in 63% yield. Cyclization of **23** with the ditosylate of polyethylene glycol, followed by oxidation with ceric ammonium nitrate (CAN) and treatment with dinitrophenylhydrazine, affords the desired chiral azophenol crowns **4n**.

Fig. 7. Absorption spectra of dye **4**-norpseudoephedrine in chloroform.

Table I. Wavelengths (nm) and absorbances of absorption maxima in chloroform.

	4		28	
	RRRR	**SSSS**	**RRRR**	**SSSS**
(R)-29 Me⟨Ph / NHz	567.9 0.5105	571.9 0.3605	573.9 0.5548	576.9 0.445
(S)-29 Me⟨Ph / NHz	571.9 0.3860	567.0 0.5631	578.9 0.4666	573.9 0.5248
(R)-30 Me⟨1-Naph / NHz	566.9 0.4190	569.9 0.2530	569.9 0.5012	577.0 0.3744
(S)-30 Me⟨1-Naph / NHz	572.0 0.2475	566.9 0.4454	576.9 0.3984	569.8 0.4864
27 Ph⟩—⟨Me / HO NHz	561.0 0.2566	553.0 0.5820	570.0 0.4341	553.0 0.5726

4

28

Of several active monoalkylamines and ethanolamines, a selective coloration with norpseudoephedrine, $Ph(OH)CH-CH(NH_2)CH_3$, is described as a typical example. The examination with CPK molecular models shows that norpseudoephedrine **27** having the R,R-configuration may form a more stable complex with $(SSSS)$-dye **4**, compared to that with $(RRRR)$-dye **4**. In fact, $(RRRR)$-**4** in chloroform was kept yellow by addition of the amine in a range of concentrations, 4.8×10^{-7} to 1.2×10^{-6} M, whereas $(SSSS)$-**4** revealed a color change of the solution from yellow to reddish violet with the same concentration of the amine (Figure 7). We also observed a few examples of enantiomer selective coloration with optically active amines and some chiral dyed crowns as shown in Table I. Further study with various amines is now in progress.

References

1. For review: S. Misumi and T. Kaneda: *Mem. Inst. Sci. Ind. Res., Osaka Univ.* **44**, 29 (1987).
2. (a) T. Kaneda, K. Sugihara, H. Kamiya, and S. Misumi: *Tetrahedron Lett.* **22**, 4407 (1981); (b) K. Sugihara, T. Kaneda, and S. Misumi: *Heterocycles* **18**, 57 (1982); (c) K. Nakashima, S. Nakatsuji, S. Akiyama, T. Kaneda, and S. Misumi: *Chem. Lett.* 1781 (1982): *Chem. Pharm. Bull.* **34**, 168 (1986).
3. (a) I. Tanigawa, K. Tsuemoto, T. Kaneda, and S. Misumi: *Tetrahedron Lett.* **25**, 5327 (1984); (b) K. Nakashima, S. Nakatsuji, S. Akiyama, I. Tanigawa, T. Kaneda, and S. Misumi: *Talanta* **31**, 749 (1984); (c) K. Nakashima, Y. Nagaoka, S. Nakatsuji, T. Kaneda, I. Tanigawa, K. Hirose, S. Misumi, and S. Akiyama: *Bull. Chem. Soc. Jpn.* **60**, 3219 (1987).
4. T. Kaneda, S. Umeda, H. Tanigawa, S. Misumi, Y. Kai, H. Morii, K. Miki, and N. Kasai: *J. Am. Chem. Soc.* **107**, 4802 (1985).
5. Y. Kai, N. Kasai, T. Kaneda, and S. Misumi: Abstract of 56th Annual Meeting of Jpn. Chem. Soc. Tokyo (1988), II, p. 1609.

Journal of Inclusion Phenomena and Molecular Recognition in Chemistry 7 (1989), 91–97.

Molecular Recognition of Hydrophobic Ammonium Substrates by a Cationic Octopus Cyclophane Bearing Noncovalently Bound Pyridoxal-5'-phosphate: A Vitamin B₆-dependent Holoenzyme Model

YUKITO MURAKAMI*, JUN-ICHI KIKUCHI, and OSAMU HAYASHIDA
Department of Organic Synthesis, Faculty of Engineering, Kyushu University, Fukuoka 812, Japan

(Received: 1 February 1988)

Abstract. The inclusion behavior of the octopus cyclophane constructed with a rigid macrocyclic skeleton and eight hydrocarbon chains was studied in aqueous media by means of fluorescence and electronic absorption spectroscopy. Both hydrophobic and electrostatic interactions came into effect in the host–guest complexation process. The cyclophane acted as an effective apoenzyme model for constitution of an artificial vitamin B₆-dependent holoenzyme by simultaneous incorporation of pyridoxal-5'-phosphate and a hydrophobic alkylammonium substrate into the host cavity to give the Schiff-base species, showing the substrate selectivity.

Key words. Cyclophane, pyridoxal-5'-phosphate, Schiff-base, vitamin B₆, hydrophobic interaction, enzyme model.

1. Introduction

Naturally occurring hosts such as enzymes and receptors selectively recognize guest molecules through various interaction modes of both rigid and flexible characters as explained on the basis of the lock-and-key and induced-fit concepts, respectively. In order to mimic extensive functions exerted for molecular recognition in biological systems, cyclophanes having an intramolecular hydrophobic cavity have been widely utilized as artificial host molecules [1]. While most of the cyclophanes previously designed provide a relatively rigid recognition site for hydrophobic guest molecules, we have recently developed so-called octopus cyclophanes, which are capable of providing a large and hydrophobic binding site constructed with a rigid macrocyclic skeleton and flexible hydrocarbon chains in aqueous media [2].

The following characteristic inclusion behavior of the octopus cyclophane has been clarified [2, 3]. (i) The host exercises molecular discrimination toward guests through hydrophobic and electrostatic interactions and strongly binds hydrophobic guest molecules. (ii) The hydrophobic binding site of the octopus cyclophane is highly apolar and acts to repress the molecular motion of guest molecules. (iii) Formation of both 1:1 and 1:2 host–guest complexes is remarkably favored due to the induced-fit binding capability of the host molecule. (iv) The host–guest complex formed with the octopus cyclophane and a hydrophobic vitamin B₁₂ derivative acts as an effective vitamin B₁₂-dependent holoenzyme model.

---

* Author for correspondence.

In order to get further insight into the inclusion behavior of the octopus cyclophane, we now prepared a new water-soluble octopus cyclophane and studied its guest-binding behavior for constitution of a vitamin B_6-dependent holoenzyme model.

2. Experimental

2.1. MATERIALS

Preparation of an octopus cyclophane, N,N',N'',N'''-tetrakis[3-(N,N-ditetradecyl-carbamoyl)-3-(trimethylammonio)acetamidopropanoyl]-2,11,20,29-tetraaza[3.3.3.3]-paracyclophane tetrabromide (**1**), is to be reported elsewhere [4]; a pale yellow solid, m.p. 136–138°C. Anal. Calcd. for $C_{180}H_{324}Br_4N_{16}O_{12} \cdot 2\,H_2O$: C, 66.31; H, 10.14; N, 6.87%. Found: C, 66.35; H, 9.90; N, 6.85%. All of the guest molecules employed in this work were obtained from commercial sources and used without further purification, as guaranteed reagents. N-Phenyl-1-naphthylamine (PNA) was recrystallized from methanol–water (4 : 1 v/v), m.p. 61–62°C.

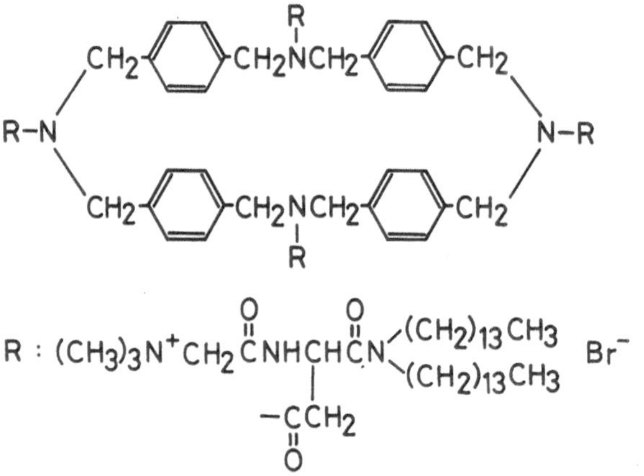

2.2. MEASUREMENTS

Elemental analyses were performed at the Microanalysis Center of Kyushu University. Fluorescence and electronic absorption spectra were taken on a Hitachi 650–40 fluorescence spectrometer and a Hitachi 220A spectrophotometer, respectively. The critical aggregate concentration (cac) was determined by surface tension measurements on a Shimadzu ST–1 surface tensometer assembled by the Wilhelmy principle: 2.5×10^{-4} mol dm^{-3} for **1** in aqueous media at room temperature. Thus, the cyclophane concentration was maintained in a range below the cac value for all measurements on the host–guest interactions.

3. Results and Discussion

3.1. CONSTITUTION OF VITAMIN B$_6$-DEPENDENT HOLOENZYME MODEL

The octopus cyclophane (**1**) showed inclusion behavior toward hydrophobic fluorescent guests in aqueous media, in a similar manner as observed with other octopus cyclophanes previously prepared [2]. For example, the anionic 8-anilinonaphthalene-1-sulfonate (ANS) was incorporated into **1** in a 1:1 stoichiometry with a large binding constant: K_1, 5.3×10^5 dm^3 mol^{-1} in an aqueous 2-[4-(2-hydroxyethyl)piperazin-1-yl]ethanesulfonate (HEPES) buffer [0.01 mol dm^{-3}, pH 8.0, μ 0.10 (KCl)] at 30.0°C. The microenvironmental polarity of the cyclophane cavity, as evaluated from the fluorescence maximum of ANS (λ_{max}, 467 nm), is equivalent to that provided by 2-propanol; $E_T(30)$, 49 kcal mol^{-1} [2]. While N-phenyl-1-naphthylamine (PNA) as a nonionic guest was also incorporated into the hydrophobic cavity of **1** [K_1, 1.0×10^5 dm^3 mol^{-1}; $E_T(30)$, 34 kcal mol^{-1} (λ_{max}, 406 nm)], cationic **1** does not show any binding affinity toward a cationic guest, 1-dimethylaminonaphthalene-5-sulfonamidoethyltrimethylammonium. The results indicate that **1** is a potent hydrophobic host exhibiting molecular discrimination toward guests as originated in the electrostatic effect.

On these grounds, a host–guest complex is expected to be formed between **1** and pyridoxal-5'-phosphate (PLP) under the conditions that PLP behaves as an anionic species. Such a host–guest interaction was examined by electronic absorption spectroscopy in an aqueous HEPES buffer (0.02 mol dm^{-3}, pH 7.0) at 30.0°C.

In the absence of the octopus cyclophane, PLP (1.0×10^{-4} mol dm^{-3}) shows an absorption maximum at 388 nm which originates from the dianionic species (A in Scheme 1) [5]. Addition of **1** to this solution resulted in a red shift of the absorption maximum along with concomitant decrease in its intensity. The isosbestic point was observed at 400 nm by changing the concentration of **1**, and the maximum wavelength was shifted to 392 nm at the host concentration of 2.0×10^{-4} mol dm^{-3}. This spectral change reflects the binding capability of the cationic host toward PLP. We have clarified that an analogous cationic octopus cyclophane is in favor of forming a 1:2 host–guest complex with nonionic guest molecules, but not so with anionic ones. This is due to mutual electrostatic repulsion between the latter guest molecules exercised in the hydrophobic cavity of the host molecule [2]. Thus, a binding constant for the interaction of **1** with PLP was evaluated on the basis of the 1:1 complex formation with reasonable reliability; 6.5×10^4 dm^3 mol^{-1}. Since the complexation was inhibited as the ionic strength of the solution was increased, the electrostatic effect is predominant in the present host–guest interaction. In addition, the octopus cyclophane provides a relatively

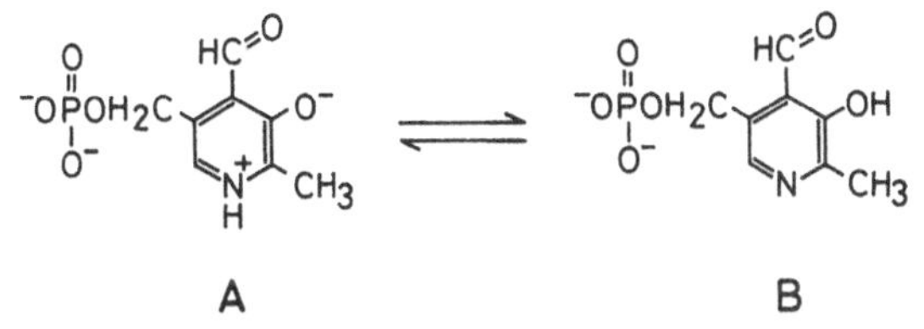

Scheme 1.

polar binding site for the hydrophilic PLP molecule because an absorption band
due to the uncharged tautomer with respect to the pyridine ring (B in Scheme 1),
which is expected to appear at ca. 350 nm [5], was not detected.

3.2. MOLECULAR RECOGNITION THROUGH SCHIFF-BASE FORMATION

We examined the substrate-recognition capability of the present vitamin B_6-depen-
dent holoenzyme model composed of the octopus cyclophane and PLP. Although
several types of reactions are catalyzed by vitamin B_6-dependent enzymes, all these
reactions are claimed to proceed through formation of a Schiff-base intermediate
derived from PLP and a substrate. Thus, our attention was focused here on
molecular recognition in the Schiff-base forming process. Alkylamines having
various hydrophobic chains were used as substrates in place of α-amino acids for
evaluation of the hydrophobic effect on the Schiff-base forming equilibrium. In an
aqueous HEPES buffer at pH 7.0, these amines are present as cationic ammonium
species. Concentrations of both **1** and PLP were maintained constant,
1.0×10^{-4} mol dm^{-3} each, for the following measurements.

Upon addition of the amine (AA) to the aqueous solution containing **1** and PLP,
the extent of the Schiff-base (SB) formation was monitored by electronic absorption
spectroscopy. A typical example of the spectral change is shown in Figure 1. Clear
isosbestic points were observed for all the measurements. The apparent SB formation constants (K_{SB}) as defined by equation (1) were evaluated according to the
Benesi–Hildebrand relationship in a manner that has been reported elsewhere [6].
The K_{SB} values in the absence of the octopus cyclophane were also evaluated. The

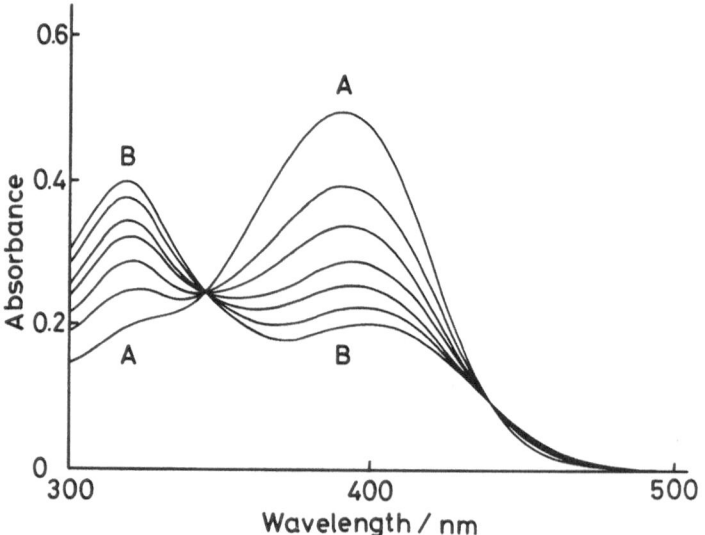

Fig. 1. Electronic absorption spectra for the Schiff-base system derived from PLP ($1.0 \times$
10^{-4} mol dm^{-3}) and varying amounts of hexylamine in the presence of **1** (1.0×10^{-4} mol dm^{-3}) in an
aqueous HEPES buffer (0.02 mol dm^{-3}, pH 7.0) at 30.0°C. Amine concentrations in mmol dm^{-3}: 0, 0.2,
0.4, 0.7, 1.1, 1.7, and 2.4 (read from A to B).

Table I.   Apparent formation constants (K_{SB}) for Schiff-bases derived from PLP and alkylammonium species at 30.0°C.[a]

$CH_3(CH_2)_{n-1}NH_3^+$	$K_{SB}/dm^3\,mol^{-1}$	
	Cyclophane system[b]	Without cyclophane
$n = 4$	240	200
$n = 5$	580	200
$n = 6$	2100	290
$n = 8$	120000	230

[a] In an aqueous HEPES buffer; 0.02 mol dm^{-3}, pH 7.0.
[b] 1, 1.0×10^{-4} mol dm^{-3}.

results are listed in Table I. While the K_{SB}

$$K_{SB} = [SB]/([PLP][AA]) \tag{1}$$

values for the various amines are nearly identical with each other in an aqueous phase, the K_{SB} value is markedly dependent on the chain length of AA in the presence of 1. Thus, the holoenzyme model composed of the octopus cyclophane and PLP clearly discriminates the ammonium substrates through the hydrophobic interaction. In addition, it is noteworthy that the present cationic host shows substrate recognition toward the guests having the same cationic charge when another guest molecule having the opposite charges, i.e. PLP, is concomitantly bound to the octopus cyclophane.

As is apparent from Figure 1, SB shows two absorption bands with maxima at 400 and 318 nm, which are assigned to two tautomeric isomers, C and D in Scheme 2, respectively [7]. Since the relative intensities of these bands sensitively vary depending on the medium polarity, the microenvironmental polarity provided by the octopus cyclophane can be estimated. Figure 2 shows electronic absorption spectra of the SB species formed with PLP and octylamine in the presence and absence of 1. While SB is present exclusively in form C in an aqueous phase, form D is remarkably favored in the presence of 1; the microenvironmental polarity in the cavity of 1 is roughly equivalent to that provided by 2-propanol. Accordingly, it is apparent that PLP bound to the hydrophilic site of the octopus cyclophane moves into a more hydrophobic domain of the host molecule via formation of SB with the hydrophobic alkylammonium substrate as schematically illustrated in Figure 3.

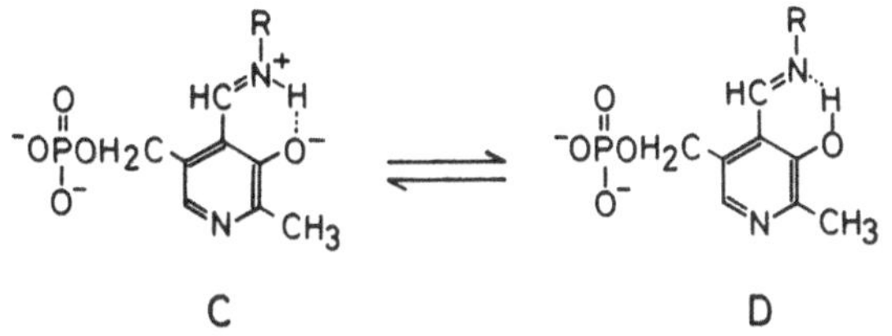

C   D

Scheme 2.

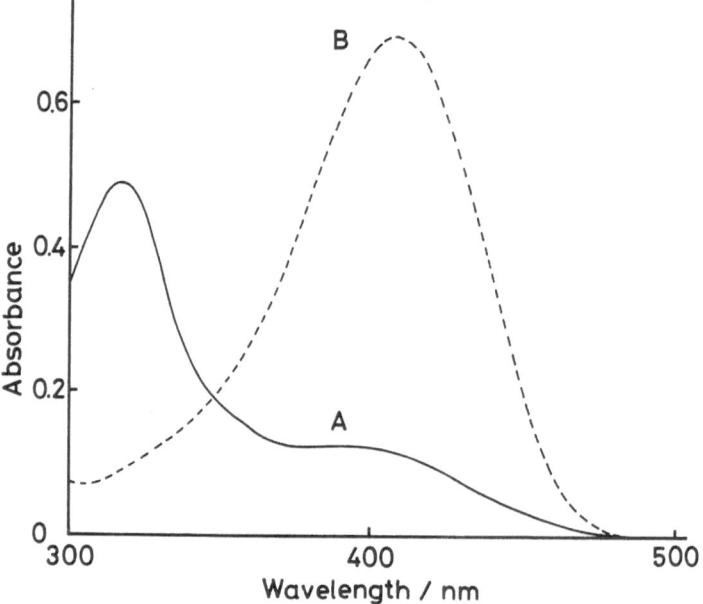

Fig. 2. Electronic absorption spectra of Schiff-base formed with PLP $(1.0 \times 10^{-4}\,\mathrm{mol\,dm^{-3}})$ and octylamine in the presence (A) and absence (B) of **1** $(1.0 \times 10^{-4}\,\mathrm{mol\,dm^{-3}})$ in an aqueous HEPES buffer $(0.02\,\mathrm{mol\,dm^{-3}},\ \mathrm{pH}\ 7.0)$ at 30.0°C. Amine concentrations in $\mathrm{mol\,dm^{-3}}$: A, 2.1×10^{-4}; B, 1.2×10^{-2}.

In conclusion, it became clear that the octopus cyclophane can be utilized as an effective apoenzyme model for constitution of an artificial vitamin B_6-dependent holoenzyme. The ternary complex is formed with **1**, PLP, and a substrate in the initial reaction stage, and then the latter two species bound to **1** undergo Schiff-base formation. Molecular recognition is exercised by the octopus cyclophane in favor of hydrophobic substrates.

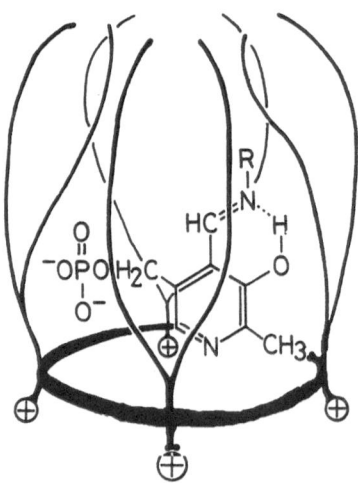

Fig. 3. Schematic representation for hydrophobic incorporation of the PLP Schiff-base into the octopus cyclophane.

Acknowledgement

The present work was partially supported by a Grant-in-Aid for Special Project Research No. 62101002 from the Ministry of Education, Science, and Culture of Japan.

References

1. I. Tabushi and K. Yamamura: *Cyclophanes I* (Topics in Current Chemistry No. 113, Ed. F. Vögtle), pp. 145–182. Springer-Verlag (1983); Y. Murakami: *Cyclophanes II* (Topics in Current Chemistry No. 115, Ed. F. Vögtle), pp. 107–155. Springer-Verlag (1983); K. Odashima and K. Koga: *Cyclophanes* (v. 2, Ed. P. M. Keehn and S. M. Rosenfeld), Chapter 11, Academic Press (1983).
2. Y. Murakami, J. Kikuchi, M. Suzuki, and T. Takaki: *Chem. Lett.* 2139 (1984); Y. Murakami, J. Kikuchi, M. Suzuki, and T. Matsuura: *J. Chem. Soc., Perkin Trans. 1,* 1289 (1988).
3. Y. Murakami, Y. Hisaeda, J. Kikuchi, T. Ohno, M. Suzuki, and Y. Matsuda: *Chem. Lett.* 727 (1986); Y. Murakami, Y. Hisaeda, J. Kikuchi, T. Ohno, M. Suzuki, Y. Matsuda, and T. Matsuura: *J. Chem. Soc., Perkin Trans. 2,* 1237 (1988).
4. Y. Murakami, J. Kikuchi, and O. Hayashida: *J. Chem. Soc., Perkin Trans. 1,* to be submitted.
5. C. M. Harris, R. J. Johnson, and D. E. Metzler: *Biochim. Biophys. Acta* **421**, 181 (1976).
6. Y. Murakami, J. Kikuchi, K. Akiyoshi, and T. Imori: *J. Chem. Soc., Perkin Trans. 2* 1445 (1986).
7. J. Llor and M. Cortijo: *J. Chem. Soc., Perkin Trans. 2* 1111 (1977).

Journal of Inclusion Phenomena and Molecular Recognition in Chemistry 7 (1989), 99–106.
© 1989 by Kluwer Academic Publishers.

Molecular Recognition and Host–Guest Interactions in Complexes of O-Bistren, C-Bistren, and Bisdien

ARTHUR E. MARTELL
Department of Chemistry, Texas A&M University, College Station, Texas 77843-3255, U.S.A.

(Received: 1 February 1988)

Abstract. The host–guest relationships of the complexes of the cryptand ligands O-BISTREN and C-BISTREN, and of the analogous macrocyclic ligand BISDIEN are compared. The affinities of their binuclear copper(II) complexes for the bridging ligands (OH^-, F^-) as 'cascade' type guests (i.e., guests of guests) are reported. The ability of the dicobalt(II)-BISDIEN complex to coordinate dioxygen and an additional bifunctional guest simultaneously, leads to the possibility of a new type of catalysis occurring within the cavity of a macrocyclic complex.

Key words. Cryptand, copper(II) complexes, catalysis, cobalt(II) complexes.

1. Introduction

The ability of macrocyclic and macrobicyclic ligands to function as hosts in binding metal ions and anions as guests depends on their structures, in particular their degree of preorganization [1], and on the nature of the donor atoms, as well as on the nature of guests. In this paper the macrobicyclic (cryptand) polyamine ligands O-BISTREN, **1**, and C-BISTREN, **2**, and the macrocyclic ligand BISDIEN, **3**, are compared with respect to how protonation and metal ion coordination influence the degree of molecular recognition of the host for various donor ions and molecules as

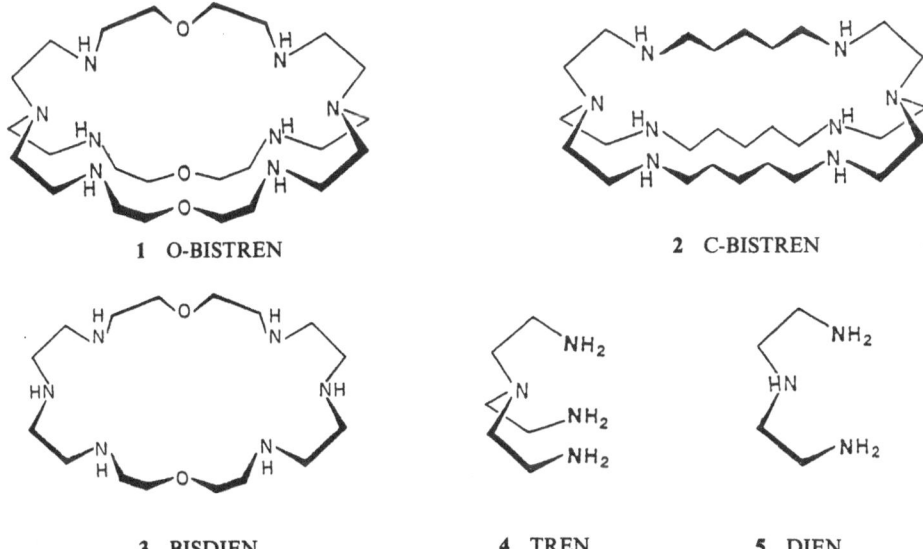

1 O-BISTREN   2 C-BISTREN

3 BISDIEN   4 TREN   5 DIEN

guests. The formulas of the free bases of **1–3** are compared with the formulas of their coordinating components, TREN and DIEN, represented by **4** and **5**, respectively.

2. Discussion

2.1. BASICITIES OF THE LIGANDS

The protonation constants of ligands **1–5**, listed in Table I, show interesting trends related to their conformations. The non-cyclic polyamines TREN, **4**, and DIEN, **5**, generally have higher second and third protonation constants than do the corresponding moieties in the cryptand and macrocyclic ligands because of reduced charge repulsions resulting from the fact that their conformations in solution are not restricted by ring formation, and assume more extended structures than those

Table I. Protonation constants and metal ion stability constants of O-BISTREN, C-BISTREN, TREN, BISDIEN and DIEN; $\mu = 0.100$ M (NaClO$_4$), $t = 25.0°$C.

Equilibrium quotient Q	Log Q				
	O-BISTREN [2, 3, 4]	C-BISTREN [5]	TREN [2]	BISDIEN [6]	DIEN [6]
$\dfrac{[HL^+]}{[H^+][L]}$	9.99	10.35	10.12	9.65	9.84
$\dfrac{[H_2L^{2+}]}{[H^+][HL^+]}$	9.02	9.88	9.41	8.92	9.02
$\dfrac{[H_3L^{3+}]}{[H^+][H_2L^{2+}]}$	7.98	8.87	8.47	8.30	4.23
$\dfrac{[H_4L^{4+}]}{[H^+][H_3L^{3+}]}$	7.20	8.38		7.64	
$\dfrac{[H_5L^{5+}]}{[H^+][H_4L^{4+}]}$	6.40	8.14		3.81	
$\dfrac{[H_6L^{6+}]}{[H^+][H_5L^{5+}]}$	5.76	7.22		3.26	
$\dfrac{[CuL^{2+}]}{[Cu^{2+}][L]}$	17.59	15.39		16.46	15.9
$\dfrac{[Cu_2L^{4+}]}{[CuL^{2+}][Cu^{2+}]}$	10.73	13.37		10.84	
$\dfrac{[Cu_2(OH)L^{3+}][H^+]}{[Cu_2L^{4+}]}$	−3.8	−7.59		−6.51	
$\dfrac{[CoL^{2+}]}{[Co^{2+}][L]}$	11.20			9.73	8.0
$\dfrac{[Co_2L^{4+}]}{[CoL^{2+}][Co^{2+}]}$	5.60			2.7	
$\dfrac{[Co_2(OH)L^{3+}][H^+]}{[Co_2L^{4+}]}$	−7.20				

shown in formulas **1–5**. The lower basicities of the amino groups of O-BISTREN relative to C-BISTREN are due to the electron-withdrawing effects of the three ether oxygens in the bridging groups of the former.

2.2. BINUCLEATING TENDENCIES OF O-BISTREN, C-BISTREN, AND BISDIEN

Because of the greater flexibility of BISDIEN, its binucleating tendencies with various metal ions are much weaker than those of the macrobicyclic ligands. The stability constants of the 1 : 1 Cu(II) and Co(II) complexes of BISDIEN are considerably higher than those of DIEN, indicating that more than three amino nitrogens of BISDIEN are coordinated to the metal ion, and the ligand probably has folded conformations in which from four (for Cu(II)) to six (for Co(II)) amino groups are coordinated [6]. Therefore it is understandable that while there is a considerable tendency for coordination with a second metal ion to form a dinuclear complex in the case of Cu(II), which effectively coordinates strongly only four amino groups, the analogous tendency for divalent transition metal ions which form octahedral complexes is very weak ($10^{2.7}$ for Co(II), and much lower for Ni(II) [6]). Binuclear complexes of BISDIEN are strongly stabilized by appropriate coordinating bridging groups, such as ethylenediamine, hydroxide ion, imidazolate ion (for Cu(II)), and dioxygen (for Co(II)) [6].

The macrobicyclic ligands O-BISTREN and C-BISTREN, on the other hand, show strong binucleating tendencies, with or without suitable coordinating bridging groups. This characteristic tendency to function as hosts for two metal ions as guests is considered to be due to preorganization which provides two tetramine cavities at opposite ends of the macrobicyclic octamine cryptand. It has been pointed out [5] that the cavity in the C-BISTREN structure seems to be less preorganized than that of O-BISTREN, probably because of the tendency of the pentamethylene bridges to self-associate through hydrophobic bonding.

In aqueous solution, in the absence of bifunctional coordinating bridging groups, the dicopper(II) complexes of both O-BISTREN and C-BISTREN polarize an internal coordinated water molecule to produce a μ-hydroxo bridge. The corresponding hydrolysis constants are very much higher (higher OH$^-$ affinity; lower pK) than those of analogous mononuclear complexes, reflecting the very strong polarizing power of two metal ions on a bridging water molecule.

There is an interesting reversal in the relative magnitudes of the successive Cu(II) binding constants of O-BISTREN and C-BISTREN. The latter has more basic donor groups and would be expected to form more stable metal complexes; however its 1 : 1 complex with Cu(II) is considerably weaker than that of O-BISTREN. On the other hand its second metal binding constant is much stronger than that of O-BISTREN with the result that the overall binding constant ($\beta = [M_2L]/[M]^2[L]$) is somewhat larger for C-BISTREN. The interpretation given to this behavior [5] involves differences in the preorganization of the ligands. Coordination of the first metal ion by C-BISTREN would require opening up of the cryptand cavity and disruption of the hydrophobic bonding associations of the hydrocarbon bridges. Thus the ligand would be prepared (preorganized) for coordination of the second metal ion, which would take place with a relatively high stability constant. Overall, formation of the dinuclear copper(II) C-BISTREN cryptate complex involves two effects operating in opposite directions, the breaking up of hydrophobic bonding, which cost energy, and the greater basicity of the amino groups, which favors higher stability, with the latter effect predominating.

2.3. REACTIONS OF BINUCLEAR CU(II) COMPLEXES WITH COORDINATING BRIDGING GROUPS

The binding constants of (isoelectronic) hydroxide and fluoride ions to the binuclear Cu(II) complexes of O-BISTREN and C-BISTREN are presented in Table II.

Table II.   Binding constants of hydroxide and fluoride ions with dinuclear Cu(II) complexes of O-BISTREN and C-BISTREN: $\mu = 0.100$ M (NaClO$_4$), $t = 25.0°$C.

Equilibrium quotient, Q	Log Q	
	O-BISTREN	C-BISTREN
$\dfrac{[Cu_2(OH)L^{3+}]}{[Cu_2L^{4+}][OH^-]}$	10.0	6.19
$\dfrac{[Cu_2FL^{3+}]}{[Cu_2L^{4+}][F^-]}$	4.5[a]	3.3[a]

[a] $\mu = 0.100$ M (0.090 M NaClO$_4$ + 0.010 M NaF).

The higher basicities of the amino groups in C-BISTREN would be expected to produce stronger Cu—N coordinate bonds, and therefore weaker coordinate bonds to secondary ligands such as F$^-$ and OH$^-$. The fluoride ion is seen to behave normally, with about a factor of ten difference in its binding constants, with Cu$_2$O-BISTREN^{4+} having the higher affinity for the fluoride anion (Table II, formulas 6 and 7).

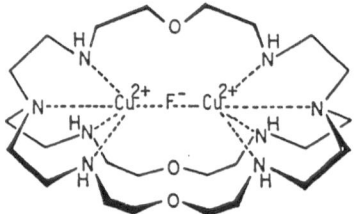   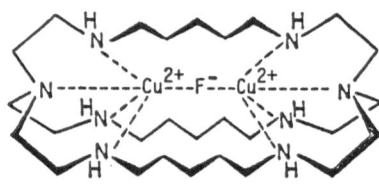

6 [Cu$_2$LF]$^{3+}$ L = O-BISTREN   7 [Cu$_2$L'F]$^{3+}$ L' = C-BISTREN

This interpretation, however, does not explain the fact that the affinity of Cu$_2$O-BISTREN^{4+}, 8, for OH$^-$ ion is nearly four orders of magnitude greater than that of Cu$_2$C-BISTREN^{4+}, 9. Therefore, it was concluded that the former has a structure which provides stabilization not available to the latter. It was suggested [2] that the high stability of the hydroxo bridge of the dicopper(II) complex of O-BISTREN is due to hydrogen bonding to one of the ether bridging groups, an interaction which is not possible with the fluoride-bridged complex, as indicated by

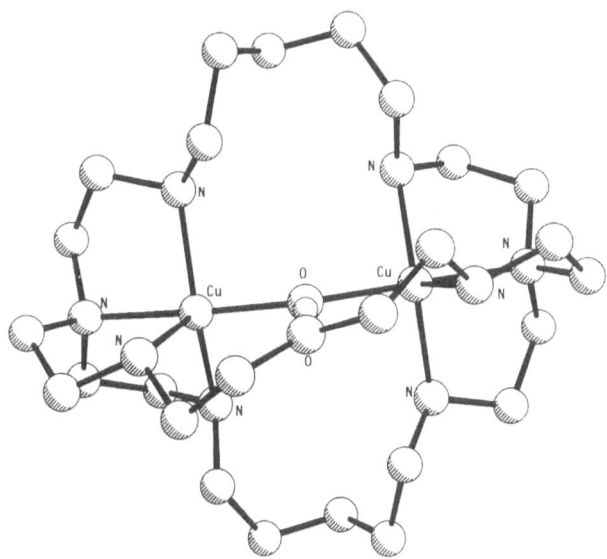

Fig. 1. Diagram illustrating structure of hydroxo-bridged dicopper(II)-O-BISTREN cryptate showing bent Cu—O—Cu bridge and proximity of OH to an ether oxygen of the ligand.

formulas **6** and **8**. Recently, it has been found possible to crystallize the hydroxo bridged complex and determine its structure by X-ray crystallography [7]. Figure 1, which is a drawing of the structure based on the crystallographic parameters obtained, confirms the hydrogen bonding of the hydroxo bridge to an ether oxygen, with a bent Cu-OH-Cu angle.

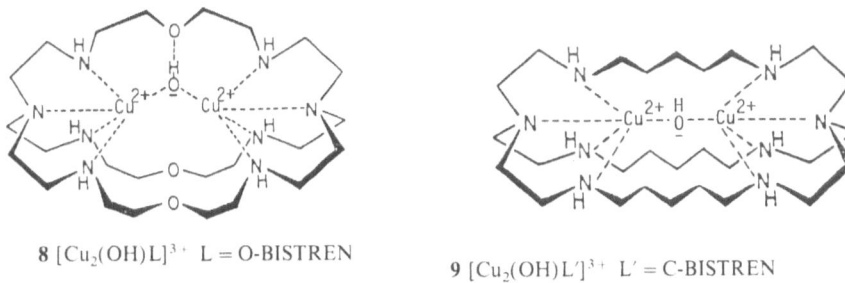

8 $[Cu_2(OH)L]^{3+}$ L = O-BISTREN

9 $[Cu_2(OH)L']^{3+}$ L' = C-BISTREN

The unusually high stability of the hydroxo-bridged dinuclear cryptate of Cu(II)-O-BISTREN has a profound influence on the nature of the complex species formed in solution. Figure 2 is a distribution diagram of the individual complex species formed as a function of pH in a solution containing a 1 : 1 ratio of the molar concentrations of O-BISTREN and Cu(II) ion. It is seen that in spite of the unfavorable concentration ratio, the binuclear complex predominates in its hydroxo-bridged form over most of the pH range. In this solution, uncoordinated ligand in various stages of protonation are seen to also be present.

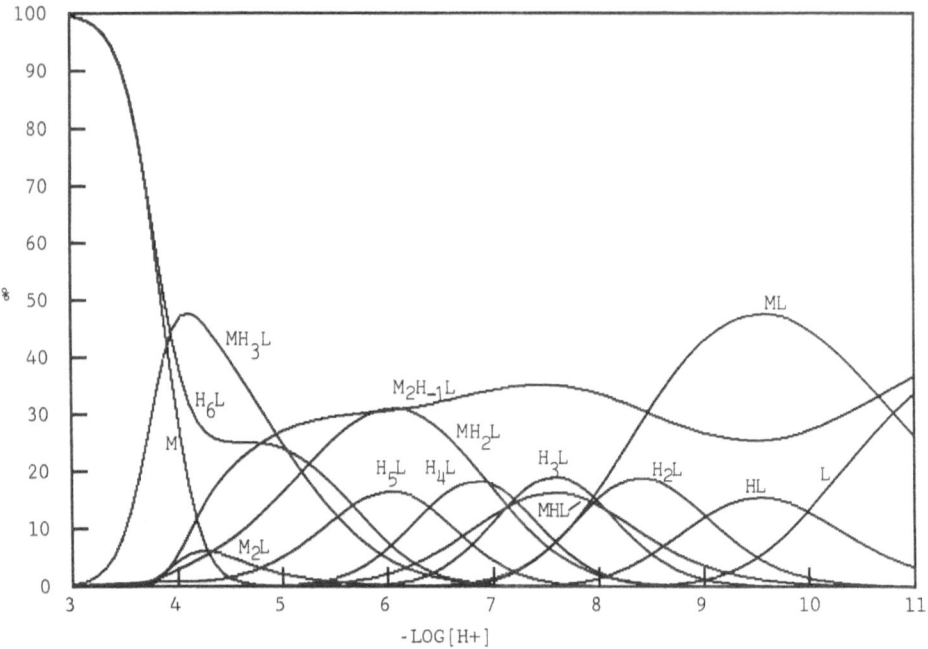

Fig. 2. Distributions of species as a function of p[H] in a solution containing a 1 : 1 molar ratio of O-BISTREN to copper(II) in 0.100 M $NaClO_4$ and 25.0°C. Component concentrations, $H_6Bistren^{6+}$ 1.00×10^{-3} M; Cu^{2+}, 9.98×10^{-4} M.

2.4. DIOXYGEN COMPLEXES

The differences in stabilities of the dioxygen adducts of the dinuclear cobalt complexes of O-BISTREN and BISDIEN also deserve comment. The higher basicities of the eight amino groups of O-BISTREN over those of the six amino groups of BISDIEN would lead to the expectation of much higher dioxygen affinity of the former over the latter [8]. It turned out, however, that the oxygenation constant associated with the formation of **10** is about three orders of magnitude lower than that of **11**. The explanation offered for this interesting reversal of the expected relative magnitudes of the oxygenation constants was based on possible steric crowding in the cavity of **10**, which interferes with metal—dioxygen bond formation. This complex turns out to be of interest for oxygen separation processes because of its rapid reversibility at moderate temperatures [9].

It is noted that the dioxygen complex of dicobalt BISDIEN, **11**, indicates only five coordinated bonds between each metal ion and the donor groups indicated. Because of the Co(III)-like nature of the metal ion, it is expected to be 6-coordinate, so that two water molecules (one for each metal) are probably present in the coordination spheres of the metal ions in **11**. This interpretation has been confirmed [6] by the hydrolysis of this complex in two steps to give di- and trihydroxo dioxygen complexes with pK's of 8.25 and 9.36 [6]. In addition to hydrolysis (hydrogen ion dissociation) it should be possible to coordinate additional donor groups, including bifunctional bridging groups, to the aquo sites. This concept has

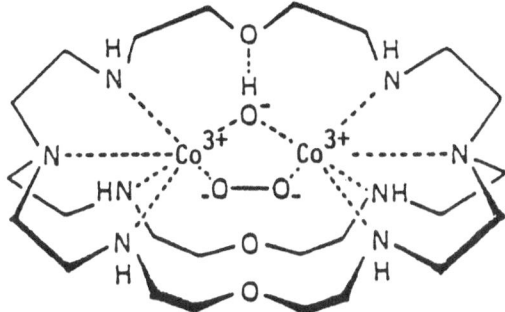

10 Suggested arrangement of coordinate bonds in dibridged dioxygen complex formed from dinuclear Co(II) O-BISTREN cryptate

recently been confirmed by the coordination of a reducing bridging group, the oxalate ion, giving an intermediate dioxygen complex, indicated by **12**, which undergoes facile electron transfer from oxalate to coordinated peroxo to give carbon dioxide and water [10]. The activation of coordination dioxygen in the cavity of the dinuclear macrocyclic complex **12** so as to react with a coordinated reductant in the same complex molecule is the first such reaction to be reported. Further studies in this laboratory are being directed toward catalysis of redox reactions involving coordinated dioxygen and coordinated substrates within the cavities of dinuclear macrocyclic complexes.

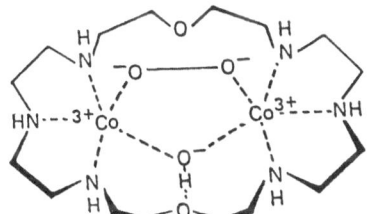

11 μ-Hydroxo-μ-peroxodicobalt BISDIEN chelate

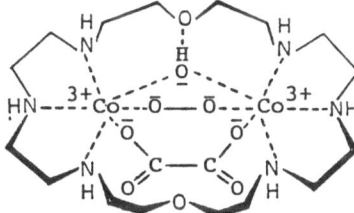

12 μ-Hydroxo-μ-oxalato-μ-peroxo dicobalt BISDIEN chelate

3. Conclusions

The results of the research described above makes possible the following general conclusions.

1. Hydroxide ion is more strongly bound than fluoride ion as bridging donor groups (guests) in hosts consisting of dinuclear copper(II) cryptates of O-BISTREN and C-BISTREN.
2. The cavities of binucleating cryptands seem to be less preorganized with hydrocarbon bridges between the coordinating moieties.

3. The p1 sence of an electronegative atom in the bridging linkages of binucleating cryptands greatly increases the binding of a bridging hydroxide ion through hydrogen bonding.
4. The greater flexibility of a binucleating macrocyclic ligand relative to an analogous macrobicyclic cryptand makes possible stronger binding of bifunctional bridging guests such as imidazolate and dioxygen.
5. The simultaneous coordination of two potentially reactive bridging guests in binuclear host complexes may produce strong catalytic effects on the guest molecules or ions.

Acknowledgement

This research was supported by a grant, No. A-259, from the Robert A. Welch Foundation.

References

1. R. D. Hancock and A. E. Martell: *Comments Inorg. Chem.* **6**, 237 (1988).
2. R. J. Motekaitis, A. E. Martell, J. M. Lehn, and E. I. Watanabe: *Inorg. Chem.* **21**, 4253 (1982).
3. R. J. Motekaitis, A. E. Martell, B. Dietrich, and J. M. Lehn: *Inorg. Chem.* **23**, 1588 (1984).
4. R. J. Motekaitis, A. E. Martell, and I. Murase: *Inorg. Chem.* **25**, 938 (1986).
5. R. J. Motekaitis, A. E. Martell, I. Murase, J. M. Lehn, and M. W. Hosseini: *Inorg. Chem.* (1988) **27**, 3630.
6. R. J. Motekaitis, A. E. Martell, J. P. Lecomte, and J. M. Lehn: *Inorg. Chem.* **22**, 609 (1983).
7. R. J. Motekaitis, A. E. Martell, P. Rudolf, and A. Clearfield: *Inorg. Chem.* in press.
8. A. E. Martell: *Acc. Chem. Res.* **15**, 155 (1982).
9. R. J. Motekaitis, and A. E. Martell: *J. Chem. Soc., Chem. Commun.*. (1988) 1020.
10. A. E. Martell and R. J. Motekaitis: *J. Chem. Soc., Chem. Commun.* (1988) 915.

Journal of Inclusion Phenomena and Molecular Recognition in Chemistry 7 (1989), 107–115.

The Interactions of Surfactant Viologens with Cyclodextrins

ANGEL E. KAIFER*, PABLO A. QUINTELA, and JODI M. SCHUETTE
Department of Chemistry, University of Miami, Coral Gables, FL 33124, U.S.A.

(Received: 1 February 1988)

Abstract. N-ethyl-N'-hexadecyl-4,4'-bipyridinium bromide ($C_{16}VBr_2$) and N-ethyl-N'-octadecyl-4,4'-bipyridinium bromide ($C_{18}VBr_2$) were used as electroactive probes to assess the interactions between surfactants and cyclodextrins. Cyclic voltammetry, visible spectroscopy, fluorescence spectroscopy and surface tension techniques were used to detect the formation of complexes between the surfactant viologen probes and α- and β-cyclodextrins. The voltammetric results suggest the formation of inclusion compounds in which the hydrophobic tail of the surfactant viologens penetrate the cyclodextrin cavity. The dimerization of the viologen cation radicals is essentially suppressed by the presence of α-cyclodextrin (ACD) while no effects are observed in the presence of β-cyclodextrin (BCD). The observed results are best explained by the relative solubility in aqueous media of each of the inclusion complexes in the several accessible viologen oxidation states.

Key words. Cyclodextrin, viologens, surfactant, voltammetry.

1. Introduction

Cyclodextrins [1–3] are glucopyranose cyclic oligomers having a characteristic toroidal shape. These compounds are soluble in aqueous media because of the hydrophilic nature of the outer surface of the torus. The inner surface is more hydrophobic and thus hydrated cyclodextrins represent a high energy state that can readily accept guest hydrophobic molecules in place of the inner water molecules [1]. Indeed, as expected for inclusion complexes, the better the fit of the guest molecule in the inner cavity of the cyclodextrin the more stable the host–guest complex will be [1–3]. β-Cyclodextrin (BCD, 7 glucopyranose units) forms complexes with many organic molecules because its cavity is perfectly suited to include substituted phenyl rings and many other commonly found groups. α-Cyclodextrin (ACD, 6 glucopyranose units) does not form so many complexes because of its smaller cavity diameter. However, the examination of CPK models reveals that ACD is well suited to bind compounds having long, linear alkyl chains such as surfactants.

Scattered reports are available in the chemical literature concerning the interactions of surfactants and cyclodextrins. For instance, Ise and coworkers [4] have reported on the interaction of colloidal electrolytes and cyclodextrins. They observed that the apparent critical micelle concentrations (cmc) of sodium dodecyl sulfate (SDS) and cetyltrimethylammonium bromide (CTABr) increase upon the addition of ACD and BCD. They concluded that the cyclodextrins form 1:1 complexes with the surfactants. More recently Satake *et al.* [5] determined the association constants of ACD with several ionic surfactants. For 1-alkanesulfonate

---

* Author for correspondence.

ions with 5–12 (n) carbon atoms in the alkyl chain, the association constant was found to increase regularly with n and become abruptly constant at $n = 10$. The association constants with sodium 1-dodecanesulfonate, dodecylammonium chloride, dodecyltrimethylammonium chloride, and 1-dodecylpyridinium chloride were all similar in magnitude.

Thomas and coworkers [6] have investigated the formation of cyclodextrin-surfactant-aromatic fluorophore ternary complexes. Quite recently, Kusumoto et al. [7] have also reported on the interaction of pyrene with BCD in aqueous surfactant solutions. Yasuda et al. [8] have studied the interactions between several rather hydrophobic viologens and BCD aiming at the development of a practical electrochromic display system. Willner and Eichen [9] have found that, a complex between octylviologen and BCD acts efficiently as a photoelectron collector on the surface of CdS and TiO_2 colloidal particles. It is then clear that (1) the interactions between surfactants and cyclodextrins are rather unexplored, and (2) the use of surfactant viologens for this purpose appears as very appropriate not only because of the electroactivity imparted by the viologen group (which enables the use of electrochemical techniques for the characterization of the interactions) but also because the properties of previously known viologen-cyclodextrin complexes appended an additional interest to the study.

In this work we report the interactions of ACD and BCD with the following two surfactant viologens:

$CH_3-(CH_2)_{14}-CH_2-^+N$⟨⟩⟨⟩$N^+-CH_2-CH_3$ 2 Br$^-$

$C_{16}VBr_2$

$CH_3-(CH_2)_{16}-CH_2-^+N$⟨⟩⟨⟩$N^+-CH_2-CH_3$ 2 Br$^-$

$C_{18}VBr_2$

2. Experimental

The surfactant viologen bromides, $C_{16}VBr_2$ and $C_{18}VBr_2$, were synthesized as reported elsewhere [10]. ACD and BCD were obtained from Fluka or Aldrich and used without further purification. No difference in behavior was detected with cyclodextrins purchased from different suppliers. Pyrene (Sigma) was recrystallized from ethanol. All solutions were freshly prepared with distilled water that had been further purified by passage through a pressurized, four-cartridge Barnstead Nanopure system.

The equipment for the electrochemical experiments has been already described. [11]. Absorption spectra were measured with a Hewlett Packard 8452A spectrophotometer. Steady-state emission spectra were recorded with an Aminco Bowman spectrophotofluorimeter. Surface tension measurements were performed on a Fisher Model 20 tensiometer using the du Nuoy method with platinum-iridium rings.

A two-compartment cell was used to obtain the cyclic voltammograms. All solutions were thoroughly deoxygenated by purging with purified nitrogen. A nitrogen blanket was maintained above the solution during the electrochemical experiments. Glassy carbon and platinum working electrodes (Bioanalytical Systems, Indiana) were regularly polished following standard procedures prior to use. All potentials were measured against the sodium chloride saturated calomel electrode (SSCE).

Samples of the viologen cation radicals were prepared by controlled potential electrolysis of solutions containing adequate concentrations of the parent dications. These electrolyses were directly performed in a spectrophotometer cuvette in which a glass strip covered with SnO_2 (transmittance 80%, Delta Technologies) was affixed to one of the optical windows so that the conductive surface faced the solution in the cuvette, thus allowing its use as the working electrode. The auxiliary and reference electrodes were situated out of the optical pathway of the spectrophotometer. This arrangement allowed the recording of the visible spectra for the reduced viologen species either in deposited films at the electrode surface or in solution.

3. Results and Discussion

Several asymmetric viologens with surfactant properties have been previously reported in the chemical literature. For instance, Gratzel and coworkers have published the synthesis and aggregation properties of a series of asymmetric, amphiphilic viologens [12]. We have already reported the synthesis and aggregation properties of $C_{16}V^{2+}$ and $C_{18}V^{2+}$ [10]. Both of these viologen derivatives form micelles and show well defined cmc values in pure water (2.8 and 1 mM, respectively). It was of interest to this work to obtain cmc values in aqueous 50 mM NaCl solutions since this is the medium commonly used for cyclic voltammetry experiments. Figure 1 shows the surface tension of 50 mM NaCl aqueous solutions containing variable concentrations of $C_{16}V^{2+}$. The shape of the plot is characteristic of micelle formation with an apparent cmc slightly below 0.4 mM. Under the same experimental conditions the cmc of $C_{18}V^{2+}$ was determined to be 0.03 mM.

The association of the surfactant viologen with the cyclodextrins was initially assessed by using the emission spectral characteristics of pyrene. To this end, a saturated solution of pyrene in 50 mM NaCl was prepared by stirring several pyrene crystals in the solution for about 12 h, followed by filtration to remove undissolved pyrene crystals. According to reported accounts [13], the final pyrene concentration in this saturated solution is about 5×10^{-7} M. Enough $C_{16}VBr_2$ was then added to this solution to make the concentration of the surfactant viologen dication 1.0 mM. The characteristic emission spectrum of pyrene was barely observable under these conditions (excitation wavelength: 332 nm). This is probably the result of effective quenching of the excited state pyrene molecules by the viologen acceptors. If one accepts the current view of solubilization of hydrophobic molecules by ionic surfactant micelles, the pyrene molecules must reside near the Stern layer in very close proximity to the polar viologen groups. This spatial proximity would explain satisfactorily the highly efficient level of quenching observed. Addition of ACD to this solution appears to diminish the efficiency of the quenching process. In the

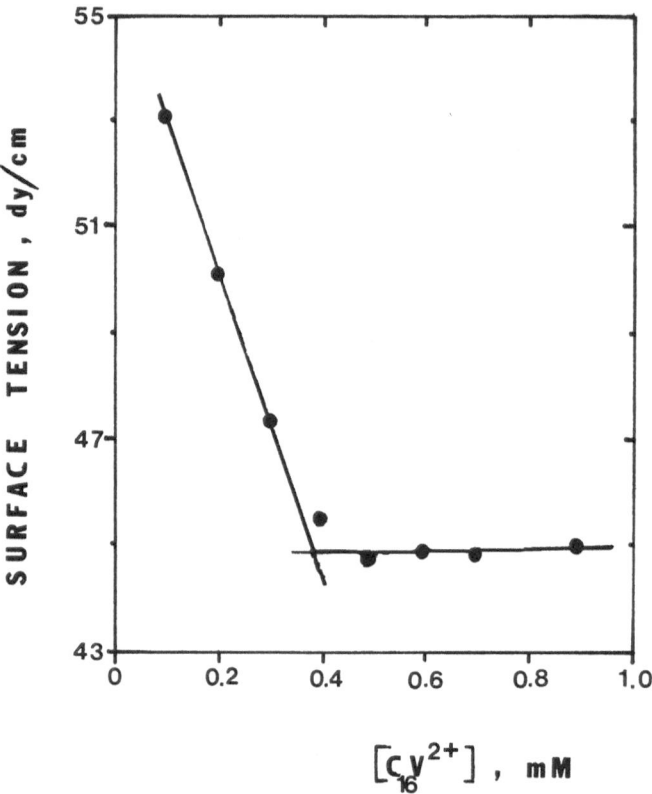

Fig. 1. Concentration dependence of the surface tension of an aqueous solution containing 50 mM NaCl and variable concentrations of $C_{16}VBr_2$.

presence of 1 mM ACD, the emission intensity increases substantially. This cyclodextrin-induced increase in the observed fluorescence intensity quickly levels off at ACD concentrations above 3 mM. It has been previously established that the ACD cavity is too small to form host–guest complexes with pyrene [7]. Thus, these results can only be explained by the formation of a rather stable complex between the surfactant viologen and ACD. The cyclodextrin host associates with the surfactant viologen molecules, causing the concentration of free viologen to decrease below the cmc (in effect destroying the micelles) and releasing the pyrene molecules to the bulk solution where the quenching process proceeds more slowly at diffusion-controlled rates. The slower quenching rate explains the observed increase in the fluorescence intensity. In agreement with this interpretation, the relative intensities of the vibronic bands in the fluorescence spectrum of pyrene after the addition of ACD correspond to those expected for an isotropic aqueous solution [13], thus supporting the disruption of the micellar aggregates by the cyclodextrin host.

The complexation of the surfactant viologens by the cyclodextrins was also clearly demonstrated with surface tension data. Figure 2 shows the surface tension of 50 mM NaCl aqueous solutions also containing 1.0 mM $C_{16}VBr_2$ as a function of the concentration of added ACD. In the absence of ACD, the surface tension is

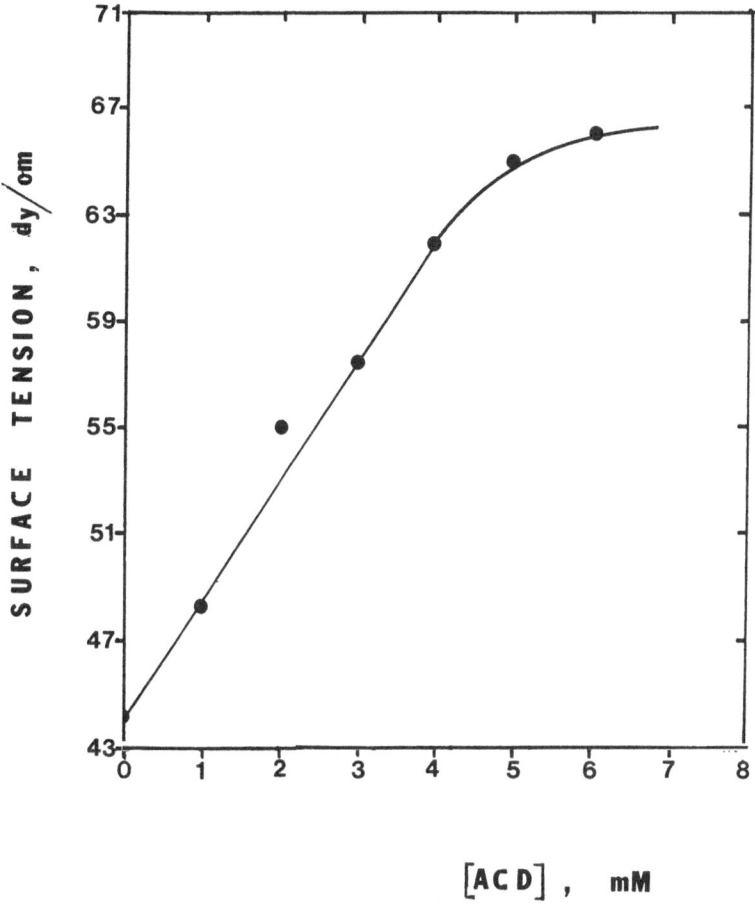

Fig. 2. Effect of ACD additions on the surface tension of a solution containing 50 mM NaCl and 1.0 mM $C_{16}VBr_2$.

44 dyne/cm. However, the addition of ACD causes a quick increase of the surface tension. When the ACD concentration is increased to the 4–5 mM level, the surface tension reaches values characteristic of surfactant-free solutions. Similar observations were made with 1.0 mM solutions of $C_{18}V^{2+}$. Since ACD does not have any surface activity [14], the increase in the surface tension values is attributable to the formation of a surfactant-cyclodextrin complex. Hence, as the concentration of the cyclodextrin increases, the concentration of free viologen in solution decreases. In turn, this diminishes the concentration of surfactant viologen ions at the air-solution interface, thus increasing the surface tension of the solution. Therefore, the addition of cyclodextrin has the overall effect of removing surfactant viologen ions from the air-solution boundary via complexation of the free viologen in the solution.

The redox behavior of the surfactant viologens is represented in Scheme I, exemplified by $C_{16}V^{2+}$. As any viologen, these surfactants can be reduced in two consecutive monoelectronic steps to yield, first, a cation radical, and, second, an

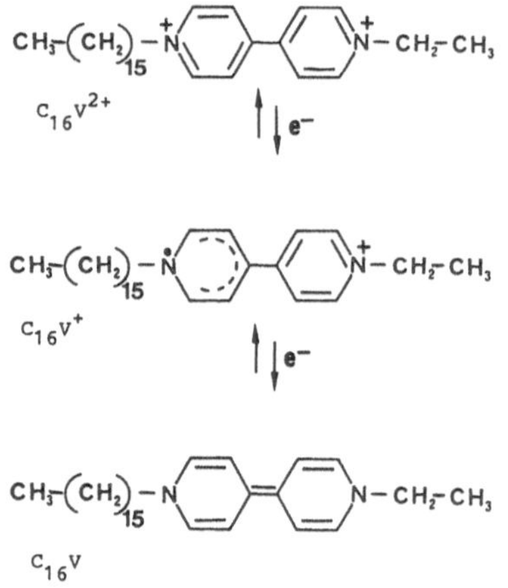

Scheme I.

uncharged, hydrophobic species. Both reductions are chemically reversible. However, the voltammetric behavior of these surfactants is highly complicated by precipitation of the two reduced forms, $C_{16}V^+$ and $C_{16}V$, at the electrode surface. These precipitation reactions have been previously reported with other hydrophobic viologens, like the commercially available heptylviologen [14], and form the basis for the proposed use of viologens as active components in electrochromic systems.

In contrast to this, the first reduction of $C_{16}V^{2+}$ in the presence of a ten-fold excess of ACD is free of these precipitation effects as shown in the voltammogram of Figure 3. The potential separation between the reduction and the oxidation peaks is about 60 mV for moderate scan rates, and a plot of reduction peak current vs the square root of scan rate was found to be linear in the range 20–400 mV/s. Both of these pieces of evidence clearly indicate that the first reduction is a reversible, diffusion-controlled electrochemical process which was assigned to the reduction of the ACD–$C_{16}V^{2+}$ complex, that is

$$\text{ACD–}C_{16}V^{2+} + e \rightleftharpoons \text{ACD–}C_{16}V^{2+}, \qquad E^{0'} = -0.66 \text{ V.}$$

The half-wave potential, as obtained from the average of the reduction and oxidation peak potentials, is very close to that reported for the more hydrophilic methylviologen (-0.69 V vs the same SSCE reference electrode) [15]. This probably reveals that the electroactive 4,4'-bipyridinium moiety is essentially unaffected by the complexation, arguing in favor of an inclusion complex between the lipophilic tail of the surfactant and the cyclodextrin host.

In the presence of a ten-fold excess of ACD, if the negative potential scan is extended to -1.0 V vs SSCE (using glassy carbon electrodes), a second reduction wave is observed. However, its shape is strongly distorted by precipitation processes indicating that the electrogenerated species (ACD–$C_{16}V$) is no longer soluble in the electrolysis medium. The first reduction couple of $C_{16}V^{2+}$ in the presence of ACD at concentration levels below 10 mM is also distorted by precipitation of the residual uncomplexed $C_{16}V^{2+}$. It is thus necessary to add an excess of cyclodextrin to the surfactant solution in order to decrease substantially the concentration of free

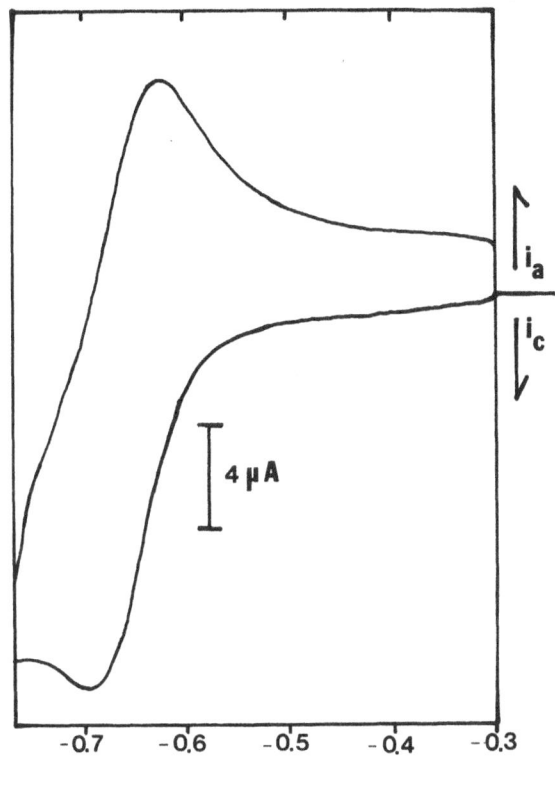

E , V vs SSCE

Fig. 3. Cyclic voltammogram on Pt of 1.0 mM $C_{16}VBr_2$ also containing 50 mM NaCl and 10 mM ACD. Scan rate = 20 mV/s.

surfactant viologen and allow a clean observation of the ACD–$C_{16}V^{2+}$/ACD–$C_{16}V^+$ couple.

The voltammetric behavior of $C_{18}V^{2+}$ in the presence of a ten-fold excess of ACD is close to that observed for the shorter chain analog. However, at slow scan rates a small desorption spike can be seen in the oxidative scan probably as a reflection of the greater hydrophobic character of this surfactant. Nonetheless, the voltammetric response at scan rates faster than 100 mV/s is essentially diffusion-controlled.

A ten-fold excess of BCD also changes the voltammetric behavior of either $C_{16}V^{2+}$ or $C_{18}V^{2+}$. However, the first reduction couple does not exhibit reversible behavior at the scan rates surveyed (up to 1000 mV/s). In both cases, the reduction current appears to be kinetically controlled, perhaps by the dissociation of the BCD complexes, but quantitative analysis of the voltammetric data is hindered by precipitation of the reduced products at the electrode surface.

An important aspect of the aqueous chemistry of viologen cation radicals is their tendency to dimerize [16, 17]. The extent of the dimerization reaction can be easily estimated by visible spectrophotometry of the cation radical solutions. Figure 4 shows the visible spectrum recorded after reduction of a 1mM solution of $C_{16}V^{2+}$

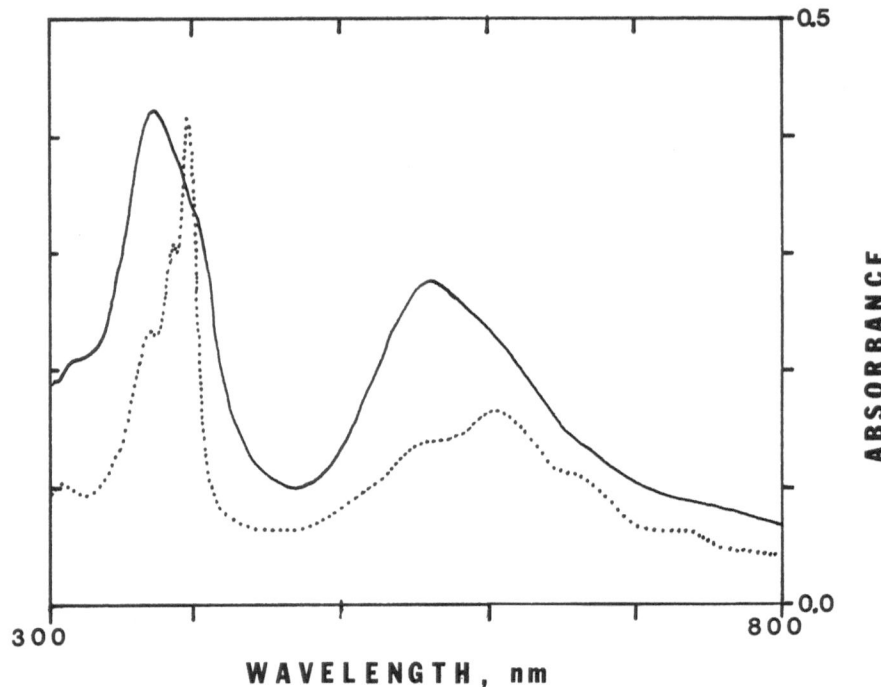

Fig. 4. Absorption spectra of reduced surfactant viologens. (——) 1.0 mM $C_{16}V^+$ in 50 mM NaCl (this spectrum corresponds to the film deposited on the electrode surface). (·····) 1.0 mM $C_{16}V^+$ + 10 mM ACD in 50 mM NaCl (this spectrum corresponds to an homogeneous solution of the cation radical species).

in the presence of 10 mM ACD (dotted line). The spectrum is characteristic of the monomeric cation radical [16, 17] and does not show any hints of dimer absorption. This is a solution spectrum since the purple color of the cation radical was observed to spread through the solution. In contrast to this, the reduction of $C_{16}V^{2+}$ in the absence of ACD yielded an insoluble purple film on the optically transparent electrode whose visible spectrum is also given in Figure 4 (continuous line). This spectrum shows all the absorptions characteristic of the dimer, and only hints of those corresponding to the monomer. The insoluble blue film is then mostly composed of dimeric $C_{16}V^+$. Interestingly, this behavior was exactly reproduced by the same surfactant viologen in the presence of a ten-fold excess of BCD. Reduction of $C_{18}V^{2+}$ followed an identical pattern, that is, soluble monomeric $C_{18}V^+$ in the presence of a ten-fold excess of ACD, and dimeric deposits in the presence of BCD or in the absence of cyclodextrin hosts. In summary, only ACD appears to be capable of preventing the extensive dimerization and precipitation of these surfactant cation radicals.

Evans and coworkers have published an interesting study of the electrochemical behavior of a ferrocene-carboxylic acid derivative in the presence of BCD [18] and a summary of general guidelines on the electrochemical methodology that can be used to assess the complexation of redox-active molecules by cyclodextrin hosts [19]. However, the amphiphilic character of the surfactant viologens used in this work prevented us from applying most of these methods because the voltammetric parameters (reduction potentials and diffusion coefficients) of the surfactant viologens in the absence (or at low levels) of cyclodextrins were not accessible due to the precipitation processes observed upon reduction. Therefore, the obtention of complexation equilibrium constants from purely voltammetric data is impossible. We have reported elsewhere [10] the determination of these association constants by a conductometric technique. The results indicate only small differences among the several cyclodextrin-surfactant viologen complexes, with the ACD complexes showing slightly larger association constants. Thus, the thermodynamic parameters of complexation do not appear to be the crucial factors to explain the observed differences in voltammetric and cation radical dimerization behavior. With the experimental information available at this point, it seems adequate to interpret these differences as the result of varying solubilities in the aqueous medium. In this way, the $ACD-C_{16}V^{2+}$ complex is slightly more soluble than the $ACD-C_{18}V^{2+}$ complex because of the slightly shorter alkyl chain of the surfactant. Moreover, BCD complexes must show less solubility as a reflection of the lower aqueous solubility of BCD as compared to ACD. These arguments can explain all the dimerization results and the electrochemical data for ACD complexes but do not explain fully the voltammetric data for BCD complexes where some kinetic effects were clearly detected.

Our research with these systems is still underway in an attempt to better define the thermodynamic and kinetic parameters that determine the behavior of these complexes. We are also currently looking at similar host systems that could be used to control the chemistry of rather reactive species, such as the viologen cation radicals.

Acknowledgement

This research was supported by a Bristol-Myers Company Grant of Research Corporation.

References

1. J. Szejtli: *Cyclodextrin and their Inclusion Complexes*; Akademiai Kiado: Budapest, 1982.
2. M. L. Bender and M. Komiyama: *Cyclodextrin Chemistry*; Springer-Verlag: Berlin, 1978.
3. W. Saenger: *Angew. Chem., Int. Ed. Engl.* **19**, 344 (1980).
4. T. Okubo, H. Kitano, and N. Ise: *J. Phys. Chem.* **80**, 2661 (1976).
5. I. Satake, T. Ikenoue, T. Takeshita, K. Hayakawa, and T. Maeda: *Bull. Chem. Soc. Jpn.* **58**, 2746 (1985).
6. S. Hashimoto and J. K. Thomas: *J. Am. Chem. Soc.* **107**, 4655 (1985).
7. Y. Kusumoto, M. Shizuka, and I. Satake: *Chem. Phys. Lett.* **125**, 64 (1986).
8. A. Yasuda, H. Kondo, M. Itabashi, and J. Seto: *J. Electroanal. Chem.* **210**, 265 (1986).
9. I. Willner and Y. Eichen: *J. Am. Chem. Soc.* **109**, 6862 (1987).
10. A. Diaz, P. A. Quintela, J. M. Schuette, and A. E. Kaifer: *J. Phys. Chem.* **92**, 3537 (1988).
11. P. A. Quintela and A. E. Kaifer: *Langmuir* **3**, 769 (1987).
12. M. P. Pileni, A. M. Braun, and M. Gratzel: *Photochem. Photobiol.* **31**, 423 (1980).
13. K. P. Ananthapadmanabhan, E. G. Goddard, N. J. Turro, and P. L. Kuo: *Langmuir* **1**, 352 (1985).
14. R. J. Jasinski: *J. Electrochem. Soc.* **124**, 637 (1977).
15. A. E. Kaifer and A. J. Bard: *J. Phys. Chem.* **89**, 4876 (1985).
16. C. L. Bird and A. T. Kuhn: *Chem. Soc. Rev.* **10**, 49 (1981).
17. E. M. Kosower and J. L. Cotter: *J. Am. Chem. Soc.* **86**, 5524 (1964).
18. T. Matsue, D. H. Evans, T. Osa, and N. Kobayashi: *J. Am. Chem. Soc.* **107**, 3411 (1985).
19. T. Matsue, T. Osa, and D. H. Evans: *J. Incl. Phenom.* **2**, 547 (1984).

Journal of Inclusion Phenomena and Molecular Recognition in Chemistry 7 (1989), 117–124.
© 1989 *by Kluwer Academic Publishers.*

Artificial Hydrolase Using Modified Dimethyl-β-cyclodextrin

HIROSHI IKEDA, RYOICHI KOJIN, CHUL-JOONG YOON, TSUKASA IKEDA,
and FUJIO TODA*
*Department of Bioengineering and Bioscience, Faculty of Engineering, Tokyo Institute of Technology,
O-okayama, Meguro-ku, Tokyo 152, Japan*

(Received: 1 February 1988)

Abstract. The first successful method for modification of dimethyl-β-cyclodextrin (β-DMCD) was demonstrated by the synthesis of a new artificial hydrolase (**2**) and the enzymatic activities of **2** were investigated. **2** caused an 1100-fold increase in the rate of hydrolysis of *p*-nitrophenyl acetate at pH 7.2, whereas unmodified β-DMCD depressed the reaction. The kinetic pK_a of **2** was 7.2, and the K_m of **2** was independent of pH values. **2** had *para*-selectivity for the hydrolysis of nitrophenyl acetate isomers.

Key words. Cyclodextrin, artificial enzyme, dimethyl-β-cyclodextrin, chymotrypsin.

1. Introduction

Cyclodextrins (CDs) can form inclusion complexes with a number of molecules. For this reason, biomimetic reactions using cyclodextrins and their derivatives have been actively studied [1–2]. We have also prepared successful artificial enzymes by modifications of cyclodextrins [6–7]. On the other hand, dimethylcyclodextrins (DMCDs) are a series of cyclic oligomers consisting of α-1,4-linked 2,6-di-*O*-methyl-D-glucopyranose units, and have quite unique and different properties from cyclodextrins [2–3]. β-DMCD has a deeper hydrophobic cavity than β-CD. Stezowski and his co-workers estimated the height of the torus of β-DMCD at 10–11 Å [4]. The most different character between them is solubility [3]. The solubility of β-DMCD in water is 55 g/100 ml water at 25°C, whereas that of β-CD is only 1.8 g/100 ml water. As temperature rises, the solubility of β-DMCD decreases, whereas that of β-CD increases (Figure 1). β-DMCD is highly soluble in many organic solvents, whereas β-CD is insoluble in most organic solvents. For example, β-DMCD is soluble in alcohol, acetone, chloroform, benzene, acetonitrile, THF, DMF, DMSO, and so on. β-CD is slightly soluble only in acetonitrile, DMF, DMSO and so on. Furthermore, the behavior of inclusion complex formation is different between CD and DMCD [3].

For these reasons, if β-DMCD were used as an enzyme model, a new unique enzyme model might be obtained. In this paper, we wish to provide the first successful method for modification of β-DMCD and to investigate the enzymatic activities of a new artificial hydrolase using modified β-DMCD.

---

* Author for correspondence.

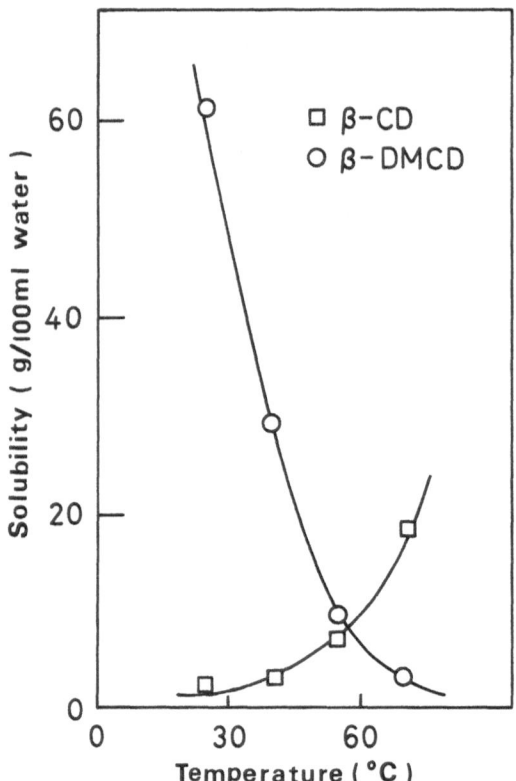

Fig. 1. Plots of the temperature dependence of solubilities of β-cyclodextrin and dimethyl-β-cyclodextrin.

2. Experimental

2.1. MATERIALS

Dimethyl-β-cyclodextrin was purchased from Toshin Chemical Co. Ltd., and recrystallized from water and a mixed solvent of chloroform and hexane several times. Its purity was checked by HPLC and NMR before use [5]. Nitrophenyl acetate isomers were prepared by a common method and purified by recrystallization from hexane.

3-[2-(4-imidazolyl)ethyl]dimethyl-β-cyclodextrin (**2**). A solution of dimethyl-β-cyclodextrin (5 g, 3.8 mmol) in dry THF (40 ml) was treated with NaH (631 mg, 13 mmol) at 0–40°C for 5 hours. To this reaction mixture was added a solution of 2-(4-imidazolyl)ethyl chloride (736 mg, 5.6 mmol) in dry THF (10 ml) at 0°C. The resulting mixture was stirred at room temperature for 10 h. The product was isolated by column chromatography on silica gel (Wakogel C-300) (elution with CHCl$_3$/MeOH = 35:1). (1.2 g, 20% yield)

Found: C, 48.58; H, 7.00; N, 1.75%, Calcd. for C$_{61}$H$_{104}$O$_{35}$N$_2$ · CHCl$_3$: C, 48.20; H, 6.85; N, 1.80%. ^{1}H-NMR (500 MHz, C$_6$D$_6$) δ = 4.32 (1H, bs, C$_{3'}$-H), 4.43 (6H, bs, C_3-H), 4.99 (6H, bs, C_3-OH), 5.26 (1H, bs, $C_{1'}$-H), 5.41 (6H, bs, C_1-H), 6.59 (1H, s, $CHCl_3$), 7.08 (1H, s, imidazole), 7.69 (1H, s, imidazole).

2.2. KINETICS

Hydrolysis reactions were followed by monitoring the appearance of nitrophenol spectrophotometrically using a HITACHI 220A spectrophotometer. The reaction was conducted in a quartz cell in the water-jacketed cell holder of the 220A. Temperature was maintained at 25°C by a HAAKE F3 circulating water bath. The reaction was initiated by adding a stock solution of ester in acetonitrile to a buffer solution in the quartz cell.

The pH of a reaction mixture did not change during the course of the reaction.

3. Results and Discussion

3.1. SYNTHESIS OF ARTIFICIAL HYDROLASE

β-DMCD (1) was treated with 3.5 equiv. of NaH at 0–40°C for 5 h and 1.5 equiv. of imidazolylethyl chloride at room temperature for 10 h in dry THF under argon (Scheme I). By chromatography on silica gel ($CHCl_3$/MeOH), the product (2) was isolated as a colorless solid in 20% yield. Elemental analysis data and estimation by peak area of the ^{1}H-NMR spectrum confirmed that 2 had only one imidazolylethyl group. The ^{1}H-NMR spectrum of 2 shows peaks at $\delta = 7.69$ and 7.08 ppm for the imidazolyl group, and small peaks at 5.26 and 4.32 for the up-field shift of $C_{1'}$-H and up-field shift of $C_{3'}$-H, respectively [8].

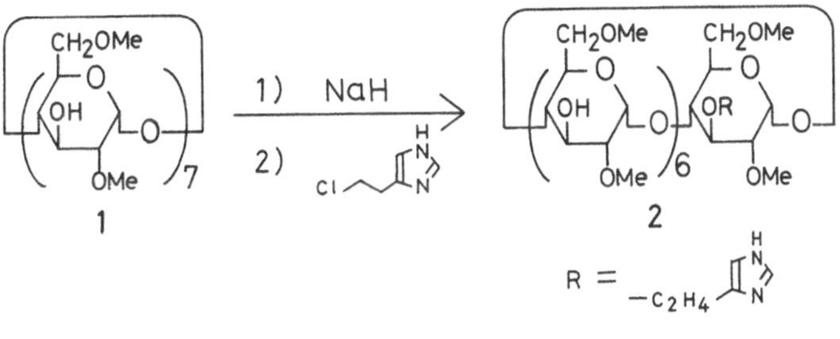

Scheme I.

3.2. HYDROLYSIS OF p-NITROPHENYL ACETATE BY 2

The initial rate of hydrolysis of p-nitrophenyl acetate (PNPA) was measured at 25°C in pH 7.2 phosphate buffer (1/15 M) in the presence of dimethyl-β-cyclodextrin bearing the imidazolylethyl group (2), dimethyl-β-cyclodextrin (β-DMCD) (1) or imidazole, and in the absence of them [8]. The conditions of large excess of PNPA were used and concentrations of PNPA were changed from 10-fold to 100-fold over those additives except 1. The reaction was followed by monitoring the released p-nitrophenol at 400 nm.

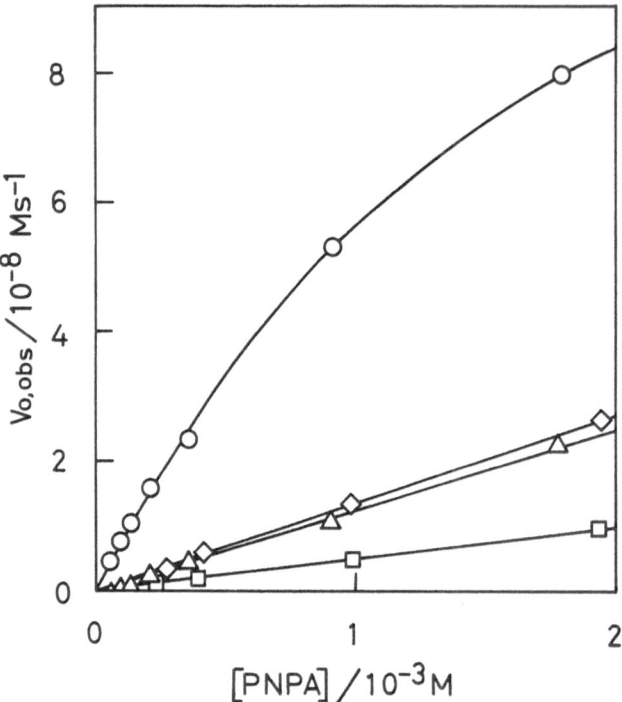

Fig. 2. Hydrolysis of p-nitrophenyl acetate in pH 7.2 phosphate buffer (1/15 M) at 25°C; $\bigcirc$ in the presence of **2** (1.10×10^{-5} M); $\diamond$ in the presence of imidazole (1.10×10^{-5} M); $\square$ in the presence of β-DMCD (9.67×10^{-3} M); and $\triangle$ in the absence of them.

The result is shown in Figure 2. Only 1 mol% of **2** caused a 5-fold increase in the rate of hydrolysis of PNPA (10^{-3} M), compared with the condition of absence of **2**, whereas 5 equiv. of β-DMCD (**1**) caused 60% depression of the reaction. The reaction was scarcely accelerated by imidazole in the same concentration as **2**. Only the binding site (β-DMCD) or active site (imidazole) was not effective for the acceleration of the hydrolysis reaction. The combined action of the binding site and active site of **2** caused a large acceleration of the reaction.

By the plotting kinetic data for **2** in the form of $1/(V_{0,obs} - V_{un})$ vs. $1/[\text{PNPA}]$ (Lineweaver–Burk plot), a straight line was obtained (Figure 3). It suggests that this reaction by **2** proceeds by the Michaelis–Menten mechanism as do cyclodextrins and their derivatives [1, 2, 6, 9]. k_{cat} and K_m were calculated by a Michaelis–Menten treatment (Table I). k_{cat} is $1.44 \times 10^{-2} \text{ s}^{-1}$, K_m is 2.6×10^{-3} M and k_{cat}/K_m is $5.54 \text{ M}^{-1} \text{ s}^{-1}$. k_{cat} and k_{cat}/K_m of **2** are nine times larger than those of β-cyclodextrin bearing a histaminyl group at the C-6 position (**3**) [6]. K_m of **2** is nearly equal to that of **3** [6]. These indicate that the binding ability of **2** is nearly the same as that of **3**, whereas the ability to accelerate the reaction of **2** is much larger than that of **3**. The ability of rate acceleration of **2** can be estimated by the ratio k_{cat}/k_{un}. k_{cat}/k_{un} of **2** is over 1100.

Also, around the optimum pH of α-chymotrypsin or pH 8, the rate of the hydrolysis reaction was measured. k_{cat} of **2** is over twice that of α-chymotrypsin (Table I) [10].

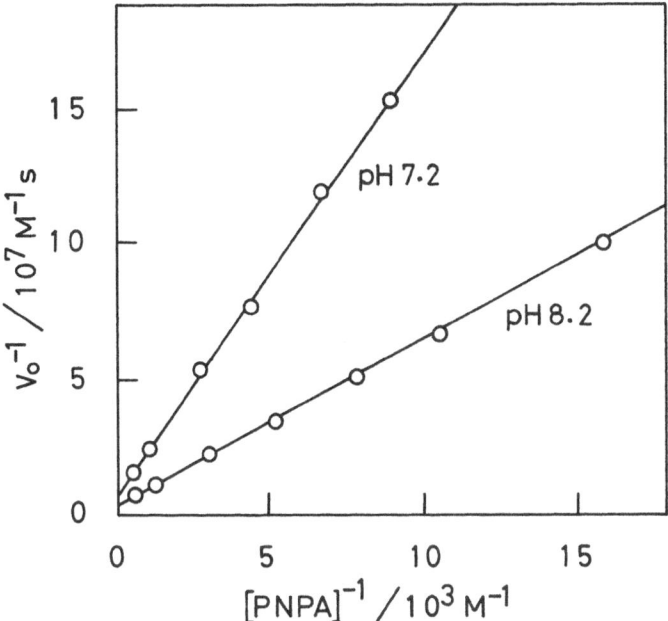

Fig. 3. Lineweaver–Burk plots for hydrolyses of *p*-nitrophenyl acetate by **2** (1.10×10^{-5} M) in pH 7.2 and by **2** (1.75×10^{-5} M) in pH 8.2 phosphate buffer (1/15 M) at 25°C.

Table I. Kinetic parameters for hydrolysis of *p*-nitrophenyl acetate.

	pH	k_{cat}	K_m	k_{cat}/K_m
		10^{-2} s^{-1}	10^{-3} M	M^{-1} s^{-1}
2	7.2	1.44	2.60	5.54
	8.2	2.67	2.90	9.21
3	7.2	0.165	2.63	0.627
α-chymotrypsin[a]	8.0	1.1	0.04	275

[a] Ref. 10.

3.3. pH DEPENDENCE OF ENZYMATIC ACTIVITY OF **2**

The pH dependence of the enzymatic activity of **2** was examined. In phosphate buffers (1/15 M) of various pH values, the accelerations of the hydrolysis reaction of *p*-nitrophenyl acetate (PNPA) by **2** were measured at 25°C. Kinetic parameters were calculated by a Michaelis–Menten treatment. As **2** has only one ionizing group, i.e., the imidazolyl group, the pH dependence of k_{cat}, K_m and k_{cat}/K_m can be represented by Eqs. 1, 2, and 3, respectively [11].

$$k_{cat} = \frac{(k_{cat})_{max}K_{ES}}{K_{ES} + [H^+]} \qquad (1)$$

$$K_m = \frac{K_S K_{ES} + [H^+] K_S K_{ES}/K_E}{K_{ES} + [H^+]} \tag{2}$$

$$k_{cat}/K_m = \frac{(k_{cat}/K_m)_{max} K_E}{K_E + [H^+]} \tag{3}$$

The pH dependence of k_{cat} of the hydrolysis reaction by **2** is shown in Figure 4a. pK_{ES} and $(k_{cat})_{max}$ of **2** were calculated by the nonlinear-least-squares fitting of

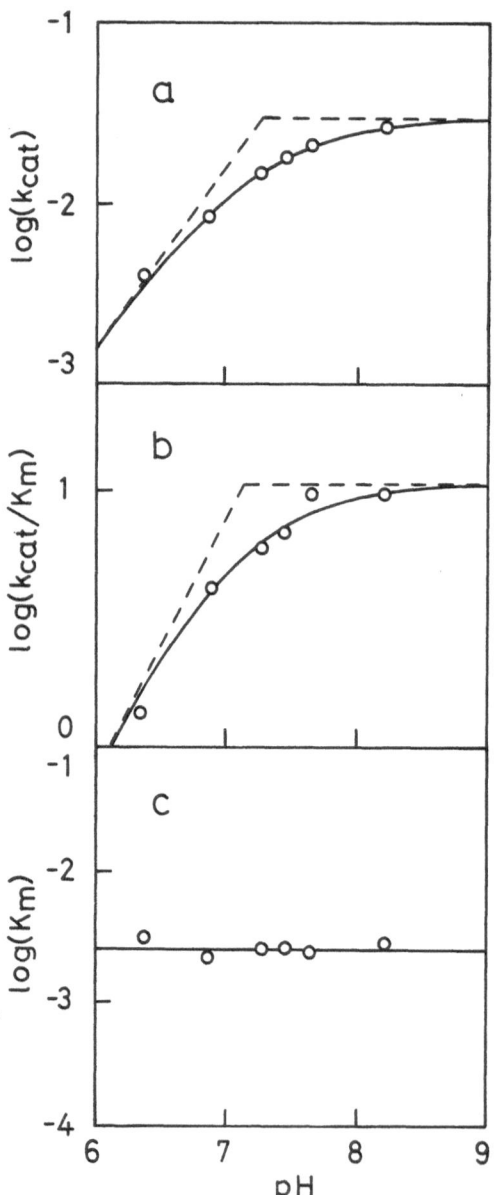

Fig. 4. Plots of the pH dependence of kinetic parameters for hydrolysis of *p*-nitrophenyl acetate by **2**.

kinetic data to Eq. 1. pK_{ES} is 7.28. It indicates that the imidazolyl group may play an important role in the rate determining step and that **2** can cause sufficient catalytic activity around a neutral pH condition. $(k_{cat})_{max}$ is $3 \times 10^{-2} s^{-1}$. It is larger than that of α-chymotrypsin. The pH dependence of k_{cat}/K_m is shown in Figure 4b. pK_E and $(k_{cat}/K_m)_{max}$ was calculated by the nonlinear-least-squares fitting of kinetic data to Eq. 3. pK_E is 7.12. $(k_{cat}/K_m)_{max}$ is $10.17 M^{-1} s^{-1}$.

pK_{ES} is the ionization constant of **2** which is including a substrate. pK_E is the ionization constant of **2** which is not including a substrate. pK_{ES} of **2** is near to pK_E. It indicates that an included substrate might rarely influence the ionization equilibrium. If pK_{ES} is equal to pK_E, it is obvious from Eq. 2 that K_m is independent of pH value. The pH dependence of K_m of **2** is shown in Figure 4c. K_m is almost constant in this pH range and its average is $2.57 \times 10^{-3} M$. This is consistent with the relationship that pK_{ES} is nearly equal to pK_E.

3.4. SUBSTRATE SPECIFICITY OF HYDROLYSIS REACTION BY 2

The regioselectivity for the hydrolysis of nitrophenyl acetate isomers by **2** was demonstrated. The reactions were carried out under the conditions of excess of substrate in pH 8.2 phosphate buffer (1/15 M) at 25°C. Kinetic parameters are shown in Table II. Both k_{cat} and k_{cat}/K_m for PNPA are larger than for the others. k_{cat}/K_m for PNPA is 6 times larger than that for o-nitrophenyl acetate (ONPA). k_{cat}/k_{un} for PNPA is also largest (Table II). These are contrary to the case of unmodified cyclodextrins which have $meta$-selectivity [9]. This may be due to the difference of the geometry of the inclusion complex, especially the distance between reaction centers.

The K_m values indicate that **2** mostly tends to form an inclusion complex with m-nitrophenyl acetate (MNPA) and it is also different from unmodified cyclodextrins [9].

k_{cat} for ONPA is 1.1 times larger than that for MNPA and K_m for the former is 2.5 times larger than that for latter. Therefore k_{cat}/K_m for the latter is 2.3 times larger than that for the former. It shows that the ability to form an inclusion complex is important for accelerating the rate of the over-all reaction.

Table II. Kinetic parameters for hydrolysis of nitrophenyl acetate isomers at 25°C in pH 8.2 phosphate buffer by **2** ($1.75 \times 10^{-5} M$)

Substrate	k_{cat}	K_m	k_{cat}/K_m	k_{un}	k_{cat}/k_{un}
	$10^{-3} s^{-1}$	$10^{-3} M$	$M^{-1} s^{-1}$	$10^{-5} s^{-1}$	—
ONPA	4.96	3.22	1.54	1.93	257
MNPA	4.48	1.27	3.54	1.40	320
PNPA	26.7	2.90	9.20	2.90	921

4. Conclusion

The first successful method for modification of β-DMCD was demonstrated as a synthesis of the new artificial hydrolase (**2**). This artificial hydrolase caused a

1100-fold increase in the rate of hydrolysis of p-nitrophenyl acetate (PNPA) at pH 7.2, whereas unmodified β-DMCD depressed the reaction. k_{cat} for the hydrolysis of PNPA of 2 is larger than that of the natural enzyme α-chymotrypsin. The kinetic pK_a of 2 was about 7.2 (that is, the average of pK_E and pK_{ES}) and K_m of 2 was independent of pH. 2 had *para*-selectivity for the hydrolysis of nitrophenyl acetate isomers.

References

1. M. L. Bender and M. Komiyama: *Cyclodextrin Chemistry*, Springer-Verlag (1978).
2. J. Szejtli: *Cyclodextrins and Their Inclusion Complexes*, Akademiai Kiado (1982).
3. J. Szejtli: *J. Incl. Phenom.* **1**, 135 (1983).
4. M. Czugler, E. Eckle, and J. J. Stezowski: *J. Chem. Soc., Chem. Commun.*, 1291 (1981).
5. K. Koizumi, Y. Kubota, T. Utamura, and S. Horiyama: *J. Chromatogr.* **368**, 329 (1986).
6. T. Ikeda, R. Kojin, C-J. Yoon, H. Ikeda, M. Iijima, K. Hattori, and F. Toda: *J. Incl. Phenom.* **2**, 669 (1984).
7. C-J. Yoon, H. Ikeda, R. Kojin, T. Ikeda, and F. Toda: *J. Chem. Soc., Chem. Commun.*, 1080 (1986).
8. H. Ikeda, R. Kojin, C-J. Yoon, T. Ikeda, and F. Toda: *Chem. Lett.*, 1495 (1987).
9. R. L. VanEtten, T. F. Sebastian, R. C. Clowes, and M. L. Bender: *J. Am. Chem. Soc.* **89**, 3242 (1967).
10. V. T. D'Souza, K. Hanabusa, T. O'Leary, R. C. Gadwood, and M. L. Bender: *Biochem. Biophys. Res. Commun.* **129**, 727 (1985).
11. A. Fersht: *Enzyme Structure and Mechanism*, 2nd ed., Freeman (1985).

Journal of Inclusion Phenomena and Molecular Recognition in Chemistry **7** (1989), 125–126.
© 1989 *by Kluwer Academic Publishers.*

James J. Christensen 1931–1987

This issue of the Journal is dedicated to the memory of the late Professor James J. Christensen. Professor Christensen died suddenly at his home on September 5, 1987. He was 56 years old.

Jim's life was devoted to service; service to his profession of chemical engineering, to his University, to his students, to his wife and family and to his community.

During his professional career, he made significant contributions in several fields including precision calorimeter design and construction, determination of thermodynamic data for ligand interactions with protons and cations, compilation of thermodynamic data, and organization of international symposia. He was co-organizer of the first Symposium on Macrocyclic Chemistry. This Symposium is now held on an annual basis.

The precision calorimeters which he designed and constructed made possible the early measurement of equilibrium constants for cation-macrocycle interaction. These values demonstrated the remarkable selectivities these compounds have for a variety of cations. His interest in compiling thermodynamic data made these data available to workers in the macrocyclic chemistry field. Advances in this field were aided significantly by having these data available.

Jim was the recipient of numerous local and national awards based on his professional accomplishments. The latest of these was the prestigious American Association of Engineering Education 3-M Award in Chemical Engineering. He received this Award and presented a lecture on the subject of creativity just one month before his death.

He was a prolific writer and gave many invited lectures at national and international meetings. He has over 300 publications, including journal articles, review articles, book chapters, books, and patents. He involved students in all of his scholarly activities. These students became co-authors of his publications and many have achieved prominence in their own right. Through his own professional accomplishments and those of his students, he has left an important legacy to science and to macrocyclic chemistry, in particular.

Jim was a native of Salt Lake City, Utah. He received B.S. and M.S. degrees in chemical engineering from the University of Utah and the Ph.D. degree in chemical engineering from Carnegie-Mellon University in 1958. His Ph.D. research involved heat transfer to coils, which served as an excellent base for his later work in the design of precision calorimeters.

He married Virginia Bills and they are the parents of five children. One of these children, Scott, is following in his father's footsteps in that he is expecting to receive his Ph.D. in chemical engineering at the University of Delaware in late 1988. Jim and Virginia took their families on sabbatical leaves to Oxford, England and the Mexico Polytechnical Institute in Mexico City. He and Virginia traveled widely.

Jim had a genuine interest in Brigham Young University and in his students, undergraduate and graduate. He was the first chairman of the Chemical Engineering Department. BYU honored him several times with outstanding research and teaching awards. He was a popular teacher and research director. He took great delight in collecting 'toys' that demonstrated the principles of thermodynamics and using them to interest his students. His favorite courses were thermodynamics and creativity and he exhibited marked creativity in his teaching and research programs.

Jim will be missed by those who knew him well. However, his life was unique in many ways and his influence will continue to be felt in the lives of his family, students, and friends.

REED M. IZATT
Department of Chemistry
Brigham Young University
Provo, Utah
May, 1988

Journal of Inclusion Phenomena and Molecular Recognition in Chemistry 7 (1989), 127–136.
© 1989 *by Kluwer Academic Publishers.*

Stable Silica Gel-Bound Crown Ethers. Selective Separation of Metal Ions and a Potential for Separations of Amine Enantiomers*

JERALD S. BRADSHAW**, REED M. IZATT**, JAMES J. CHRISTENSEN,
KRZYSTOF E. KRAKOWIAK‡, BRYON J. TARBET, and RONALD L. BRUENING
*Departments of Chemistry and Chemical Engineering, Brigham Young University,
Provo, UT 84602, U.S.A.*

S. LIFSON**
Department of Chemical Physics, Weizmann Institute of Science, Rehovot 76100, Israel.

(Received: 27 April 1988)

Abstract. Silica gel-bound crown ethers and aza macrocycles have been synthesized with the attaching arm connected to the carbon framework of the macrocycles. The interactions of these bound macrocycles with cations are almost identical to those involving the analogous free macrocycles. This has allowed for predictable cation separation, concentration, and removal processes to be performed on a small scale. Quantum mechanical calculations and NMR measurements indicate that similarly bound chiral macrocycles will be capable of use in separating chiral organic amines.

Key words. Macrocyclic compounds, silica gel-bound macrocycles, metal ion separation, chiral macrocycles.

1. Introduction

Selectivity in chemical interactions is the basis for some of the most remarkable phenomena in living and non-living systems. Enzyme catalysis, antibody-antigen interactions, involvement of specific metals in metalloproteins, enantiomeric selectivity, and metal catalysis in industrial reactions provide a partial listing of systems where selectivity is found. The obvious success and efficiency of such systems encourage the identification and investigation of underlying principles which make recognition possible in chemical interactions. Such understanding could lead to the design of new host molecules capable of interacting with guest species in a predetermined, selective manner.

Macrocycles have several features which are desirable in host molecules for an investigation of this type. These molecules are preorganized to allow through synthetic design the fitting of host to guest in a complementary fashion. Donor atoms, aromatic rings, substituent groups and chirality can be designed into these hosts in such a manner as to enhance selectivity for similar chemical guest species.

* Dedicated to the memory of Professor James J. Christensen who died on 5 September 1987.
** Authors for correspondence.
‡ Permanent address for K.E.K., Department of Chemical Technology, School of Medicine, 90145 Lodz, Poland.

Finally, by appropriate synthetic design, particular macrocycles can be attached to a solid support to form a material capable of making quantitative separations of like chemical species in a chromatographic column.

During the past two decades, much attention has been given to the design of marcocycles capable of selective recognition of guest species. Pedersen [1] first reported the crown ether compounds and recognized that they showed selective interaction with alkali metal ions. Lehn and Cram and their coworkers [2, 3] have designed numerous crown ether-type macrocycles and have studied their selective interaction with cations. The significance of the work of these individuals resulted in their receipt of the 1987 Nobel Prize in Chemistry [4]. An extensive compilation of equilibrium constant (K) data is available [5] from which selectivities in a given solvent can be calculated. It is evident from these K data that use of the proper macrocycle can result in selectivity for nearly any metal cation.

We have made use of the cation-selective properties of macrocycles to design liquid membrane [6], supported liquid membrane [6], and macrocycle-bonded silica gel [7] systems for specific cation separations. The selective transport of Ag^+ [8, 9], Pb^{2+} [8], K^+ [10] and Li^+ [11] was observed in a bulk liquid membrane system using various macrocyclic ligands. In the supported liquid membrane (thin sheet and hollow fiber) systems, selective and predicted separations of Na^+/K^+, Cd^{2+}/Hg^{2+}, and others have been achieved [12, 13].

A major problem with the separation of metal ions using extraction or membrane systems is the slow but steady loss of the expensive macrocyclic compounds from the organic membrane or layer. To circumvent this problem, we have attached various macrocyclic compounds to silica gel using a stable hydrocarbon-ether linkage [7, 14, 15]. Log K values for the interaction of these silica gel-bound macrocycles towards various metal ions were found to be the same ($\pm 10\%$) as those for the analogous unbound macrocycles toward the same cations in water. This paper describes the synthesis of some of these silica gel-bound macrocycles, their metal ion complexation properties, their use in the separation and concentration of certain cations from cation mixtures and the potential use of silica gel-bound chiral macrocycles for the separation of enantiomeric ammonium salts.

2. Synthesis of Silica Gel-Bound Macrocycles

The silica gel-bound macrocycles were prepared as shown in Scheme I. The allyloxymethyl-substituted macrocycle, a catalytic amount of chloroplatinic acid and an excess of diethoxymethysilane were refluxed in benzene for 8 to 24 hours. The benzene and excess diethoxymethysilane were removed under reduced pressure to give the crude silane product. The IR and NMR spectra of these materials were examined to determine the extent of the hydrosilylation reaction. An absence of an IR band at about 2170 cm^{-1} and an NMR peak at about 4.6 δ indicated that there were no free Si—H functions in the mixture. Also loss of the peaks at 4.5–6.0 δ in the NMR indicated a complete reaction of the vinyl-substituted macrocycle due to the absence of peaks indicative of the vinyl group.

The crude macrocycle-containing diethoxysilane was dissolved in chloroform and added to a known amount of 60–200 mesh silica gel so that the macrocycle to silica gel ratio was about 1 : 10 (wt). The solvent was removed on a rotary evaporator to

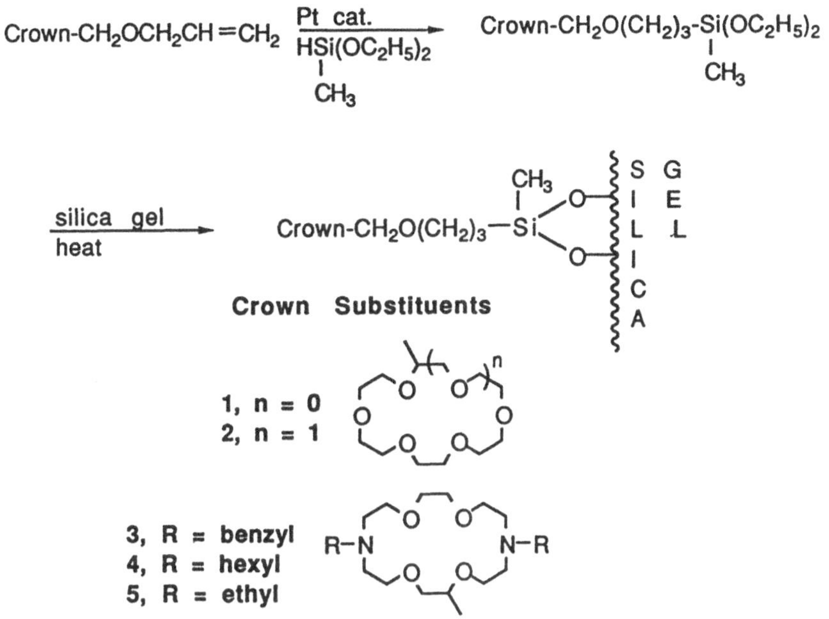

Crown-CH$_2$OCH$_2$CH=CH$_2$ $\xrightarrow[\substack{\text{HSi(OC}_2\text{H}_5)_2 \\ | \\ \text{CH}_3}]{\text{Pt cat.}}$ Crown-CH$_2$O(CH$_2$)$_3$-Si(OC$_2$H$_5$)$_2$
$\qquad\qquad\qquad\qquad\qquad\qquad\qquad\qquad\qquad\qquad\qquad\quad$ |
$\qquad\qquad\qquad\qquad\qquad\qquad\qquad\qquad\qquad\qquad\qquad\quad$ CH$_3$

$\xrightarrow[\text{heat}]{\text{silica gel}}$ Crown-CH$_2$O(CH$_2$)$_3$—Si ... SILICA GEL

Crown Substituents

1, n = 0
2, n = 1

3, R = benzyl
4, R = hexyl
5, R = ethyl

Scheme I. Preparation of silica gel-bound crown compounds.

obtain an even coating of the diethoxysilane on the silica gel. The coated silica gel was then heated at 120°C in a Kugel-rohr apparatus for about 24 hours to form the covalent linkage. A chloroform wash of the silica gel-bound macrocycle materials gave little or no residue.

Triethoxysilane was also used in many of the hydrosilylation reactions (Scheme I, first reaction). The resulting macrocycle-substituted triethoxysilane could then form a three-bonded linkage to the silica gel. The final bonding of the diethoxy- or triethoxysilane material to silica gel could also be carried out in refluxing toluene.

The starting allyloxymethyl-18-crown-6 and 15-crown-5 compounds needed to prepare 1 and 2 were synthesized as reported [16]. A 3-butenyl substituted 18-crown-6 was prepared in the same manner. The allyloxymethyl-substituted diaza-18-crown-6 compounds needed to prepare 3 5 were prepared by five different synthetic sequences [15]. The most often used procedure to prepare the diazacrowns is through cyclic *bis*-amide compounds like 6 and 7 but without the allyloxymethyl substituents and where R = hydrogen (Scheme II) [17–19]. This sequence to form 6 and 7 requires the preparation of allyloxymethyl-substituted triglycolic acid 10 [19]. In the cyclization step, the diacid chloride and the diamine must be added simultaneously to a large amount of toluene. The reduction of 6 and 7 gave a mixture of desired products 3 and 4 as well as the reduced hydroxymethyl analogs 8 and 9. Products 8 and 9 can easily be reacted with allyl bromide to give 3 and 4. In actual practice, the crude reduction product containing both 3 and 8 or 4 and 9 can be treated with allyl bromide to attain the maximum yields of 3 and 4. Because of the many steps in this reaction leading to an overall yield of 12% [15], a different synthetic route to the allyloxymethyl-substituted diazacrown compounds has been developed.

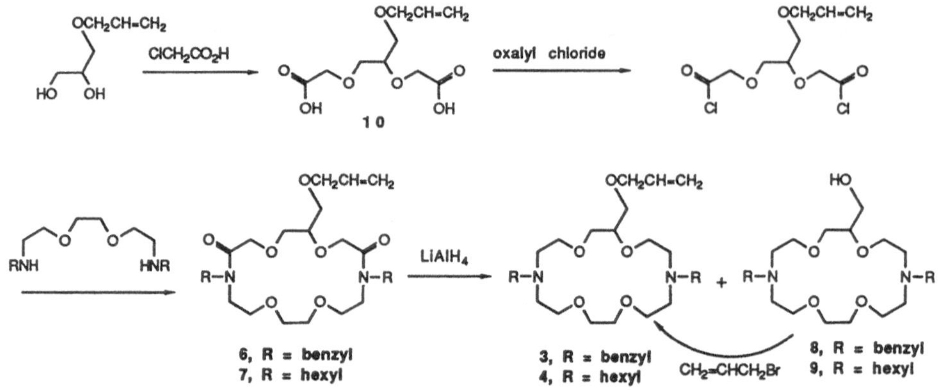

Scheme II. Allyloxymethyl diaza-18-crown-6 compounds from allyloxymethyl-substituted triglycolic acid.

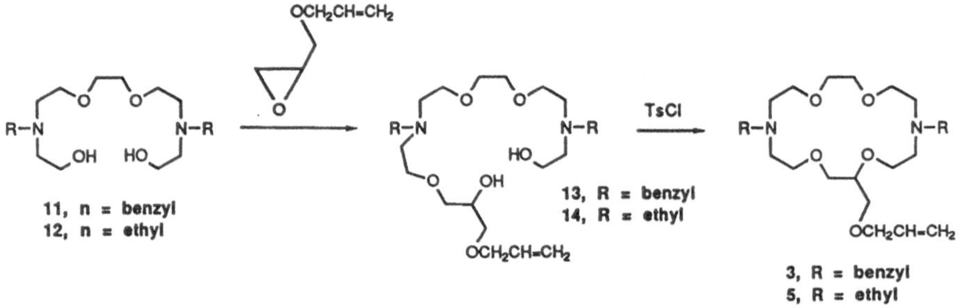

Scheme III. Allyloxymethyl diaza-18-crown-6 from N,N'-dialkyldiazaoligoethylene glycol.

Scheme III shows a convenient three step synthesis of the allyloxymethyl-substituted diazacrowns 3 and 5 [15]. The diazapentaethylene glycols 11 and 12 were prepared in good yields from triethylene glycol dichloride and N-benzyl- or N-ethylethanolamine [20]. The reaction of 11 or 12 with the allyloxymethyl-substituted epoxide gave a good yield of the allyloxymethyl-substituted diazahex-aethylene glycol 13 or 14. Compound 14 was purified by distillation. The Okahara ring closure reactions of 13 and 14 with tosyl chloride [21] gave excellent yields of 3 and 5, respectively. The overall yields for this process was 34% for 3 and 35% for 5 [15].

3. Determination of Log K Values for the Silica Gel-Bound Macrocycles

Log K values for the interaction of the silica gel-bound crown compounds with various cations have been determined [7]. A small amount of the silica gel material was placed in a chromatography column and the column was equilibrated with known concentrations of the cations studied. The cation binding properties of silica gel itself were determined by making blank measurements with plain silica gel. Binding of the cations of interest to the silica gel sites was made negligible by including excess concentrations of a cation (e.g., Mg^{2+}) which does not complex with the crown but competes effectively with the cations of interest for plain silica gel as measured in plain silica gel measurements. The equilibrium expression for 1 : 1 cation-macrocycle interaction is given by Equation 1:

$$K = \frac{F(1 + K_1[H^+] + K_1 K_2 [H^+]^2)}{(1 - f)[M^{n+}]} \tag{1}$$

where f = the fraction of ligand sites containing bound cations, K_1 and K_2 are the protonation constants applicable to 3, and $[M^{n+}]$ and $[H^+]$ are the equilibrium molar free cation and proton concentrations, respectively. The quantities $[M^{n+}]$ and $[H^+]$ are taken to be the effluent M^{n+} and H^+ concentrations as determined by atomic absorption spectroscopy (AA) and pH measurements when these concentrations are equal to the input concentrations. The total number of moles of ligand sites is known from the organic synthesis and was checked by quantitatively loading every macrocycle site with a strongly interacting cation (i.e., for 3, Ag^+ at concentrations $\geqslant 1 \times 10^{-5}$ M and pH > 8.0 were found to be sufficient to load the column with Ag^+). After equilibrium was reached, the column was stripped using either pure water, a complexing agent such as EDTA or, for 3, an acidic solution such as an acetic acid-acetate buffer. The resulting solution of known volume was analyzed for cation concentration by AA. The fraction of ligand sites containing the cation can now be calculated as moles of bound cation/mole of ligand. The pK_a values for 3 were determined by repeating the above log K experiments for Ag^+ and Cd^{2+} at several pH values and curve fitting the results according to Equation 1.

In Table I, log K values for the association of several cations with the silica gel-bound and analogous free crown compounds are compared. The agreement between the two sets of log K values is generally good. Four of the free crown log K values presented in the table were measured in methanol. Log K values for unbound 18-crown-6 interaction with Cd^{2+} and Ni^{2+} have only been obtained in methanol. Log $K(H_2O)$ values [5] for unbound 15-crown-5-Sr^{2+} (1.95) and unbound 18-crown-6-Ca^{2+} (<0.5) interaction have been measured but are unreliable since the heats of reaction used to make the calorimetric measurements were small. Values for the same reactions in methanol shown in Table I were also measured calorimetrically, but had large heats of reaction. Log K values for cation-macrocycle interactions have been shown to be 2 to 3 log K units higher in methanol than in water [22, 23]. Hence, the degree of interaction for metal ions with the silica gel-bound macrocycles for these four cases also appears to be close to that with the free macrocycles in water.

The similarity of the log K values for the silica gel-bound and unbound crown-cation interactions suggests that both crown entities are effectively solvated by the aqueous solution. Thus, the silica gel-bound macrocycles form complexes in the same manner as do the free crowns in water. On the other hand, the bonding of macrocycles to hydrocarbon polymers, such as polystyrene, causes a considerable modification in metal ion binding properties in both organic solvents and aqueous solutions. In particular, aqueous solutions cannot be treated effectively because the hydrocarbon polymers are not wetted by water [24, 25].

Table I.   Log K values at particular ionic strength (I) values for the interaction of M^{n+} with silica gel-bound and the analogous free (in parentheses) crown compounds (1-3)[a]

Cation	Log K		
	1	2	3
$H^+(1)$	–	–	8.9 ± 0.2 ($I = 0.5$) $(9.08^b - I = 0.1)$
$H^+(2)$	–	–	7.5 ± 0.2 ($I = 0.5$) $(7.94^b - I = 0.1)$
Sr^{2+}	0.57 ± 0.05 ($I = 3$) (2.63^c)	2.83 ± 0.01 ($I = 3$) (2.72)	2.4 ± 0.2 ($I = 0.5$) $(2.57^b - I = 0.1)$
Ba^{2+}	–	3.56 ± 0.01 ($I = 3$) (3.87)	–
Cd^{2+}	–	0.39 ± 0.13 ($I = 3$) (3.0^c)	5.0 ± 0.2 ($I = 0.5$) $(5.25^b - I = 0.1)$
Ca^{2+}	–	1.03 ± 0.10 ($I = 3$) (3.9^c)	–
Pb^{2+}	–	3.96 ± 0.05 ($I = 3$) (4.27)	–
Tl^+	1.38 ± 0.01 ($I = 3$) (1.23)	2.01 ± 0.06 ($I = 3$) $(2.2 - I = 0.1)$	–
Ni^{2+}	–	<0.2 ($I = 3$) (2.9^c)	–
K^+	–	1.75 ± 0.03 ($I = 1$) 2.10 ± 0.03 ($I = 0$)[d] (2.03)	
Ag^+	0.90 ± 0.15 ($I = 1$) (0.94)	1.61 ± 0.09 ($I = 0.5$) $(1.50, 1.60)$	8.2 ± 0.2 ($I = 0.5$) $(7.8^b - I = 0.1)$

[a] Ref. [5].
[b] The values are for diaza-18-crown-6 in Ref. [5].
[c] Values valid in methanol. See text for comparison to H_2O values.
[d] Value adjusted to 0 M ionic strength using the 1 M ionic strength activity coefficient for KNO_3 (0.443) from Ref. [23].

4. Removal, Separation and Concentration of Metal Ions Using the Silica Gel-Bound Macrocycles

The similarity of the log K values for the bound macrocycles to those involving the unbound macrocycles suggests that prediction of metal separations using silica gel-bound macrocyclic ligands should be possible. Thus, one has a powerful means to predict separations using available data compilations [5]. Such separations have been studied. For example, 18-crown-6 gel material **2** was used to separate 0.001 M concentrations of the alkaline earth cations [14]. The log K values for the association of unbound 18-crown-6 with Sr^{2+} and Ba^{2+} in water are 2.72 and 3.87, respectively [5]. The log K value for 18-crown-6-Ca^{2+} interaction should be between 0.9 and 1.9 as described in Section 3. Magnesium ions do not complex with 18-crown-6. All the heavier alkaline earth cations were separated from Mg^{2+} by their being retained on the gel **2** column. Sr^{2+} and Ba^{2+} were selectively retained on the column over Ca^{2+} by factors of 54 and 339, respectively, and Ba^{2+} was retained over Sr^{2+} by a factor of 10 [14]. These selectivity numbers are similar to the predicted selectivities based on the relative values of the respective metal ion-unbound macrocycle association constants [14]. All of these separations were performed in 1 M HNO_3 so that alkaline earth cation-blank silica gel interaction would be negligible.

In each of the separation experiments mentioned above, the metal ion that was retained on the column was stripped off the column in a concentrated form by an aqueous EDTA solution. The desired ion can then be recovered. This process can also be used to determine trace amounts of various cations in an aqueous solution. A large volume (i.e. 1 L) of water containing ppb levels of Hg^{2+} and Pb^{2+} and also containing several hundred ppm of Ca^{2+}, Mg^{2+}, Na^+, and K^+ as the nitrate salts was rapidly passed through 2–4 g of gel 3, 4, or 5. The metal ions formed complexes with the diazacrown and were removed quantitatively from solution. A much smaller amount of either 1 M aqueous HCl or EDTA was used to remove the metal ions from the column. Hence, a more concentrated solution of Hg^{2+} and Pb^{2+} was obtained and the solution was easily analyzed for Hg^{2+} and Pb^{2+} content by flame or inductively coupled plasma atomic absorption spectrophotometry.

5. Potential Separations of Enantiomeric Amines Using Silica Gel-Bound Chiral Macrocyclic Ligands

Chiral compounds 15, 16, 18, and 20 shown in Figure 1 have been synthesized and their selective complexation with the enantiomers of organic ammonium salts have been studied using the temperature dependent NMR technique [26–29]. In many instances, excellent chiral recognition by the chiral ligand for one of the enantiomers of a chiral organic ammonium salt was observed. Table II shows the differences in $\Delta G^{\ddagger}$ values as obtained by the temperature dependent NMR technique for the interaction of the chiral ligands with (R)- and (S)-1-(1-naphthyl)ethyl ammonium perchlorate. Also in Table II are differences of the calculated values of the conformational strain energies of the same (R) and (S) complexes. The strain energy is the main component of $\Delta G^{\ddagger}$. Furthermore the other components, that are temperature and solvent dependent, are to a large extent the same for the (R) and the (S) complex, and are therefore mostly canceled out in the difference [30, 31].

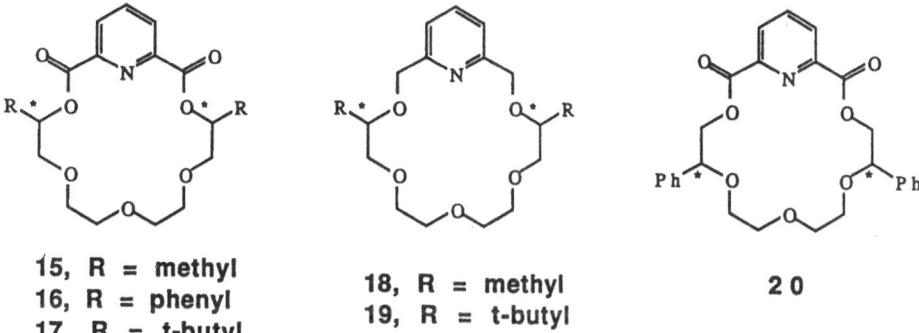

15, R = methyl
16, R = phenyl
17, R = t-butyl

18, R = methyl
19, R = t-butyl

20

Fig. 1. Chiral crown compounds.

Table II. Differences in free energy of activation values ($\Delta G^{\ddagger}$, kcal mol^{-1}) for the interaction of chiral macrocyclic ligands with (R)- and (S)-1-(1-naphthyl)ethyl ammonium perchlorate as determined experimentally (NMR) and as calculated from empirical energy functions.[a]

Ligand	$\Delta\Delta G^{\ddagger b}$	
	Observed	Calculated[a]
(S,S)-15	1.1[c]	0.7
(S,S)-16	0.7[c]	2.5
(S,S)-17	[d]	2.2
(S,S)-18	1.6[c]	1.7
(S,S)-19	[d]	2.2
(S,S)-20	0[c]	0.1

[a] Calculation method is given in Refs. [30] and [31].
[b] $\Delta\Delta G^{\ddagger} = \Delta G^{\ddagger}_R - \Delta G^{\ddagger}_S$
[c] Reference [28].
[d] Compounds have not yet been prepared.

Consequently, the calculated conformational strain energy difference represents approximately $\Delta\Delta G^{\ddagger}$. The calculated values for these interactions are based on the empirical functions of bond lengths, bond angles, torsional angles and interatomic coulombic and Lennard-Jones interactions. The calculations were performed by the Empirical Force Field method that has been described in detail [30, 31]. The method yields the equilibrium conformations for which the total energy is at a local minimum and determines their energy. In the present macrocyclic molecules, the number of such local low-energy minima is limited because the 'conformational space' is very restricted by the conditions of ring closure. A thorough scan of the conformational space produced all low energy minima of both enantiomers, including the most stable (R) and (S) conformations, whose energy difference is given in the last column of Table II.

The reliability of this method in making theoretical predictions that were borne out by experiment has been checked and confirmed in many instances [30, 31]. More recently, this method has been used in the discovery of a hydration pattern in enniatin crystals that escaped detection by X-ray diffraction analysis [32] and in the design of biomimetic ferric ion carriers [33].

Three of the compounds listed in Table II [(S,S)-15, 16, and 18] exhibited chiral recognition for the (R)-form of 1-(1-naphthyl)ethyl ammonium perchlorate. These compounds also exhibited chiral recognition for the (S)-form of the hydrogen perchlorate salt of methyl phenylalaninate [28]. (R,R)-15 exhibited chiral recognition for (S)-1-(1-naphthyl)ethyl ammonium perchlorate by nearly the same magnitude ($\Delta\Delta G^{\ddagger} = 0.9$) as (S,S)-15 did for the (R)-form ($\Delta\Delta G^{\ddagger} = 1.1$). The difference may be due either to enantiomeric impurities or to the finite accuracy of the measurements. (S,S)-20, with the phenyl substituents in the more mobile polyether portion of the molecule, did not show chiral recognition for the (R)- and (S)-forms of the ammonium salt (Table II) [28].

The computer calculations, based on empirical energy functions, gave similar values for the free energy differences for the interaction of either (S,S)-15 or (S,S)-18 (both with methyl substituents) and the (R)- and (S)-forms of the ammonium salt (see Table II). The calculated $\Delta\Delta G^{\ddagger}$ values for the interaction of (S,S)-**16** (with phenyl substituents) and the (R)- and (S)-forms of the ammonium salt were much higher than those for the observed interactions. Perhaps an estimate of the other contributions to $\Delta\Delta G^{\ddagger}$, or more refinement of the calculation parameters is needed in this case. The calculated value for the interaction of (S,S)-**20**, with the phenyl substituents in the less rigid polyether portion of the molecule, and the (R)- and (S)-forms of the ammonium salt also indicated little or no energy differences. The similarity, in most cases, of the calculated and observed $\Delta\Delta G^{\ddagger}$ values has prompted us to calculate the energy differences for the interaction of other chiral bis-alkyl-substituted macrocyclic compounds to determine the chiral ligands which will give the best chiral recognition for the organic amine enantiomers. Two such calculations for the bis-t-butyl-substituted macrocycles (**17** and **19**) are shown in Table II. We are presently preparing these compounds to determine the chiral recognition factors experimentally.

We will attach the chiral ligand which displays the best recognition for the enantiomers of the chiral organic amines to silica gel in a manner similar to that given in Scheme I. Although Cram and his coworkers have attached one chiral macrocycle to silica or polystyrene gel [34], few actual separations of chiral organic amines or ammonium salts have been carried out. We expect to demonstrate enantiomeric separations of specific enantiomeric amines.

Acknowledgement

The authors are grateful for financial support from the Office of Naval Research (Grant No. N00014-88-K-0115) and the United States–Israel Binational Science Foundation. The authors also wish to thank Dr. C. E. Felder for his assistance in the Empirical Force Field calculations.

References

1. C. J. Pedersen: *J. Am. Chem. Soc.* **89**, 7077 (1967).
2. D. J. Cram: *Science* **219**, 1177 (1983).
3. J. M. Lehn: *Science* **227**, 849 (1985).
4. *C & E N*, October 19, 1987, pp. 4, 5.
5. R. M. Izatt, J. S. Bradshaw, S. A. Nielsen, J. D. Lamb, J. J. Christensen and D. Sen: *Chem. Rev.* **85**, 271 (1985).
6. R. M. Izatt, G. C. LindH, R. L. Bruening, J. S. Bradshaw, J. D. Lamb and J. J. Christensen: *Pure Appl. Chem.* **58**, 1453 (1986).
7. J. S. Bradshaw, R. L. Bruening, K. E. Krakowiak, B. J. Tarbet, M. L. Bruening, R. M. Izatt and J. J. Christensen: *J. Chem. Soc., Chem. Commun.* 812 (1988).
8. J. J. Christensen, S. P. Christensen, M. P. Biehl, S. A. Lowe, J. D. Lamb and R. M. Izatt: *Sep. Sci. Technol.* **18**, 363 (1983).
9. R. M. Izatt, G. C. LindH, R. L. Bruening, P. Huszthy, C. W. McDaniel, J. S. Bradshaw and J. J. Christensen: *Anal. Chem.* **60**, 1694 (1988).
10. R. M. Izatt, G. C. LindH, G. A. Clark, Y. Nakatsuji, J. S. Bradshaw, J. D. Lamb and J. J. Christensen: *J. Membrane Sci.* **31**, 1 (1985).
11. R. M. Izatt, G. C. LindH, R. L. Bruening, P. Huszthy, J. D. Lamb, J. S. Bradshaw and J. J. Christensen: *J. Incl. Phenom.* **5**, 739 (1987).
12. J. D. Lamb, R. L. Bruening, R. M. Izatt, Y. Hirashima, P. K. Tse, and J. J. Christensen: *J. Membrane Sci.* **37**, 13 (1988).

13. R. M. Izatt, R. L. Bruening, M. L. Bruening, G. C. LindH, and J. J. Christensen: *Anal. Chem.*, in press.
14. R. M. Izatt, R. L. Bruening, M. L. Bruening, B. J. Tarbet, K. E. Krakowiak, J. S. Bradshaw and J. J. Christensen: *Anal. Chem.* **60**, 1825 (1988).
15. J. S. Bradshaw, K. E. Krakowiak, R. L. Bruening, B. J. Tarbet, P. B. Savage and R. M. Izatt: *J. Org. Chem.* **53**, 3190 (1988).
16. I. Ikeda, S. Yamamura, Y. Nakatsuji and M. Okahara: *J. Org. Chem.* **45**, 5355 (1980).
17. H. Stetter and J. Marx: *Liebigs Ann. Chem.* **607**, 59 (1957).
18. J. M. Lehn: *Acts. Chem. Res.* **11**, 49 (1978).
19. D. A. Babb, B. P. Czech and R. A. Bartsch: *J. Heterocyclic Chem.* **23**, 609 (1986).
20. K. E. Krakowiak and J. S. Bradshaw: *J. Org. Chem.* **53**, 1808 (1988).
21. K. Kuo, M. Miki, I. Ikeda and M. Okahara: *Tetrahedron Lett.* 4273 (1978).
22. H. Frensdorf: *J. Am. Chem. Soc.* **93**, 600 (1971).
23. R. M. Izatt, R. E. Terry, D. P. Nelson, Y. Chan, D. J. Eatough, J. S. Bradshaw, L. D. Hansen and J. J. Christensen: *J. Am. Chem. Soc.* **98**, 7626 (1976).
24. See M. Takagi and H. Nakamura: *J. Coord. Chem.* **15**, 53 (1986) for a review.
25. R. S. Paredes, N. S. Valera and L. F. Lindoy: *Aust. J. Chem.* **39**, 1071 (1986).
26. J. S. Bradshaw, B. A. Jones, R. B. Davidson, J. J. Christensen, J. D. Lamb, R. M. Izatt, F. G. Morin and D. M. Grant: *J. Org. Chem.* **47**, 3362 (1982).
27. B. A. Jones, J. S. Bradshaw, P. R. Brown, J. J. Christensen and R. M. Izatt: *J. Org. Chem.* **48**, 2635 (1983).
28. R. B. Davidson, J. S. Bradshaw, B. A. Jones, N. K. Dalley, J. J. Christensen, R. M. Izatt, F. G. Morin and D. M. Grant: *J. Org. Chem.* **49**, 353 (1984).
29. J. S. Bradshaw, P. K. Thompson, R. M. Izatt, F. G. Morin and D. M. Grant: *J. Heterocyclic Chem.* **21**, 897 (1984).
30. S. Lifson, C. E. Felder and A. Shanzer: *J. Am. Chem. Soc.* **105**, 3866 (1983).
31. S. Lifson, C. E. Felder, A. Shanzer and J. Libman: in *Progress in Macrocyclic Chemistry*, Vol. 3 (Eds. R. M. Izatt and J. J. Christensen), pp. 241–308, Wiley Interscience (1987).
32. S. Lifson, C. E. Felder and M. Dobler: *Acta Crystallogr. B* **43**, 179 (1987).
33. Y. Tor, J. Libman, A. Shanzer and S. Lifson: *J. Am. Chem. Soc.* **109**, 6517 (1987).
34. G. D. Y. Sogah and D. J. Cram: *J. Am. Chem. Soc.* **101**, 3035 (1979).

Journal of Inclusion Phenomena and Molecular Recognition in Chemistry 7 (1989), 137–153.

Inclusion Chemistry for the Modeling of Heme Proteins

DARYLE H. BUSCH* and NEIL A. STEPHENSON
Chemistry Department, The Ohio State University, Columbus, Ohio 43210, U.S.A.

(Received: 11 April 1988)

Abstract. Early attention to the modeling of heme proteins is enhancing the understanding of biochemistry. Those studies are also contributing to the development of techniques for the modeling of still more intricate, multifunctional, variously selective natural systems. Selectivity in simple systems may involve the molecular capability to bind only one of a family of related species or it may mean the ability to select and control one of a number of possible functions of a given bound species. Complicated systems simultaneously combine the two kinds of simple selectivities for two or more different classes of guest, often with synergistic interrelationships. The subject is developed around examples of binary, tertiary, and quarternary complexes designed to model the behavior of monooxygenases.

Key words. Selectivity, hemoglobin, myoglobin, O_2 transport, O_2 binding, O_2 carrier, cytochrome P450, synthetic enzymes, biomimics, molecular cavity, cyclidene, superstructure, lacuna, molecular design, host/guest complex, quarternary complex.

1. Introduction

Nature manages the intricate choreography of life processes through multilayers of variously coupled, highly selective molecular events. The chemist is presently learning to produce a plodding burlesque of that elegant molecular selectively. The active sites of heme proteins provide useful motivations for early learning experiences of this kind because of the relatively obvious and simple character of some of the selectivities involved. O_2 transport and storage and electron transport are the sole functions of the heme proteins hemoglobin, myoglobin and cytochrome c and these functions represent single specific chemical processes. Consequently, their biomimics constitute relatively straightforward synthetic goals. In contrast, enzymes such as cytochromes P450, and cytochrome c oxidase, the terminal member of the mitochrondrial electron transport chain, involve sequences of separately identifiable chemical processes, making duplication of their primary chemistry a much more complicated endeavor.

1.1. SELECTIVITY IN SIMPLE SYSTEMS

The work summarized here fits into the broad subject of molecular inclusion chemistry [1–5]. Cavities in molecules are tailored electronically and geometrically to the chemical purpose at hand. In simple systems, the goal might be selective coordination of metal ions or selective binding of small ligands to previously coordinated metal ions.

---

* Author for correspondence.

Selective coordination of metal ions is most often understood to mean the action of the ligand to bind to one metal ion in preference to some other metal ion [6–14]. Such selectivity is important in separations chemistry [13–17] and in nature [18–20] and is fundamental to coordination chemistry [6, 10, 11, 17].

An equally important but different kind of selective coordination is emphasized here. It involves providing those ligand characteristics that *select the capabilities* of a particular metal ion when it is bound to the ligand [21]. This constitutes one aspect in the design of metal complexes capable of combining selectivity with such small ligand molecules as O_2, CO, H_2S, or CO_2. We discuss the requirements for O_2 carriers to illustrate these general principles.

1.2. SELECTIVITY IN COMPLICATED SYSTEMS

Complicated natural systems typify the need to implement the simultaneous selective binding of more than one species. The example of small molecule binding to previously bound metal ions constitutes the first step in that direction. Enzyme/substrate complexes illustrate the need to study the simultaneous, variously dependent or independent, binding of both metal ions and substrates within more complicated ligand systems. Homogeneous catalysis is a vast area in need of the implicit ability to organize molecules in a single multicomponent, yet specifically arranged complex. Other likely areas of interest are intricate biomimics, synthetic enzymes, molecular machines, and molecular switches.

1.3. CLASSES OF CAVITY EFFECTS

Figure 1 attempts to summarize the general concepts associated with the use of inclusion chemistry to produce selective binding for various chemical purposes. The

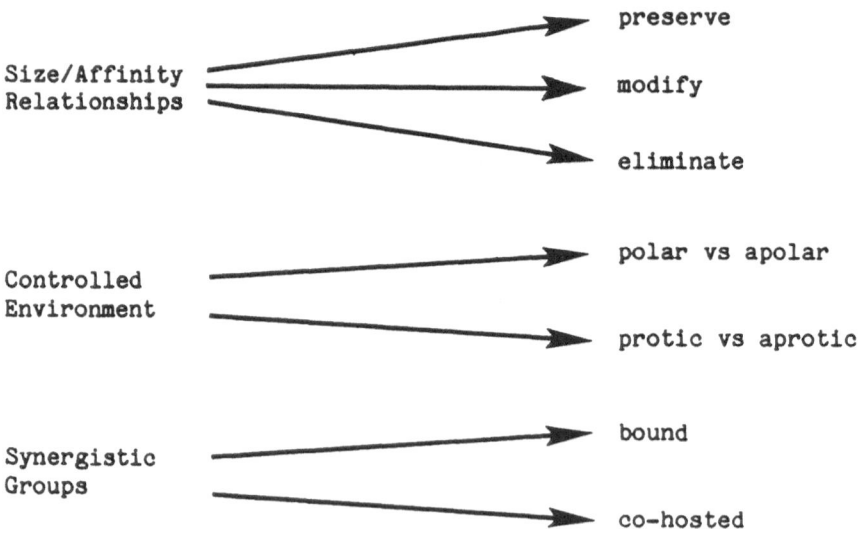

Fig. 1.   Classes of cavity effects.

affinity a given host molecule displays toward individuals within a family of guest species will depend on size, as well as other, relationships. The affinity may be preserved without change, modified because of crowding or the advantages of well placed attracting groups, or the affinity might even be eliminated. The interior of a cavity may be polar or apolar and this may have a profound effect on the affinity, the stability, or the reactivity of a particular guest. Specific groups within the cavity may exert synergistic effects and such groups may be moieties bound to the cavity wall or they may be co-hosted molecular entities. Such groups may bind directly to a guest species (e.g., hydrogen bond to bound O_2), or they might deliver or extract protons to/from a guest.

2. Molecular Recognition in Transition Metal Systems

2.1. SMALL LIGAND BINDING

As pointed out by others [22–24] and emphasized in our work [25, 26], electronic criteria can be specified for the ability of cobalt(II) or iron(II) complexes to bind to O_2. In the case of iron, this is indicated by a proximity of the electrode potential for the iron(III)/iron(II) couple of the complex in question to that of the natural oxygen carriers [27, 28]. For cobalt, oxygen affinities have been related to the potential of the cobalt(III)/cobalt(II) couple [23]. Well established families of oxygen carriers are illustrated in Figure 2. The obvious topological requirement of such a ligand is to leave a site available for the binding of O_2 while providing a hospitable electronic environment. For both iron(II) and cobalt(II) derivatives [29, 30], it has been shown that coordination of a competing sixth ligand can prevent O_2 binding.

The traditional role of ligands has been to bind to the metal ion and control its electronic and topological properties, spin state, coordination number, stereochemistry, and extent of coordination saturation [31–34]. Modern research expands the role of the ligand by appending additional moieties to the ligand [21, 35]. The added structural components are described collectively as *superstructure* [35]. For small ligand binding, the appended superstructure is used to provide a protected cavity, called a *lacuna* [36], within which the small target molecule can coordinate to the metal ion. Lacunar porphyrins [2, 37–43] present a fascinating array of structures as shown in Figure 3. These lacuna promote selective binding of the ligand of choice. Further, they serve to determine the immediate environment of the bound ligand (O_2) and to limit the interactions between the bound ligand and other molecular species. This, in turn, can have profound effects on the reactions of the bound group.

The early examples of superstructured porphyrins, the capped [40] and picket fence porphyrins [44, 45], showed that superstructures can select against the binding of large base molecules [46–48], leaving the cavity available for binding to small molecules. It has long been believed that the lesser selectivity of hemoglobin and myoglobin for CO over O_2, when compared to free porphyrins, derives from the relative steric suitability for O_2 of the vacant space within which the small ligand must bind (*vide infra*) [49, 50].

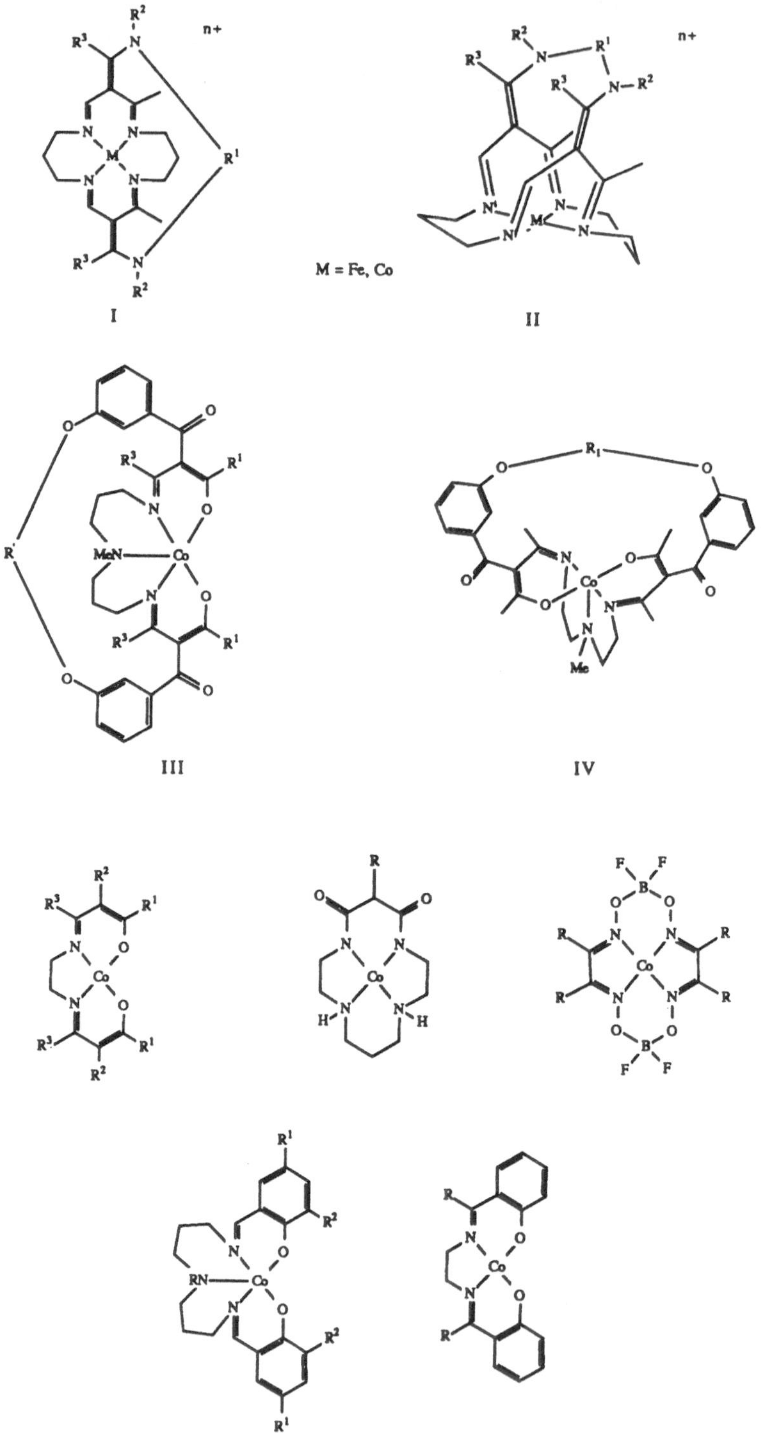

Fig. 2.   Some transition metal complexes which bind dioxygen reversibly.

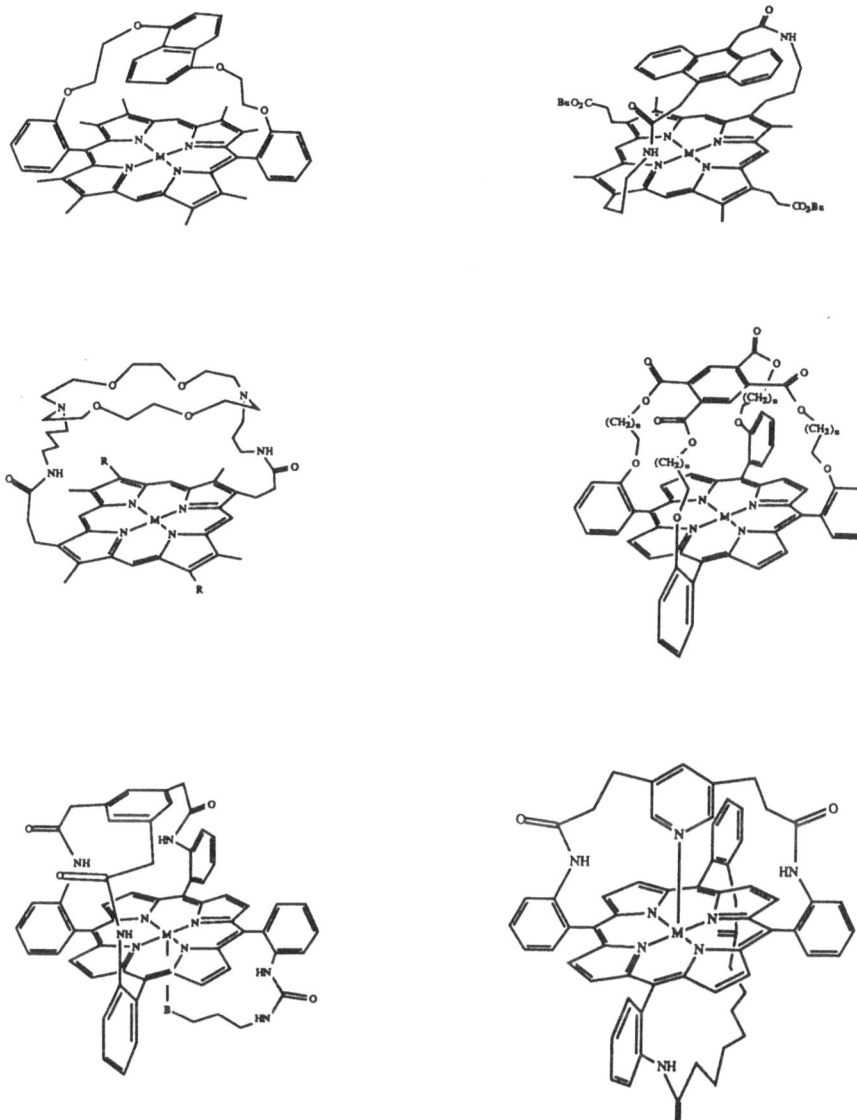

Fig. 3. Superstructured porphyrin complexes.

Many possible uses have been proposed for transition metal O_2 carriers [25, 26]. These applications require the capability of controlling certain critical properties [51], call them engineering parameters, of the oxygen carrier. For example, the partial pressures of O_2 at which the oxygen binds to or is released by the transition metal atom depends on the equilibrium constant for the binding process. Further, the rate of oxygen release from the metal complex is important in order to maximize the oxygen flux in a gaseous stream when an oxygen carrier is used to separate oxygen from air.

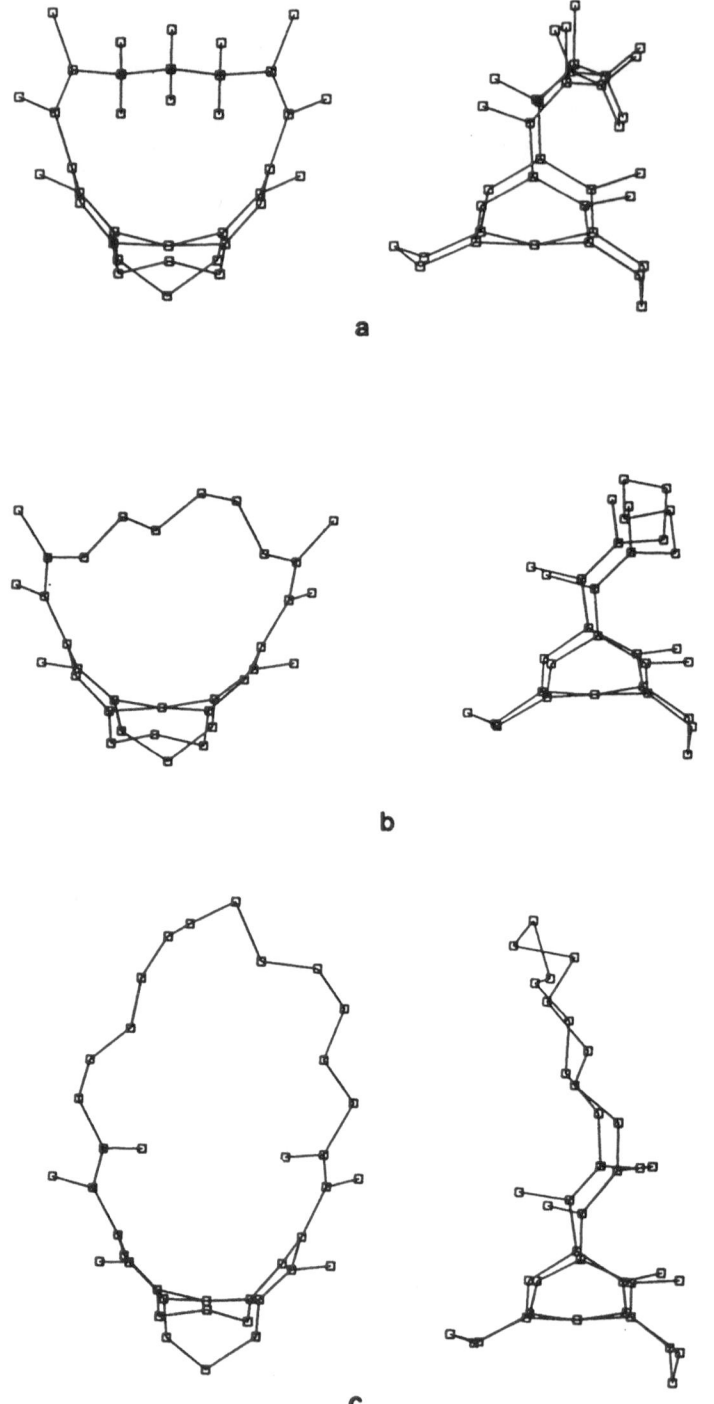

Fig. 4.   Lacunar cyclidene complexes:
(a) [Cu(Me,Me,C$_3$,cyclidene)]$^{2+}$, (b) [Co(Me,Me,C$_6$,cyclidene)]$^{2+}$, (c) [Ni(Me,Me,C$_{12}$,cyclidene)]$^{2+}$.

Both electronic and steric means can be used to modify and control the oxygen affinity of a metal complex. The lacunar cyclidene complexes [28, 52], structures **I** and **II** (Figure 2), bind O_2 in a cavity whose size is controlled by the size and orientation of a bridging group, R^1 (Figure 4) [25, 26, 28, 55]. The short trimethylene bridge so constricts the cavity that oxygen is not bound [54, 55]. Successively adding methylene groups from 4 through 7 produces a steady increase in O_2 affinity (Figure 5) [25, 53]. Thereafter, the O_2 affinity remains constant, indicating that the intrinsic affinity of the cyclidene ligands has been achieved; the smaller cavities decrease the affinity through steric constraint. Thus, it is possible to select the O_2 affinity as well as select against larger competing ligands. In fact, the cavity can be closed to all ligands, no matter how small.

Figure 6 shows the structures of lacunar complexes having small ligands in their cavities [56–59]. The O_2 adduct is easily formed because the cavity shape favors a small ligand that binds in an angular fashion (Figure 6b). In contrast, the thiocyanate ligand, that normally tends to be approximately linear when bound through nitrogen to cobalt(III), is forced into a distorted angular structure (Figure 6a).

The design characteristics of the lacunar cyclidene complexes can be incorporated into the structures of other better known oxygen carrier ligands. The Schiff base ligands derived from β-diketones [23, 60, 61] have been modified with the addition of a lacuna [62] and the optimal derivative shown in structures **III** and **IV**, Figure 2, also has a built-in axial ligand [63]. Special risers are built into the bridging group since this parent tetradentate ligand is basically planar.

The cyclidene complexes, structures **I** and **II**, Figure 2, have provided the only well established examples of nonporphyrin iron(II) dioxygen carriers [26, 28, 52]. As the examples in Figure 7 show, depending on the substituent on the bridge

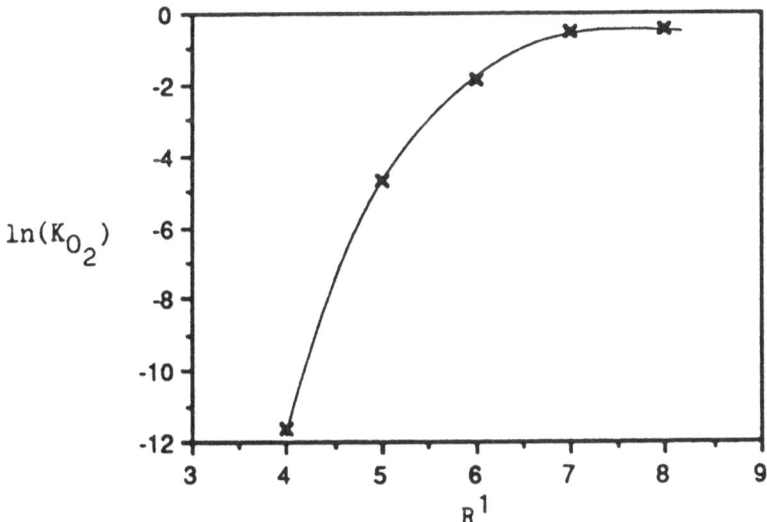

Fig. 5.   Dependence of dioxygen affinity on bridge length for [Co(Me,Me,R^1,cyclidene)]$^{2+}$ at 20°C in acetonitrile containing 1.5 M 1-MeIm.

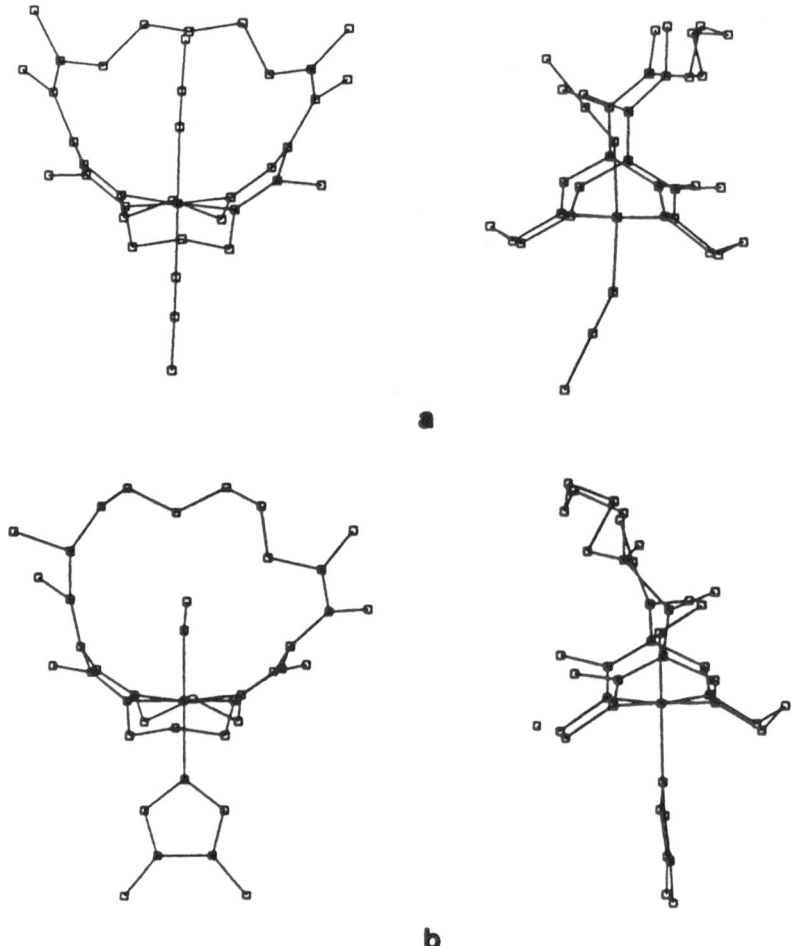

Fig. 6. Orientations of coordinated ligands for cobalt(II) cyclidene complexes:
(a) $[Co(Me,Me,C_6,cyclidene)(NCS)_2]^+$, (b) $[Co(Me,Me,C_6,cyclidene)O_2MeIm]^{2+}$.

nitrogen atoms, the cavity is either tall and narrow or short and wide. Effectively, the tall cavity accommodates the linearly coordinated CO molecule without change. However, the short, wide cavity must be greatly changed if CO is to be coordinated (Figure 8). The relative fits are reflected in the values of the equilibrium constants for CO binding to the two iron(II) complexes (Table I) [59, 64]. Obviously, the configuration of the cavity of these lacunar complexes can rearrange substantially in order to accept the guest species.

Examples discussed earlier in other contexts show how the cavity shape is changed upon entry of the small ligand (guest). The structure of the hexamethylene bridged cyclidene complex (Figure 4b) shows that the central two methylene groups of the chain fold back into the cavity, rather like the tail of a scorpion [57]. In contrast, as shown in Figure 6b, when a small ligand goes into the same cavity, this portion of the bridge swings up and away from the cavity into another favorable bridge conformation [57].

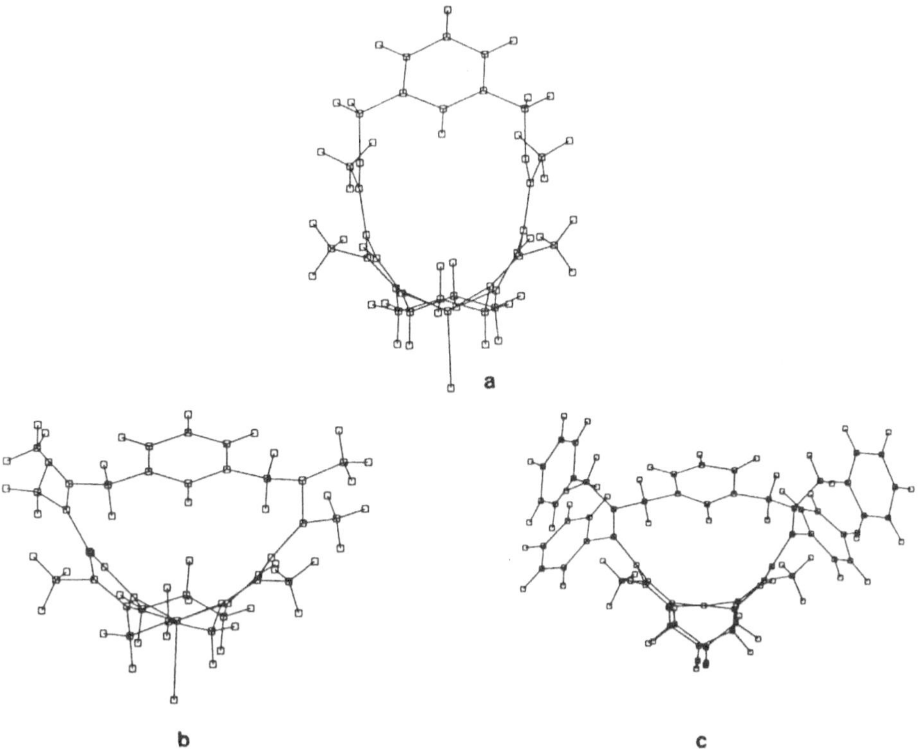

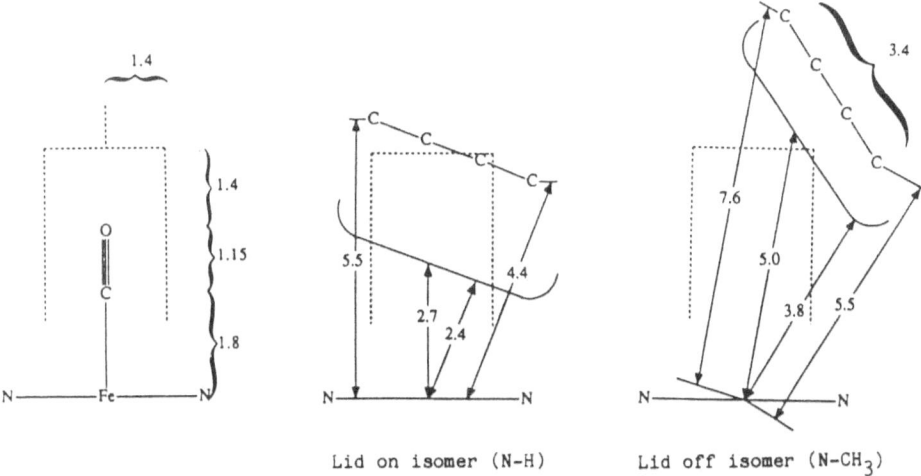

Fig. 7. Dependence of cavity shape on substituents for lacunar cyclidenes:
(a) [Fe(Me,H,*m*-xyl,cyclidene)Cl]$^+$, (b) [Fe(Me,Me,*m*-xyl,cyclidene)Cl]$^+$,
(c) [Ni(Ph,Bz,*m*-xyl,cyclidene)]$^{2+}$.

Fig. 8. Cavity sizes and the accommodation for CO in lacunar cyclidene complexes (distances reported in angstrom units).

Table I. Equilibrium constants for CO binding to iron(II) lacunar cyclidene complexes in acetonitrile at 0°C.

[Cl$^-$], M	[Fe(Me,Me,m-xyl,cyclidene)]$^{2+}$	[Fe(Me,H,m-xyl,cyclidene)]$^{2+}$
0	4.27×10^{-1}	–
8×10^{-3}	2.5×10^{-2}	0.5
1.0×10^{-3}	1.2×10^{-3}	1.0×10^{-2}

The two m-xylylene bridged iron(II) complexes having the short, wide cavities (Figure 7) provide essentially identical space for small ligand binding; however, the relative rates of oxidation by O_2 of the two complexes differ by a factor of 10^4 [26, 28, 55]. The complex having only methyl groups flanking the cavity oxidizes with a half life of about an hour at $-25°C$ while the species having the bulky phenyl and benzyl groups in the same positions oxidizes with an estimated half life of about 2 years under the same conditions. In fact, the latter species has a half life at ambient conditions of many hours. It has been proposed that the mechanism of autoxidation of these complexes involves a competition between O_2 binding and electron transfer when the O_2 molecule approaches the iron(II) complex [26, 55, 65]. In this context, the extra bulk appears to greatly impede the electron transfer process. This remarkable steric effect shows yet another way in which superstructure can help control the behavior of molecules.

2.2. SIMULTANEOUS METAL ION AND SUBSTRATE BINDING

The capability to simultaneously form inclusion complexes with organic molecules through the influence of hydrophobic interactions, was added to the superstructured cyclidene ligands by enlarging the permanent void (Figure 9) [66, 67]. The expected regiospecific mode of binding is shown in Figure 10.

The copper(II) complex was selected because it has a single unpaired electron, making it useful for the nuclear magnetic resonance technique used in these studies [68, 71]. The magnetic field due to the unpaired electron on the metal center affects the rates of relaxation of the protons of the organic *guest* molecule that invades the permanent cavity of the ligand. Because there are no bonds between the copper(II) atom and any part of the guest molecule, the rate of this relaxation process has a straightforward dependence on the distance between the metal atom and the protons in question [69, 70]. Thus, one can map the position of the guest, with respect to the metal atom, in these systems.

The data in Table II show that the α protons, and therefore the hydroxyl group, of an alcohol are furthest from the metal ion [71]. At the same time, the ω protons of the alcohol are nearest the metal center. Thus, the alcohols all enter into host/guest complexation regiospecifically. Also since the OH group is at about the same distance from the metal ion in all cases, it must remain in the solvent sheath of the host molecule. Thus, the binding arises from hydrophobic relationships involving the alkyl group of the alcohol and the hydrophobic interior of the cavity.

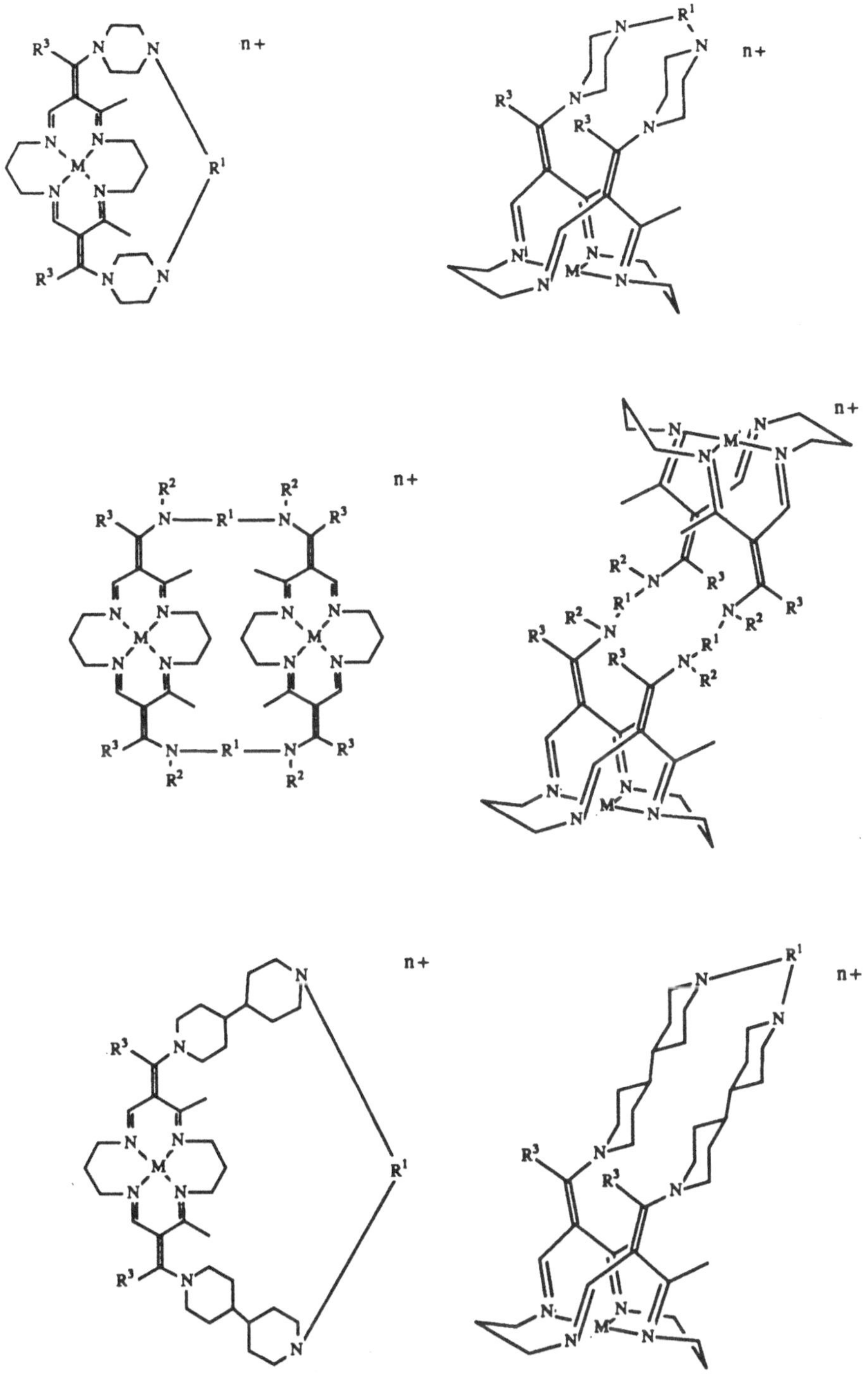

Fig. 9. Cyclidene complexes containing large cavities for guest binding.

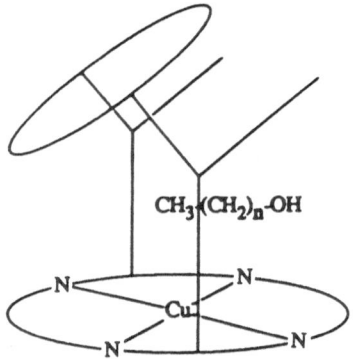

$CH_3(CH_2)_n-OH$

Fig. 10. Schematic representation of guest binding to a copper(II) lacunar cyclidene complex.

Combining the results of the solution NMR studies on the guest/host complexes with the X-ray determined coordinates for the atoms comprising the host molecule, it is possible to use modern computer graphics and molecular mechanics to produce reasonable images of these species [71]. Figure 11 shows the results which indicate that the guest molecule resides near the top of the cavity, leaving ample space for the binding of O_2 in case the metal ion were appropriate for that reaction, as well.

Corresponding measurements with phenols in place of the alcohols but using the same host molecules revealed that the cavity in this cyclidene complex is too small to accommodate a benzene ring (Figure 12) [66]. Replacing the piperazine riser in the host molecule by a bipiperidine moiety produces a much larger cavity that accommodates the phenols very easily [72].

Table II. Calculated distances for various protons in several alcohols for 'vaulted' cyclidene hosts (^{1}H NMR $v = 300$ MHz; [Host] $= 6.48 \times 10^{-5}$ M; [guest] $= 10^{-2}$ M).

	distance, Å			
alcohols	α	β	γ	δ
CH_3OH	8.6			
CH_3CH_2OH	8.4	7.5		
$CH_3CH_2CH_2OH$	9.0	8.2	7.5	
$CH_3CH_2CH_2CH_2OH$	9.0	7.7	7.4	6.6
average	8.8 ± 3	7.8 ± 3	7.5	6.6
t-$(CH_3)_3COH$		7.4		
i-$(CH_3)_2CHOH$	9.2	8.2		
$CH_2{=}C(CH_3)CH_2OH$	9.0		7.6	
		(7.2)		
overall average	8.9 ± 2	7.8 ± 3	7.4 ± 1	(6.6)

3.1 Multisite Hosts Containing Transition Metal Ions

The concept that a single ligand, of admittedly somewhat complicated design, might organize several molecular entities within a single guest/host/coordination entity is particularly intriguing. Models that exist in nature are enzyme/cofactor/substrate

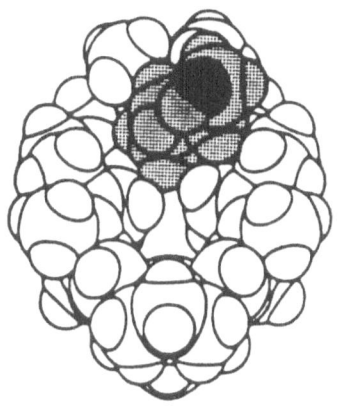
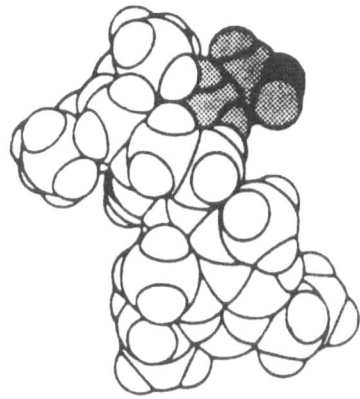

Fig. 11.   Binding of *n*-BuOH to a 'vaulted' cyclidene complex.

complexes, or the complicated clusters of proteins that are associated with some functions, e.g., cytochrome c oxidase. Relatively unimaginative examples of the various classes of species that might be involved are given in Figure 13. In fact, if chemists are to gain control over intricate molecular processes, then some control must be gained over the organization of molecules during the course of the crucial events. The coordination template effect [73] constituted the first example of such control, but it was concerned with only a single kind of organizational process. The range of possibilities for organization of molecular events within multisite guest/ host chemistry appears limitless. Again, one thinks of biomimicry, catalysis, and molecular machines, including molecular switches.

Here we consider a rare example of a quaternary complex in which four separate, distinctly different molecular species reside simultaneously within a single complex [74]. The general concept, shown in Figure 14, is a model for the so-called *ternary complex* of cytochromes P450 [75]. This is a misnomer since the complex is actually quaternary; i.e., it involves the enzyme protein, the heme prosthetic group, the substrate, and the O_2 cofactor. Our complex involves the *vaulted* cyclidene ligand, the cobalt ion, the O_2 molecule and a guest molecule [74].

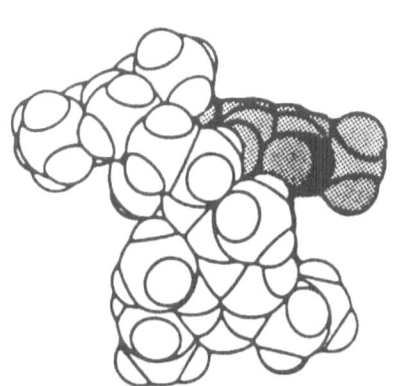

Fig. 12.   Binding of 2,5-dimethylphenol to a 'vaulted' cyclidene complex.

Probable Guest Species—
    Transition metal ion
    Substrate molecule
    Cofactor or cofactors

Examples of Metal Ions—
    Redox enzyme models: Fe, Co, Mn, Cu
    Solvolytic enzyme models: Zn, Co, Mg, Mn

Examples of Substrates—
    Select linear hydrocarbons for oxidation
    Select ester or amide groups for hydrolysis

Examples of Cofactors—
    O_2 binding for monooxygenase models
    Nucleophile binding for esterase model

Fig. 13.   Multisite hosts containing transition metal ions.

Most of the properties favorable to reversible O_2 binding have been preserved or redesigned into the ligand molecule used in this work. The critical experiments involve the same kinds of measurements used in studying guest/host complexes as described above [68, 71, 74]. However, they differ in that the dioxygen complex of the cobalt/cyclidene complex is used to define the regiospecific binding of the substrate molecule.

The cobalt(II) complexes of the cyclidene ligands are low spin and excellent O_2 carriers [25, 53]. Their O_2 adducts are typical for such cobalt derivatives and have a single unpaired electron, that is localized mainly on the O_2 moiety. Thus, the unpaired electron of the bound O_2 can be used to probe the protons of the

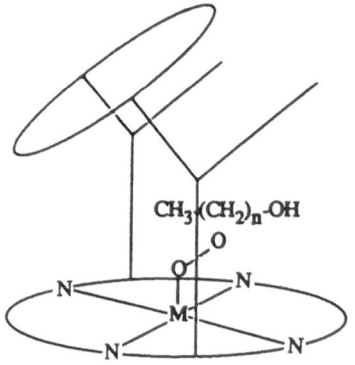

Fig. 14.   Model for the ternary complex of cytochromes P450.

guest/host molecule. A difficulty is associated with the fact that the delocalization of the electron deviates from the simple model used previously to calculate distances on the basis of magnetically accelerated relaxation rates (as described above) [69, 70]. However, the very fact that this system gives the same effect on guest proton relaxation rates is proof of the presence of the guest within the cavity of the host molecule [74].

Additional problems that had to be confronted derived from the stability of the O_2 adduct. The great stability of the cobalt/cyclidene/O_2 adducts depends on the presence of a reasonably small lacuna in the structure [25, 53]. The vaulted complexes have very large cavities and the resulting cobalt/O_2 adducts autoxidize relatively rapidly. Consequently, it was necessary to study the quaternary complex at the lowest possible temperature; 1°C for the solvent D_2O. This assured the saturation of the O_2 forming equilibrium so that only a single NMR relaxing agent was present in the solvent. At the same time, the lower temperature reduced the quality of the measurement to some extent.

Within the limits of the measurements, the results show unequivocally that the cobalt(II)/O_2 complex is serving as host for the organic guest molecule. The calculated distances between the center of electron spin density and the protons are shorter than in the cases where the electron is confined to a metal ion. This is expected since the O_2 moiety resides between the metal ion site and the guest site. The limitations described advise against quantitative interpretation of the difference [53].

Acknowledgement

The financial support of the National Institutes of Health, Grant No. GM10040 and of the National Science Foundation, Grant No. CHE8703723 is greatly appreciated.

References

1. *Inclusion Compounds*, J. L. Atwood, J. E. D. Davies and D. D. MacNicol (Eds.), Academic Press, Orlando, 1984.
2. T. J. Meade and D. H. Busch: in *Progress in Inorganic Chemistry*, S. J. Lippard (Ed.), Vol. 33, pp. 59–126, Wiley, New York, 1985.
3. D. J. Cram and J. M. Cram: *Acc. Chem. Res.* **11**, 8 (1978).
4. R. Breslow: *Adv. Chem. Ser.* **191**, 1 (1980).
5. N. K. Dalley: *Synthetic Multidentate Macrocyclic Compounds*, R. M. Izatt and J. J. Christensen (Eds.), Academic Press, New York, 1978, p. 209.
6. L. F. Lindoy: *Synthesis of Macrocycles – the Design of Selective Complexing Agents*, R. M. Izatt and J. J. Christensen (Eds.), Wiley, New York, pp. 53–92, and references therein.
7. P. G. Potyin and J-M. Lehn: *Synthesis of Macrocycles– the Design of Selective Complexing Agents*, R. M. Izatt and J. J. Christensen (Eds.), Wiley, New York, 1987, pp. 167–239.
8. P. E. Riley, K. Abu-Dari and K. N. Raymond: *Inorg. Chem.* **22**, 3940 (1983).
9. P. Muehl and K. Gloe: *Int. Solvent Extr. Conf.* [Proc.], Volume 1, Paper 80–236, 1980.
10. J. J. Christensen, D. J. Eatough and R. M. Izatt: *Chem. Rev.* **74**, 351 (1974).
11. H. Tsukube: *J. Coord. Chem.* **16**, 101 (1987).
12. D. J. Cram and S. P. Ho: *J. Amer. Chem. Soc.* **108**, 2998 (1986) and references therein.
13. J. O. Reynolds and I. V. Flavelle: *Symp. Ser.-Australas. Inst. Min. Metall.* **43**, 33 (1985).
14. Y. Anjaneyulu, P. C. Mouli, C. S. Kavipurapu and M. R. P. Reddy: *J. Indian Chem. Soc.* **64**, 377 (1987).

15. J. Rebek, Jr. and R. V. Wattley: *J. Am. Chem. Soc.* **102**, 4853 (1980).
16. M. Okahara and Y. Nakatsiyi: *Top. Curr. Chem.* **128**, 37 (1985).
17. C. J. Pederson: *Fed. Proc., Fed. Am. Soc. Exp. Biol.* **27**, 1305 (1968).
18. B. R. Byers: *NATO ASI Ser. A* **117**, 217 (1986).
19. D. A. Baldwin and T. J. Egan: *S. Afr. J. Sci.* **83**, 22 (1987).
20. M. Dobler: *Ionophores and their Structures*, Wiley, New York, 1981.
21. D. H. Busch and C. Cairns: *Synthesis of Macrocycles – the Design of Selective Complexing Agents*, R. M. Izatt and J. J. Christensen (Eds), Wiley, New York, 1987, pp. 1–52.
22. L. H. Vogt, Jr., H. M. Faigenbaum and S. E. Wilberly: *Chem. Rev.* **63**, 269 (1963).
23. M. J. Carter, D. P. Rillema and F. Basolo: *J. Am. Chem. Soc.* **96**, 392 (1974).
24. T. G. Traylor, D. K. White, D. H. Campbell and A. P. Berzinis: *J. Am. Chem. Soc.* **103**, 4932 (1981).
25. D. H. Busch: *Oxygen Complexes and Oxygen Activation by Transition Metals*, A. E. Martell and D. T. Sayer (Eds.), Plenum, New York, 1988, pp. 61–85.
26. D. H. Busch: Totally Synthetic Iron(II) Dioxygen Carriers Based on Lacunar Cyclidene Ligands, *La Transfusions del Sangue*, **33**, No. 1, 57 (1988).
27. D. H. Busch, D. G. Pillsbury, F. V. Lovecchio, M. A. Tait, Y. Hung, S. C. Jackels, M. C. Rakowski, W. P. Schammel and L. Y. Martin: *ACS Symp. Ser.* **38**, 32 (1977).
28. N. Herron, L. L. Zimmer, J. J. Grzybowski, D. J. Olszanski, S. C. Jackels, R. W. Callahan, J. H. Cameron, G. G. Christoph and D. H. Busch: *J. Am. Chem. Soc.* **105**, 6585 (1983).
29. C. J. Weschler, D. L. Anderson and F. Basolo: *J. Am. Chem. Soc.* **97**, 6707 (1975).
30. F. A. Walker: *J. Am. Chem. Soc.* **95**, 1154 (1973).
31. *The Chemistry of Coordination Compounds*, J. C. Bailar (Ed.), Reinhold, New York, 1956.
32. *Coordination Chemistry*, A. E. Martell (Ed.), Van Nostrand Reinhold, New York, 1971.
33. A. E. Martell and M. Calvin: *Chemistry of Metal Chelate Compounds*, Prentice-Hall, New York, 1954.
34. *Coordination Chemistry of Macrocyclic Compounds*, G. A. Melson (Ed.), Plenum Press, New York, 1979.
35. W. P. Schammel, L. L. Zimmer and D. H. Busch: *Inorg. Chem.* **19**, 3159 (1980).
36. D. H. Busch, D. J. Olszanski, J. C. Stevens, W. P. Schammel, M. Kojima, N. Herron, L. L. Zimmer, K. A. Holter and J. Mocak: *J. Am. Chem. Soc.* **103**, 1472 (1981).
37. J. E. Baldwin, M. J. Crossley, T. Klose, E. A. O'Rear and M. K. Peters: *Tetrahedron* **38**, 27 (1982).
38. T. G. Traylor, M. J. Mitchell, S. Tsuchiya, D. H. Campbell, D. V. Stynes and N. Koga: *J. Am. Chem. Soc.* **103**, 5234 (1981).
39. C. K. Chang: *J. Am. Chem. Soc.* **99**, 2819 (1977).
40. J. Almog, J. E. Baldwin and J. Huff: *J. Am. Chem. Soc.* **97**, 227 (1975).
41. J. P. Collman, J. I. Brauman, T. J. Collins, B. R. Iverson, G. Lang, R. G. Pettman, J. L. Sessler and M. A. Walters: *J. Am. Chem. Soc.* **105**, 3038 (1983).
42. M. Momenteau and D. Lavalette: *J. Chem. Soc., Chem. Commun.*, 341 (1982).
43. T. G. Traylor, N. Koga, L. A. Dearduff, P. N. Swepston and J. A. Ibers, *J. Am. Chem. Soc.* **106**, 5132 (1984).
44. J. P. Collman, R. R. Gagne, T. R. Halbert, T. R. Marchon and J. C. Reed: *J. Am. Chem. Soc.* **95**, 7868 (1973).
45. J. P. Collman: *Acc. Chem. Res.* **10**, 265 (1977).
46. P. E. Elis, Jr., J. E. Linard, T. Szymanski, R. D. Jones, J. R. Budge and F. Basolo: *J. Am. Chem. Soc.* **102**, 1889 (1980).
47. T. Hashimoto, J. E. Baldwin, F. Basolo, R. L. Dyer, M. J. Crossley: *J. Am. Chem. Soc.* **104**, 2101 (1982).
48. J. P. Collman, J. I. Brauman, K. M. Doxsee, J. L. Sessler, R. M. Morris and Q. H. Gibson: *Inorg. Chem.* **22**, 1427 (1983).
49. T. G. Traylor: *Acc. Chem. Res.* **14**, 102 (1981).
50. E. Antonini and M. Brunori: *Hemoglobin and Myoglobin and their Reaction with Ligands*, Elsevier, New York, 1971, p. 93.
51. J. A. T. Norman, G. P. Pez and D. A. Roberts: *Oxygen Complexes and Oxygen Activation by Transition Metals*, A. E. Martell and D. T. Sawyer (Eds.), Plenum, New York, 1988.
52. N. Herron and D. H. Busch: *J. Am. Chem. Soc.* **103**, 1236 (1981).

53. J. C. Stevens and D. H. Busch: *J. Am. Chem. Soc.* **102**, 3285 (1980).
54. N. Herron, M. Y. Chavan and D. H. Busch: *J. Chem. Soc., Dalton Trans.*, 1491 (1984).
55. L. Dickerson: Ph.D. Thesis, The Ohio State University, 1986.
56. D. H. Busch, J. C. Stevens, P. D. Jackson, D. Nosco, N. Matsumoto, M. Kojima and N. Alcock: submitted for publication.
57. J. C. Stevens, P. J. Jackson, W. P. Schammel, G. G. Christoph and D. H. Busch: *J. Am. Chem. Soc.* **102**, 3283 (1980).
58. P. J. Jackson, C. Cairns, W.-K. Lin, N. W. Alcock and D. H. Busch: *Inorg. Chem.* **25**, 4015 (1986).
59. D. H. Busch, L. L. Zimmer, J. J. Grzybowski, D. J. Olszanski, S. C. Jackels, R. C. Callahan and G. G. Christoph: *Proc. Natl. Acad. Sci., U.S.A.* **78**, 5919 (1981).
60. E. C. Neiderhoffer, J. H. Timmons and A. E. Martell: *Chem. Rev.* **84**, 137 (1984).
61. R. D. Jones, D. A. Summerville and F. Basolo: *Chem. Rev.* **79**, 139 (1979).
62. D. Ramprasad, W.-K. Lin, K. A. Goldsby and D. H. Busch: *J. Am. Chem. Soc.* **110**, 1480 (1988).
63. R. Delgado, M. W. Glogowski and D. H. Busch: *J. Am. Chem. Soc.* **109**, 6855 (1987).
64. L. L. Zimmer: Ph.D. Thesis, The Ohio State University, 1979.
65. N. Herron, L. Dickerson and D. H. Busch: *J. Chem. Soc., Chem. Commun.* 884 (1983).
66. K. J. Takeuchi and D. H. Busch: *J. Am. Chem. Soc.* **103**, 2421 (1981).
67. K. J. Takeuchi and D. H. Busch: *J. Am. Chem. Soc.* **105**, 6812 (1983).
68. W.-L. Kwik, N. Herron, K. J. Takeuchi and D. H. Busch: *J. Chem. Soc., Chem. Commun.*, 409 (1983).
69. A. S. Mildvan and R. K. Gupta: *Methods Enzymol.* **49**, 322 (1978).
70. I. Solomon and J. Bloembergen: *Chem. Phys.* **25**, 261 (1956).
71. T. J. Meade, W.-L. Kwik, N. Herron, N. W. Alcock and D. H. Busch: *J. Am. Chem. Soc.* **108**, 1954 (1986).
72. T. J. Meade, N. W. Alcock and D. H. Busch: unpublished results.
73. M. C. Thompson and D. H. Busch: *J. Am. Chem. Soc.* **86**, 3651 (1964).
74. T. J. Meade, K. J. Takeuchi and D. H. Busch: *J. Am. Chem. Soc.* **109**, 725 (1987).
75. R. E. White and M. J. Coon: *Ann. Rev. Biochem.* **49**, 315 (1980).

Journal of Inclusion Phenomena and Molecular Recognition in Chemistry 7 (1989), 155–168.
© 1989 *by Kluwer Academic Publishers.*

Progress Toward Artificial Metalloenzymes: New Ligands for Transition Metal Ions and Neutral Molecules

JOANNE F. KINNEARY, THERESE M. ROY, JEFFREY S. ALBERT,
HEUNGSIK YOON, THOMAS R. WAGLER, LUCY SHEN, and
CYNTHIA J. BURROWS*
*Department of Chemistry, State University of New York at Stony Brook, Stony Brook,
NY 11794-3400, U.S.A.*

(Received: 1 February 1988)

Abstract. New nickel catalysts have been developed for the oxidation of alkenes to epoxides, alcohols, aldehydes and ketones. Mechanistic studies indicate that the oxidation reactions are very sensitive to the nature of the catalyst; only certain ligands including salen and the macrocycles cyclam and dioxocyclam render Ni(II) effective as a catalyst. A Ni(III) or Ni(IV)-oxo species has been postulated as the catalytically active oxidant which leads to oxygen atom transfer to alkenes in a stepwise process. Both iodosylbenzene and hypochlorite have been used as terminal oxidants; both systems give high yields of epoxidation of alkenes and varying amounts of C=C bond cleavage products. In order to reach an ultimate goal of hydrocarbon oxidation within a molecular recognition system, new molecular receptors for organic substrates have been investigated. The receptors are constructed from two subunits of cholic acid and display amphophilic character – a hydrophobic exterior and a hydrophilic interior. Conformational properties in the presence of polar guests in CDCl$_3$ are described.

Key words. Metalloenzymes, catalysis, oxidation, nickel, alkene.

1. Introduction

Nature has engineered proteins with complex structural features in order to carry out sophisticated organic transformations. One example is cytochrome P-450, an enzyme capable of selectively hydroxylating a variety of hydrocarbon substrates [1]. This process is an enviable one; reagents for organic synthesis which achieve even non-selective hydroxylation of hydrocarbons are few. Due to the extraordinary interest in both metal-catalyzed oxidation of organic substrates [2] and in the molecular recognition of organic compounds [3], it is a compelling challenge to construct a mimic of an oxidative enzyme. The approach involves the assembly of an artificial enzyme possessing a reactive site oriented with respect to a binding site for the substrate. Such an approach has been undertaken by others in efforts to generate artificial hydrolases and oxidases [4]. An alternative approach is to control the steric environment around a catalytic site without incorporation of a discrete substrate binding site. Notable examples of this approach are in the area of transition metal-catalyzed olefin epoxidation [5]. It was our intention to begin by studying the two aspects of the former approach individually in order to develop both new reagents for oxidation as well as new molecular receptors.

---

* Author for correspondence.

2. Olefin Oxidation using Nickel Complexes

Numerous model systems for cytochrome P-450 have focussed upon the use of porphyrin complexes of chromium, manganese and iron in the presence of O_2, peracids, hydroperoxides, N-oxides or iodosylarenes as terminal oxidant [5, 6]. It is generally agreed that these systems generate a high valent metal-oxo species which participates in a rebound mechanism to shuttle an oxygen atom to a hydrocarbon [7].

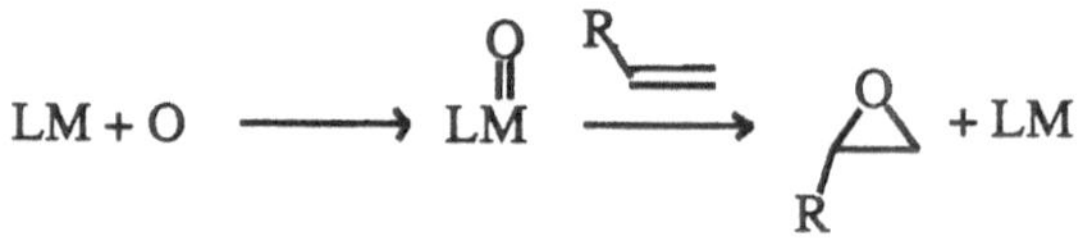

$$LM + O \longrightarrow LM\overset{O}{\parallel} \xrightarrow{\overset{R}{\diagup\!\!=}} \overset{O}{\underset{R}{\triangle}} + LM$$

L = ligand, O = oxidant

We sought macrocyclic ligands for transition metals that might offer a wider range of synthetic architectures for the construction of a substrate binding site adjacent to the metal oxidation site. There are only a few examples of non-porphyrinic metal complexes which catalyze hydrocarbon hydroxylation or olefin epoxidation [8]. Although most first row transition metals have been shown to mediate olefin epoxidation under some conditions, nickel was conspicuously absent from this list when we began this work. This was surprising in light of the vast amount of research directed toward nickel polyamine complexes [9] and the discovery of a tetraaza-macrocyclic nickel complex, Factor F_{430}, present in a redox enzyme [10]. Since the cyclam[1] ligand is well known to stabilize the Ni(III) oxidation state [11] and certain polyamine complexes have allowed generation of the Ni(IV) state [12] we undertook a program of study using tetraaza-macrocyclic nickel complexes as oxidation catalysts.

2.1. USE OF NICKEL(II) CYCLAM COMPLEXES WITH IODOSYLBENZENE

Concurrent work in our laboratory [13] and in Kochi's [14] has shown that Ni(II) cyclam complexes act as catalysts for olefin epoxidation and, to a lesser extent, alkane hydroxylation when iodosylbenzene (PhIO) is used as terminal oxidant.

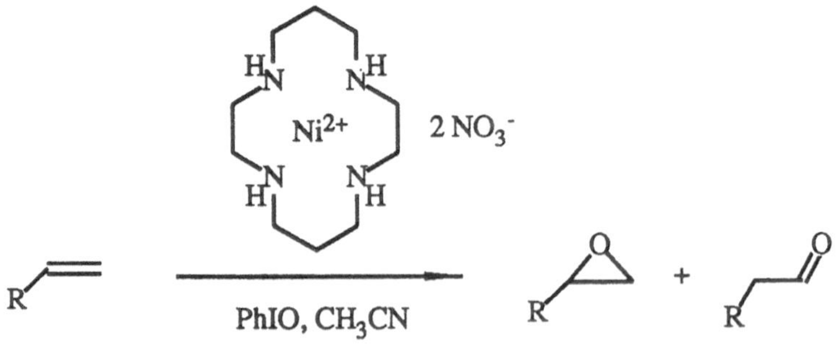

Table I. Products of Ni(cyclam)(NO$_3$)$_2$-catalyzed oxidation of phenyl-substituted alkenes using PhIO as terminal oxidant[a,b]

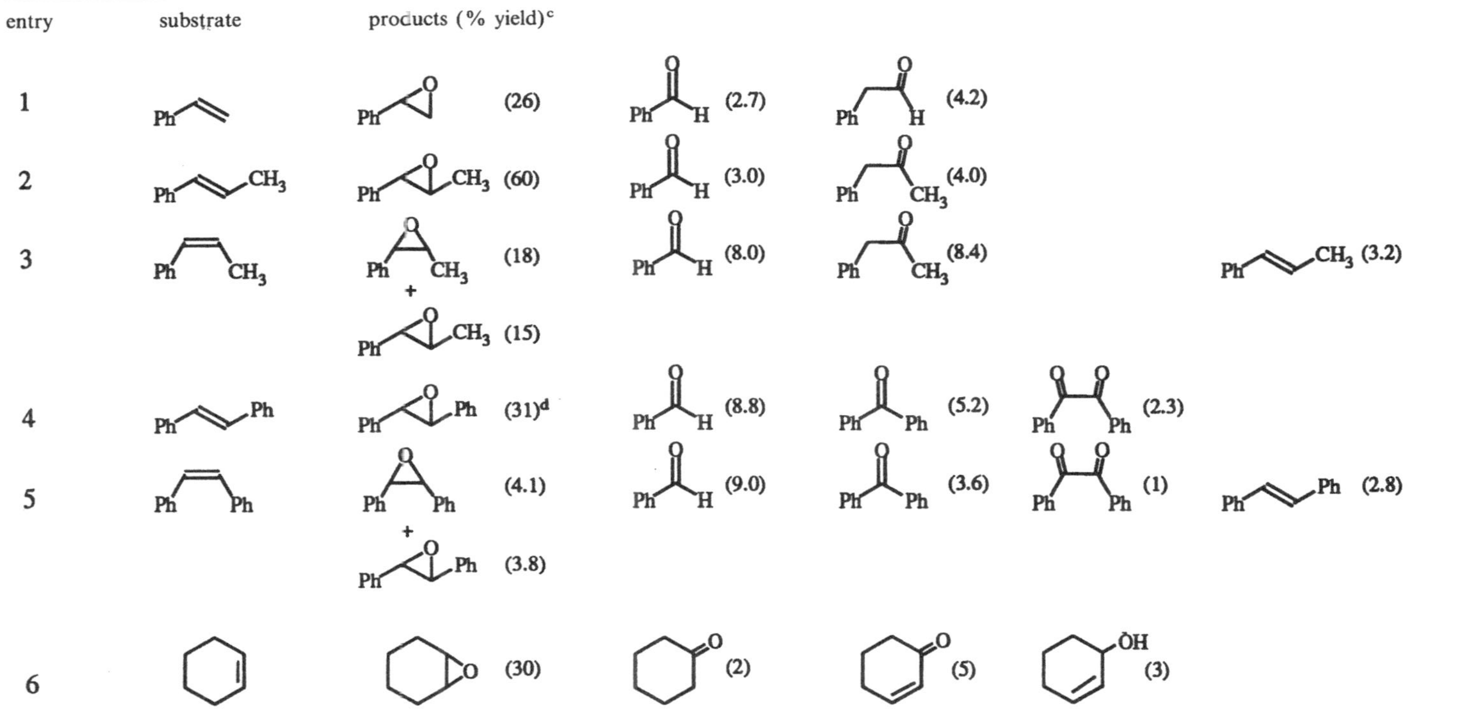

entry	substrate	products (% yield)[c]				
1	Ph–CH=CH$_2$	Ph-epoxide (26)	PhCHO (2.7)	PhCH$_2$CHO (4.2)		
2	Ph–CH=CH–CH$_3$ (trans)	Ph-epoxide-CH$_3$ (60)	PhCHO (3.0)	PhCH$_2$COCH$_3$ (4.0)		
3	Ph–CH=CH–CH$_3$ (cis)	Ph-epoxide-CH$_3$ (18) + Ph-epoxide-CH$_3$ (15)	PhCHO (8.0)	PhCH$_2$COCH$_3$ (8.4)		Ph–CH=CH–CH$_3$ (3.2)
4	Ph–CH=CH–Ph (trans)	Ph-epoxide-Ph (31)[d]	PhCHO (8.8)	PhCOPh (5.2)	PhCO–COPh (2.3)	
5	Ph–CH=CH–Ph (cis)	Ph-epoxide-Ph (4.1) + Ph-epoxide-Ph (3.8)	PhCHO (9.0)	PhCOPh (3.6)	PhCO–COPh (1)	Ph–CH=CH–Ph (2.8)
6	cyclohexene	cyclohexene oxide (30)	cyclohexanone (2)	cyclohexenone (5)	cyclohexenol-OH (3)	

[a] Reaction time = 5 hr. [b] See text for standard reaction conditions. [c] Based on olefin. [d] About 1% Z-isomer formed.

The reaction conditions are mild but an excess of oxidant is required for high yields of oxidized products. In a typical experiment, 0.1 mmol $Ni(NO_3)_2$, 0.5 mmol olefin, and 2.0 mmol PhIO in 5.0 mL dry degassed CH_3CN were allowed to react at room temperature for 2–5 hours. Most of the PhIO is consumed within the first hour of the reaction. The yields of oxidized products based on starting olefin are listed in Table I [15]. Low turnover is likely due to the competitive oxidation of CH_3CN as solvent. Nearly quantitative yields of products could be obtained if additional aliquots of PhIO were added to the reaction at intervals.

Consideration of the oxidation products of the reactions lends insight into the mechanism of the reaction. Any mechanistic proposal must account for the following facts: (i) The reaction proceeds faster with E-olefins than with Z. This is contrary to most reports of metal-catalyzed epoxidations. In addition, partial isomerization of Z to E-alkenes is observed. (ii) Only partial retention of configuration is observed. This is evident from the study of Z-stilbene and Z-β-methylstyrene. (iii) Over-oxidation to give products of C=C bond cleavage (e.g. benzaldehyde) is observed, and the yield of such products increases if O_2 is added to the reaction medium. (iv) Rearranged products such as phenylacetaldehyde are produced in small amounts. (v) Both oxygen atom transfer to alkenes (epoxidation) and C—H hydrogen atom abstraction reactions are observed (cf. cyclohexene).

If the reaction mechanism is in any way similar to that proposed for Cr(III), Mn(III) and Fe(III) porphyrin complexes, the first step of the reaction involves formation of a nickel-oxo intermediate. There are several possible formulations of such an intermediate, shown as **I–IV** below.

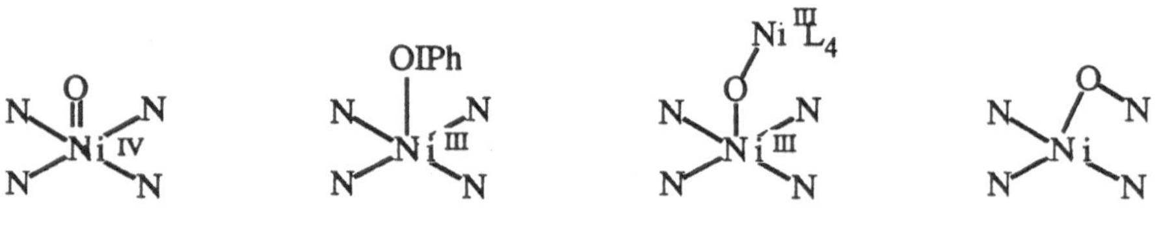

I   II   III   IV

Species **I**, a macrocyclic $L_4Ni{=}O$ structure, is drawn by analogy to iron-oxo-porphyrin complexes. The bonding scheme for the nickel complex suggests that formulation of a Ni=O double bond may not be justified. Theoretical studies of d^4 metals support the molecular orbital picture shown below [16]. A d^6 Ni(IV) species would not have empty d_{xz}, d_{yz} orbitals available for π bonding to oxygen.

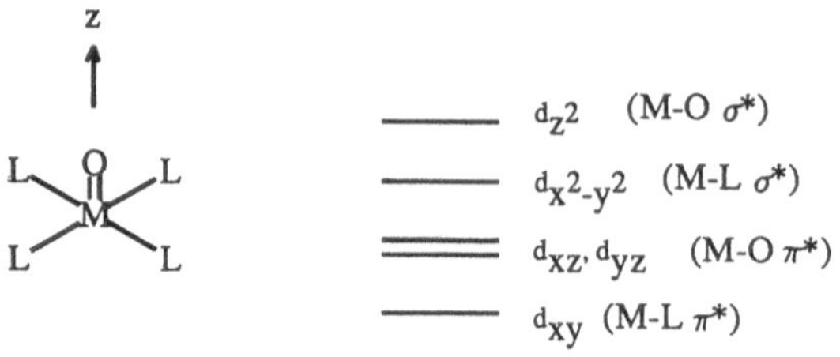

The nickel-oxo species **I** might be better formulated as the oxyanion $Ni(IV)$—O^- or a $Ni(III)$—$O\cdot$ oxy radical. No nucleophilic behavior typical of an oxyanion has been observed in this reaction. In fact, a Hammett study of p-substituted styrenes gave a linear correlation ($r^2 = 0.998$) with σ^+ [15]. The observed ρ^+ value of -0.82 is consistent with electrophilic character of the oxygen atom attack at an olefin and is in the same general range observed for metal-oxo-catalyzed olefin oxidations [17]. The results in Table I are also consistent with the formation of a nickel-oxo species with considerable radical character. Addition of a radical species to an oelfin would be expected to proceed more rapidly with, for example, E-stilbene. The intermediate carbon radical generated would be resonance stabilized. Resonance stabilization could only be achieved with the non-planar substrate Z-stilbene after C—C bond rotation, consequently, its transition state for formation would be higher in energy. Concerted formation of an oxametallacyclobutane **V** is ruled out by the observation of only partial retention of configuration in epoxide products [18]. Rather, an intermediate **VI** is implied – one with dual radical/carbocation character.[2]

V   VIa   VIb

The fate of radical **VIa** may be reductive elimination to produce an epoxide or trapping by O_2 to lead to over-oxidation products. Structure **VIb** is essentially a resonance structure of **VIa** and suggests a pathway for the migration of H or Ph to yield the products phenylacetaldehyde, phenylacetone or benzophenone.

The mechanism of C=C bond cleavage in the presence of O_2 was also studied by analyzing the ^{18}O content of products when $PhI^{18}O$ was used as terminal oxidant. In the absence of O_2, essentially 100% of the oxygen content of the epoxide products originated from $PhI^{18}O$. Examination of the ^{18}O content of benzaldehyde was more informative. When derived from E-stilbene, benzaldehyde showed a 45% incorporation of ^{18}O; the corresponding value for styrene as starting material was 2%. These results agree with intermediate **VIa** in which R_2 is always Ph and R_1 is either Ph or H. An ^{18}O atom from PhIO would always produce a label at the R_1 carbon. In these studies, dioxygen was present at very low concentration and was unlabeled. If the concentration of O_2 was increased by continuous bubbling through the reaction mixture, *the ^{18}O content of all products was diminished*. We have proposed a pericyclic mechanism of the Ni-oxo-olefin-O_2 adduct which is consistent with these observations [15]. This mechanism regenerates the nickel-oxo intermediate **I** in unlabeled form and provides further support for the existence of **I** as the first intermediate in the reaction sequence.

This scheme suggests an overall pathway for nickel cyclam-catalyzed oxidation of olefins which uses PhIO stoichiometrically to produce a high valent nickel-oxo intermediate capable of three different types of reactions. One pathway is H atom

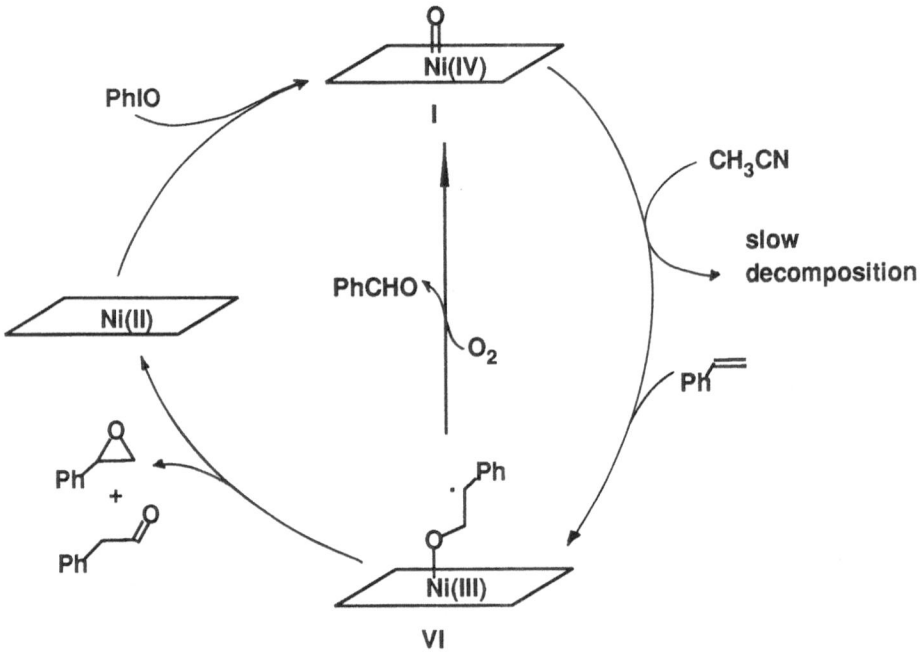

¹⁸O label shown as ●.

abstraction from hydrocarbons; this reaction may occur slowly with solvent molecules as well. A second pathway is the anticipated olefin epoxidation route. A third reaction uses O_2 stoichiometrically to cleave C=C bonds and regenerates the catalytically active oxidant. Studies are under way to exploit this third process since it is catalytic in PhIO.

Scheme I.

2.2. USE OF NICKEL(II) SALEN COMPLEXES WITH HYPOCHLORITE

A number of problems arose in the use of iodosylbenzene as terminal oxidant for olefin oxidation. The compound is a relatively expensive oxidant and was used in large excess in order to give good reaction yields. In addition, PhIO is only sparingly soluble in most solvents so that the heterogeneous nature of the reaction makes quantitative kinetic or spectroscopic analysis difficult. In a survey of other oxygen atom donor reagents, we found that tert-butyl hydroperoxide, *N,N*-dimethyl-*p*-cyanoaniline-*N*-oxide, NaIO$_4$ and H$_2$O$_2$ were unreactive as terminal oxidants. On the other hand, use of sodium hypochlorite under phase transfer conditions similar to those discovered by Meunier [19] lead to efficient olefin oxidation [20]. This reaction solves some of the problems of PhIO: NaOCl is inexpensive, only a two to four-fold excess of the oxidant is used, and the reaction is more nearly homogeneous, at least in the organic phase. The best catalyst under these conditions was Ni(II) salen.

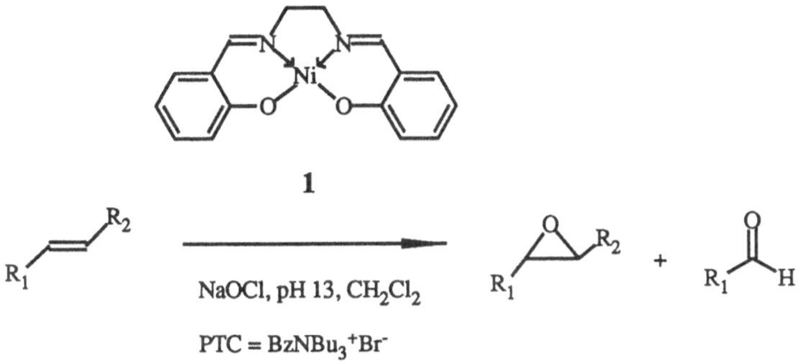

1

NaOCl, pH 13, CH$_2$Cl$_2$

PTC = BzNBu$_3$$^+Br^-$

Typical reaction conditions involved 4.0 mmol olefin, 1 mmol nickel complex and 0.15 mmol benzyltributylammonium bromide in 10 mL CH$_2$Cl$_2$ to which 20 mL 0.77 M NaOCl (pH 13) were added. A fine black precipitate is formed immediately upon mixing which may be nickel peroxide. This material was shown to be inert toward olefin oxidation and disappeared later in the reaction when all the oxidant was consumed. Table II lists the yields of styrene oxide formed from oxidation of styrene as a function of the nickel complex. The best results were obtained using Ni(II) salen as catalyst.

A summary of the reactions of various hydrocarbon substrates is given in Table III. For aryl-substituted alkenes, epoxidation is the major pathway, but substantial amounts of C=C bond cleavage to carbonyl compounds was also observed. As in the nickel cyclam-catalyzed reactions with PhIO, the reaction showed a slight preference for *E*-stilbene over the *Z*-isomer; however, the epoxide product in both cases was exclusively the *E*-isomer. This may reflect a longer-lived nickel-oxo-olefin

Table II.   Yields of styrene oxide as a function of catalyst[a]

Catalyst	% Yield
Ni(II) salen	44.3
Ni(II) cyclam (OTf)$_2$	4.7
Ni(II) TPP	0
Ni(OAc)$_2$	trace
no catalyst[b]	0.7

[a] 6 hr reaction time; see text for standard reaction conditions.
[b] 18 hr reaction time.

Table III. Percent conversion and yields of products from Ni^{II} (salen)-catalyzed oxidation of alkenes by OCl$^-$ after 5 hours

substrate	% conversion[a]	epoxide[b]	PhCHO[c]	selectivity[d]
styrene	98	44	6	45
Z-β-methylstyrene	100	84[e]	10	84
E-β-methylstyrene	100	89[e]	0	89
Z-stilbene	45	12[e]	12	27
E-stilbene	80	46[e]	0	58
cyclohexene	87	23		26
norbornene	94	30[f]		32

[a] Disappearance of starting material. [b] Based on starting alkene. [c] Remainder of product is PhCO$_2$H. [d] Epoxide yield/% conversion. [e] E-epoxide only. [f] exo-epoxide only.

intermediate capable of rapid C—C bond rotation prior to reductive elimination. The epoxidation of alkyl olefins was complicated by the production of substantial amounts of chlorinated products.

Overall, the reaction of Ni(II) salen/NaOCl with olefins bears considerable similarity to that of the Ni(II) cyclam/PhIO system. Production of C=C bond cleavage products is only modestly affected by the presence or absence of O$_2$. A mechanistic scheme is suggested below that accounts for the formation of benzaldehyde from stilbene by reaction with two equivalents of hypochlorite. In future work, it will be important to control the relative amounts of the epoxidation vs. C=C cleavage pathways and to minimize chlorination reactions.

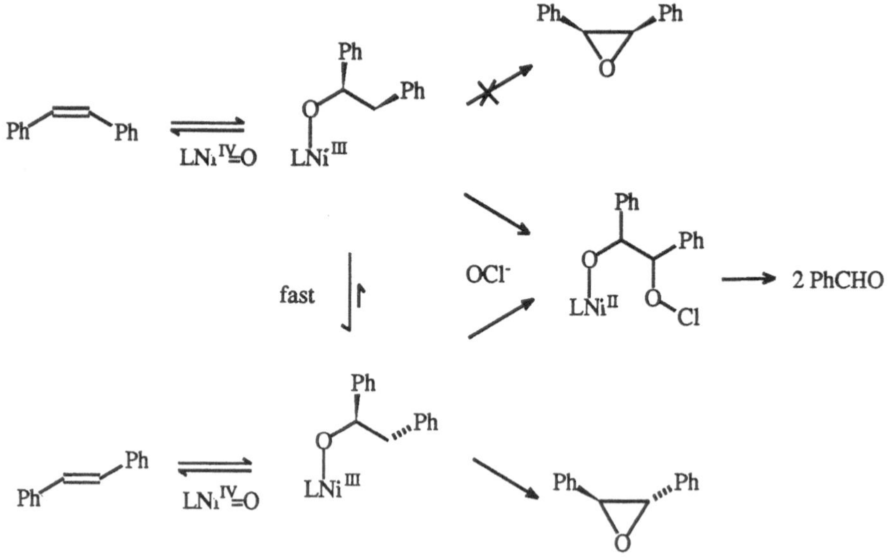

Scheme II.

2.3. USE OF NICKEL(II) DIOXOCYCLAM COMPLEXES WITH HYPOCHLORITE

From the studies described above, it is evident that certain 14-membered square planar chelating rings containing strong donor atoms render Ni(II) salts active as oxidation catalysts. To approach our ultimate goal of incorporating substrate binding sites adjacent to the catalytic site, it is necessary to use a macrocycle that possesses additional functional groups as points of attachment. Ideally, this macrocycle should also be optically active in order to avoid the complications of mixtures of diastereomeric ligands and to provide the possibility of chiral recognition. We chose to explore analogs of the dioxocyclam ligand since its ability to stabilize the Ni(III) oxidation state was well known [21]. Our unsubstituted ligands are derived from amino acids and contain two stereogenic units. We have developed a general synthetic route to the ligands **VIIa–d** from phenylalanine, tryptophan, valine and leucine.

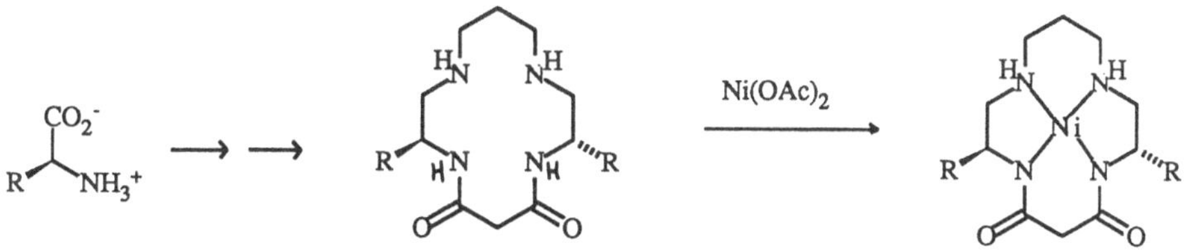

VII

a R=CH$_2$Ph

b R= 3-indolemethyl

c R=CH(CH$_3$)$_2$

d R=CH$_2$CH(CH$_3$)$_2$

Macrocycle **VIIa** readily forms the doubly deprotonated nickel complex upon addition of Ni(OAc)$_2$. The methylene chloride soluble complex was unreactive under reaction conditions similar to those employed in the nickel cyclam/PhIO system. However, oxidation of E-β-methylstyrene occurred readily when the hypochlorite phase transfer conditions were used. Epoxidation represented about 50% of the reaction pathway with the remainder yielding a mixture of benzaldehyde and other over-oxidation products.

In summary, a variety of square planar macrocyclic nickel complexes are capable of catalysis of olefin oxidation using strong terminal oxidants. With knowledge of the reaction mechanism and suitable functionalization of the periphery of the macrocycle, new catalysts might be developed which show high substrate selectivity as well as interesting regio- and stereochemistry.

3. Design of New Molecular Receptors for Neutral Molecules

For the organic chemist, synthetic reactions are most conveniently carried out in organic solvents. Reactions involving nucleophiles or bases are often more rapid in non-polar solvents due to poor solvation of polar species. For these reasons, we have focussed our attention upon the design of new molecular receptors which are soluble in non-polar solvents, but which possess a hydrogen-bonding cavity for the inclusion of polar substrates. In the design of a new receptor, we sought a structure with convergent hydrogen-bonding groups surrounded by a superstructure of overall hydrophobic character. Cholic acid was therefore an ideal building block for this purpose since it is a highly functionalized, rigid steroid with the amphophilic properties of a detergent. Our first generation receptors are diamides derived from condensation of two cholic acid molecules with simple diamines [22].

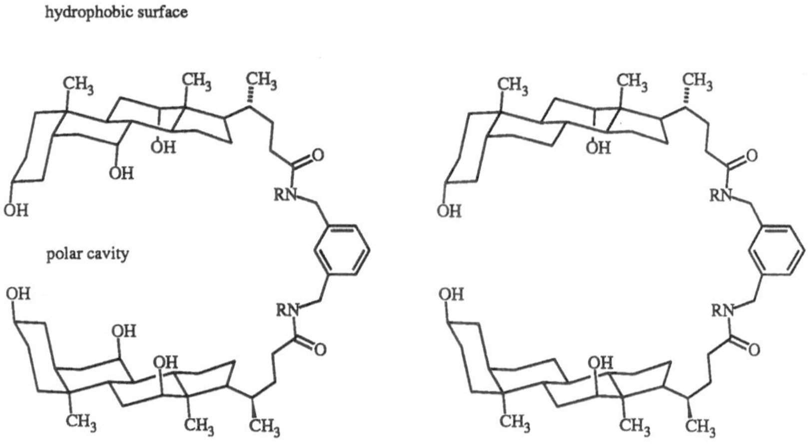

VIII   IX

The properties of the cholic acid dimers **VIII** have been compared to the deoxycholate analogs **IX** in order to give information about the importance of the hydrogen-bonding hydroxyl groups. Examination of the ¹H-NMR spectra of compounds **VII** and **IX** (R = H) under various conditions displayed interesting behavior of the two receptors. The diastereotopic benzylic hydrogens of **VIII** were doublets of doublets under all conditions, but the chemical shifts of these two hydrogens in particular were temperature and solvent dependent. At low temperature in dry CDCl$_3$ the resonances were separated by about 1 ppm. Near 65°, the peaks converged to the point of having nearly the same chemical shift. The same phenomenon was observed for the deoxycholate analog **IX**, except that the low temperature separation was smaller and the temperature of convergence was lower. These data are plotted in Figure 1.

This phenomenon of peak convergence as a function of temperature could be reproduced at room temperature by incremental addition of a hydrogen-bonding

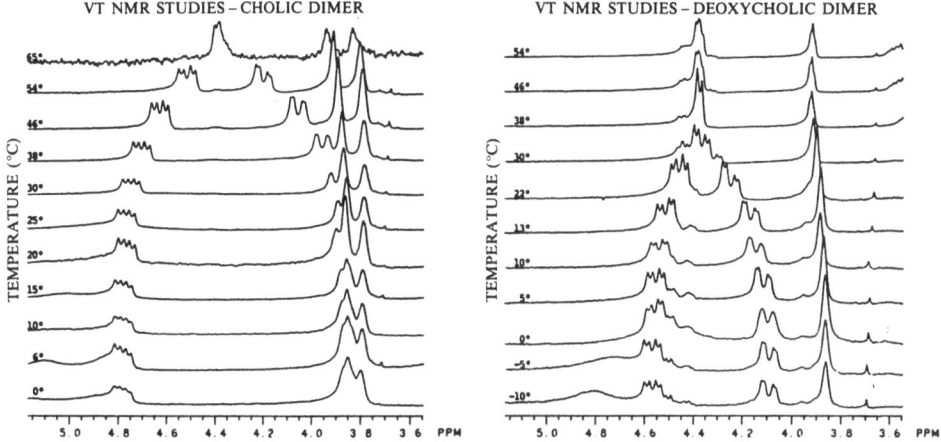

Fig. 1.

solvent such as methanol (see Figure 2). In contrast, the spectrum of *N*-benzyl-cholamide shows a collapsed multiplet at 4.4 ppm for the benzylic protons under all conditions.

An explanation consistent with these results is that the cholamide dimers may exist in two limiting conformations. In non-polar solvents and in the absence of hydrogen-bonding substrates, **VIII** may exist in a folded conformation with in-tramolecular hydrogen bonds (Figure 3). We are not currently able to detect whether or not one or two water molecules might also be present and acting as bridging groups between cholate hydroxyls as is seen in the crystal structure of cholic acid [23]. The addition of heat or methanol would be expected to break

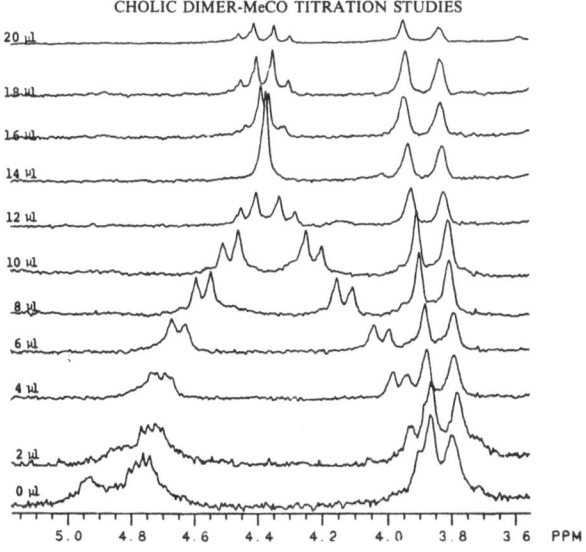

Fig. 2.

hydrogen bonds and generate an open, freely rotating species. The closed conformation would have a rigid structure with the two benzylic hydrogens in potentially quite different chemical environments. The open form would allow for near averaging of these environments through rotation.

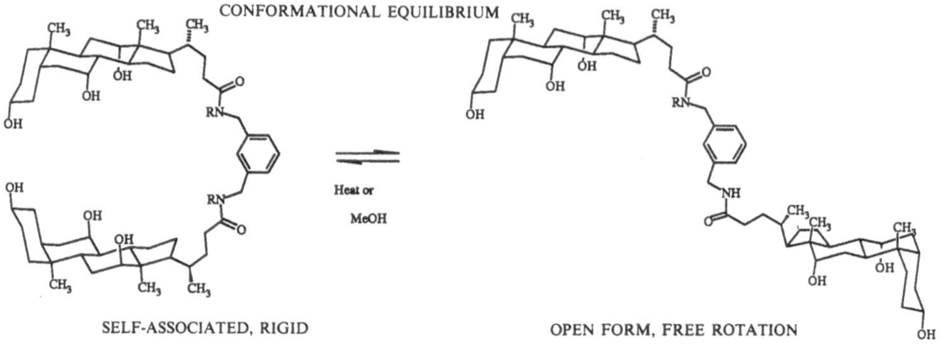

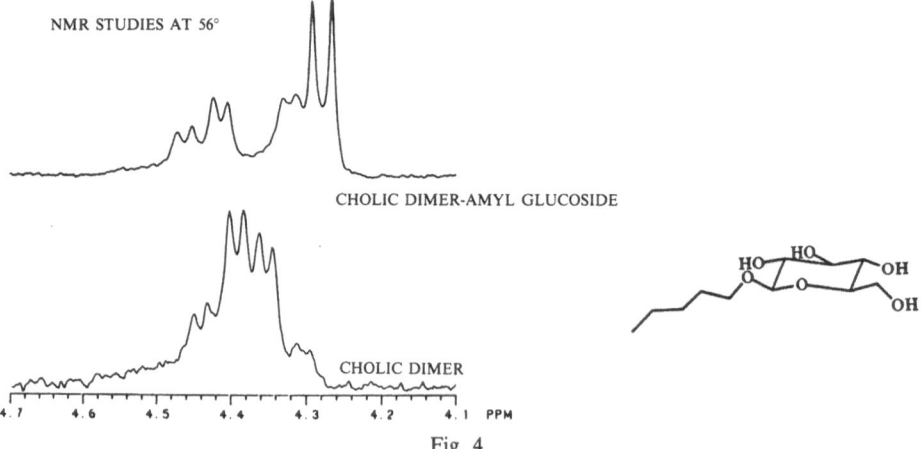

Fig. 3.

An interesting consequence of these experiments is that we can use this phenomenon of temperature convergence of signals as an indication of substrate binding. If a substrate hydrogen bonds in the central cavity between the two cholate moieties, one would anticipate a stabilization of the closed form resulting in well-separated NMR signals. This would lead to a higher temperature of convergence of the signals. A very interesting result that we have obtained in this area is such an observation using n-pentylglucopyranoside as substrate. A 1 : 1 molar ratio of **VIII** and the glucoside were studied at various temperatures in anhydrous $CDCl_3$. The results at 56° are shown in Figure 4. In the presence of substrate, convergence of signals had occurred *to a lesser extent, consistent with binding*.[3] Introduction of water or other solvent impurities would have led to the opposite result. This is an unusual example of binding of a carbohydrate derivative to a synthetic molecular receptor. By further synthetic elaboration of the receptor we hope to increase its binding strength and selectivity for polar substrates.

Fig. 4.

4. Prospects for the Future

Inclusion phenomena range from the encapsulation of single atoms to the binding of complex organic molecules. In the case of transition metal ions, the encapsulating ligand may have a profound effect on the reactivity of the metal ion particularly in the case of redox processes. For organic substrates, synthetic molecular receptors offer the possibility of dissolution in unusual media and orientation relative to approaching reagents. Progress in both the design of new catalytic species and of effective binding agents will accelerate the success of chemists in mimicry of biological catalysts.

Acknowledgement

We are grateful to the National Science Foundation (CHE-8706616) and the National Institutes of Health (GM-34841) for grants supporting this work. In addition, we thank Profs. J. S. Valentine and R. H. Holm for helpful discussions.

Notes

[1]Abbreviations used: cyclam = 1,4,8,11-tetraazacyclotetradecane; salen = N,N'-ethylenebis(salicylideneamine); TPP = 5,10,15,20-tetraphenylporphyrin.

[2]These structures are formally resonance forms only if no nuclear motion accompanies electronic reorganization.

[3]We have not eliminated the possibility of aggregation phenomena. Low solubility of **VIII** (R = H) has limited determination of solution molecular weights by vapor pressure osmometry.

References

1. P. R. Ortiz de Montellano: *Cytochrome P-450: Structure, Mechanism, and Biochemistry*, Plenum, New York (1986).
2. (a) R. A. Sheldon and J. K. Kochi: *Metal-Catalyzed Oxidations of Organic Compounds*, Academic Press, New York (1981); (b) R. H. Holm: *Chem. Rev.* **87**, 1401 (1987).
3. (a) D. J. Cram: *Science* **219**, 1177 (1983); (b) J. M. Lehn: *Science* **227**, 849 (1985); J. Rebek, Jr.: *Science* **235**, 1478 (1987).
4. For selected examples, see: (a) I. Tabushi: *Acc. Chem. Res.* **15**, 66 (1982); (b) R. Breslow, J. B. Doherty, G. Guillot, and C. Lipsey: *J. Am. Chem. Soc.* **100**, 3227 (1978); (c) T. J. Meade, K. J. Takeuchi, and D. H. Busch: *J. Am. Chem. Soc.* **109**, 725 (1987); (d) D. J. Cram, P. Y. S. Lam, and S. P. Ho: *J. Am. Chem. Soc.* **108**, 839 (1986); (e) G. Schuermann and F. Diederich: *Tetrahedron Lett.* **27**, 4249 (1986); (f) T. J. Van Bergen and R. M. Kellogg: *J. Am. Chem. Soc.* **99**, 3882 (1977); (g) S. Sasaki, M. Shionoya, and K. Koga: *J. Am. Chem. Soc.* **107**, 3371 (1985); (h) J. M. Lehn and C. Sirlin: *J. Chem. Soc., Chem. Commun.*, 949 (1978); (i) J. Wolfe, D. Nemeth, A. Costero, and J. Rebek, Jr.: *J. Am. Chem. Soc.* **110**, 982 (1988).
5. (a) T. Katsuki and K. B. Sharpless: *J. Am. Chem. Soc.* **102**, 5974 (1980); (b) J. T. Groves and R. S. Myers: *J. Am. Chem. Soc.* **105**, 5791 (1983); (c) D. Mansuy, P. Battioni, J. Renaud, and P. Guerin: *J. Chem. Soc., Chem. Commun.*, 155 (1985); (d) B. R. Cook, K. Suslick, and B. R. Reinert: *J. Am. Chem. Soc.* **108**, 7281 (1986).
6. For leading references, see: (a) J. P. Collman, T. Kodadek, and J. I. Brauman: *J. Am. Chem. Soc.* **108**, 2588 (1986); (b) T. G. Traylor, T. Nakano, R. A. Miksztal, and B. E. Dunlap: *J. Am. Chem. Soc.* **109**, 3625 (1987); (c) D. Ostovic, C. B. Knobler, and T. C. Bruice: *J. Am. Chem. Soc.* **109**, 3444 (1987); (d) J. T. Groves: *Ann. N.Y. Acad. Sci.* **471**, 99 (1984).
7. J. T. Groves and G. A. McClusky: *J. Am. Chem. Soc.* **98**, 859 (1976).

8. (a) C. Eskenazi, G. Balavoine, F. Meunier and H. Riviere: *J. Chem. Soc., Chem. Commun.*, 1111 (1985); (b) C. M. Che and W. K. Cheng: *J. Chem. Soc., Chem. Commun.*, 1443 (1986); (c) S. Krishnan, D. G. Kuhn, and G. A. Hamilton: *J. Am. Chem. Soc.* **99**, 8121 (1977); (d) N. Herron and C. A. Tolman: *J. Am. Chem. Soc.* **109**, 2837 (1987); (e) C. L. Hill and R. B. Brown, Jr.: *J. Am. Chem. Soc.* **108**, 536 (1986); (f) J. C. Dobson, W. K. Seok, and T. J. Meyer: *Inorg. Chem.* **25**, 1514 (1986); (g) R. B. VanAtta, C. C. Franklin, and J. S. Valentine: *Inorg. Chem.* **25**, 4121 (1984).

9. (a) D. H. Busch: *Acc. Chem. Res.* **11**, 392 (1978); (b) A. Buttafava, L. Fabbrizzi, A. Perotti, G. Poli, and B. Sehgi: *Inorg. Chem.* **25**, 1456 (1986); (c) S. Battacharya, R. Mukerjee, and A. Chakravorty: *Inorg. Chem.* **25**, 3448 (1986).

10. A. von Pfaltz, B. Juan, A. Faessler, A. Eschenmoser, R. Jaenchen, H. H. Gilles, G. Diekert, and R. K. Thauer: *Helv. Chim. Acta.* **65**, 828 (1982).

11. E. S. Gore and D. H. Busch: *Inorg. Chem.* **12**, 1 (1973).

12. A. G. Lappin, P. Osvath, and S. Baral: *Inorg. Chem.* **26**, 3089 (1987).

13. J. F. Kinneary, T. R. Wagler, and C. J. Burrows: *Tetrahedron Lett.* **29**, 877 (1988).

14. J. D. Koola and J. K. Kochi: *Inorg. Chem.* **26**, 908 (1987).

15. J. F. Kinneary, J. S. Albert, and C. J. Burrows: *J. Am. Chem. Soc.*, **110**, 6124 (1988).

16. R. H. Holm, reference 2(b).

17. E. G. Samsel, K. Srinivasan, and J. K. Kochi: *J. Am. Chem. Soc.* **107**, 7606 (1985).

18. J. P. Collman, J. I. Brauman, B. Meunier, S. A. Raybuch, and T. Kodadek: *Proc. Natl. Acad. Sci. U.S.A.* **81**, 3245 (1984).

19. B. Meunier, E. Guilmet, M. E. De Carvahlo, and R. Poilblanc: *J. Am. Chem. Soc.* **106**, 6668 (1984).

20. H. Yoon and C. J. Burrows: *J. Am. Chem. Soc.* **110**, 4087 (1988).

21. E. Kimura, T. Koike, R. Machida, R. Nagai, and M. Kodama: *Inorg. Chem.* **23**, 4181 (1984).

22. C. J. Burrows and R. A. Sauter: *J. Incl. Phenom.* **5**, 117 (1987).

23. L. Lessinger: *Cryst. Struct. Comm.* **11**, 1787 (1982).

Journal of Inclusion Phenomena and Molecular Recognition in Chemistry **7** (1989), 169–171.

Template Synthesis of Ligands for Highly Charged Metal Cations

KENNETH N. RAYMOND

Chemistry Department, University of California, Berkeley, California 94720, U.S.A.

(Received: 1 February 1988)

Macrocyclic and macrobicyclic ligands may enhance the stability of their metal complexes due to the inherent entropic and kinetic properties of a ligand ring or cage-type structure. In general, as the binding site becomes more encapsulated or preformed the reorganization entropy decreases, leading to a relatively higher formation constant for metal binding. Thus it would be expected that the formation constants be greatest for the series macrobicyclic > macrocyclic > exocyclic (Figure 1).

exocyclic   macrocyclic   macrobicyclic

X = bidentate chelating subunit

Fig. 1.   Possible topologies for macrocyclic ligands.

For synthetic purposes, the idealized macrobicyclic structure in Figure 1 can be viewed as a 'capped' tripod. We have explored a new binding subunit for this work, 2,3-dihydroxyterephthalamide. A highly successful example of this approach is shown in Figure 2, where an essentially one-step synthesis gives a 50% isolated yield of the macrocycle formed from six amide linkages.

In the quite different chemistry of the lanthanides we have used a conceptually similar approach to prepare cage complexes of the lanthanides from which the metal ion cannot escape without breaking a C—C or C—N bond. The proposed structure for such a ytterbium complex is shown in Figure 3. This complex is relatively hydrolytically stable. In contrast, the intermediate compounds with 1 or 2 methylene bridges are readily hydrolyzed since they allow ready exit of the metal ion. The structures of several of these intermediates have been determined.

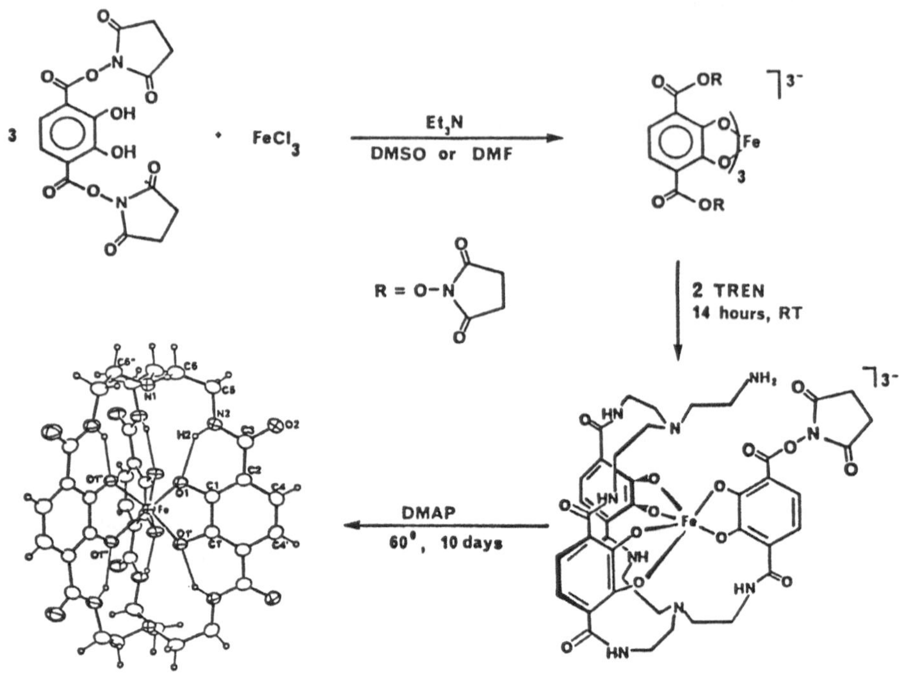

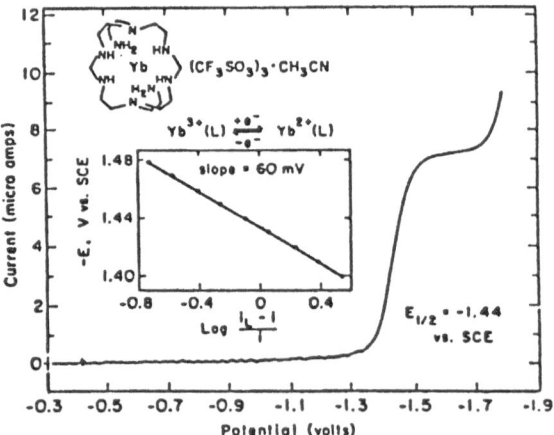

Fig. 2.   Template synthesis of ferric (bicapped TRENCAM).

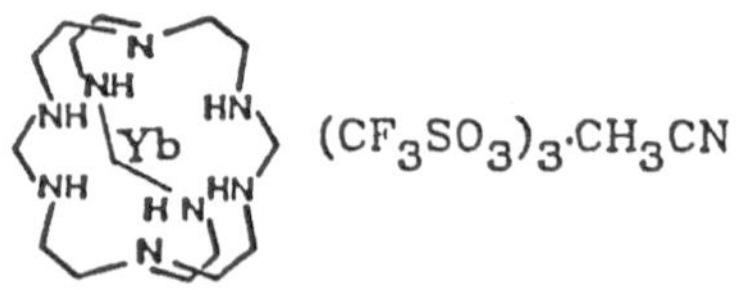

Fig. 3.   A schematic drawing of the proposed structure of the completely encapsulated ytterbium complex.

Fig. 4.   Normal pulse polarogram of Yb^{3+} (L) in 0.3 M sodium triflate acetonitrile solution. Negative potentials (V vs SCE) are plotted to the right and reduction currents are plotted upward. The inset shows a plot of $-E$ vs log $(i_L - i/i)$ for the normal pulse polarogram.

The electrochemistry for the Yb complex shown in Figure 4 (along with the electrochemical behavior) shows the $+3$ complex is stabilized by 10^{14} relative to the $+2$ complex.

Acknowledgement

I thank my coworkers, whose names appear in the publications listed below, whose efforts are summarized in my presentation.

References

1. T. J. McMurry, S. J. Rodgers, and K. N. Raymond: *J. Am. Chem. Soc.* **109**, 3451 (1987).
2. T. J. McMurry, M. W. Hosseini, T. M. Garrett, F. E. Hahn, Z. E. Reyes, and K. N. Raymond: *J. Am. Chem. Soc.* in press.
3. S. J. Rodgers, C. W. Lee, C. Y. Ng, and K. N. Raymond: *Inorg. Chem.* **26**, 1622 (1987).
4. P. H. Smith and K. N. Raymond: *Inorg. Chem.* **24**, 3469 (1985).

Journal of Inclusion Phenomena and Molecular Recognition in Chemistry **7** (1989), 173–182.

Polycarboxylate Crown Ethers: Synthesis, Complexation, Applications

PHILIP J. DUTTON, THOMAS M. FYLES*, and STEVEN P. HANSEN
Department of Chemistry, University of Victoria, Victoria, B.C., Canada, V8W 2Y2

(Received: 1 February 1988)

Abstract. Crown ethers derived from tartaric acid present a number of interesting features as receptor frameworks and offer a possibility of enhanced metal cation binding due to favorable electrostatic interactions. The synthesis of polycarboxylate crown ethers from tartaric acid is achieved by simple Williamson ether synthesis using thallous ethoxide or sodium hydride as base. Stability constants for the complexation of alkali metal and alkaline earth cations were determined by potentiometric titration. Complexation is dominated by electrostatic interactions but cooperative coordination of the cation by both the crown ether and a carboxylate group is essential to complex stability. Complexes are stable to pH 3 and the ligands can be used as simultaneous proton and metal ion buffers. The low extractibility of the complexes was applied in a membrane transport system which is a formal model of primary active transport.

Key words. membrane transport, metal cation binding, tartaric acid.

1. Introduction

The field of molecular recognition by synthetic receptors arose from the problem of alkali metal cation complexation. Initial studies by Pedersen on crown ethers [1], followed by the cryptands of Lehn [2] and more recently Cram's spherands [3] illustrate an increasingly sophisticated and successful approach to the recognition of spherical cations. To a certain degree the problem may be regarded as 'solved', as stable and selective complexes of the alkali metals and alkaline earth cations are now well described. However, with respect to recognition coupled to other supramolecular functions such as transport, the general problem remains unresolved in many important aspects. From a practical viewpoint, the synthetic complexity of some ligands precludes their use. In other applications the complexation kinetics, or the pH range of optimal complexation, may not be suitable. Thus the continued investigation of problems in spherical cation recognition appears to be justified.

From this perspective, crown ethers derived from (+)tartaric acid possess a number of appealing features as frameworks for the construction of specific complexing agents. The basic skeletons are readily assembled via reliable procedures [4, 5] and the carboxylate groups provide an easy entry into a range of derivatives [6, 7]. Secondly, tartarate-derived groups show a marked preference for occupying axial positions on the macrocycle [4, 8, 9] thereby restricting conformational mobility [9, 10]. Finally, the presence of charged groups on a macrocycle periphery leads to markedly enhanced cation complexation relative to uncharged forms [4, 10].

We have been exploring the chemistry of this class of compounds for the past few years with a focus on synthesis and on a detailed examination of cation

---

* Author for correspondence.

complexation. As well, we have investigated applications in which stable alkali metal or alkaline earth cation complexes are required, particularly in aqueous and acidic solution. This report surveys some recent progress in all these areas.

2. Synthesis of Polycarboxylate Crown Ethers

The central reaction is, of course, the Williamson ether synthesis. Early reports on the preparation of tartaric acid ethers [11], suggested that the base thallous ethoxide, (TlOEt), was essential to avoid epimerization of the chiral centers. The first syntheses thus utilized this base in dimethyformamide (DMF), and oligo-ethylenglycol diiodides for the preparation of di- and tetra-carboxylate crown ethers [4, 12]. More recently, we found that by strict control of stoichiometry, sodium hydride could be used successfully to displace tosylate without loss of chiral integrity [5]. Scheme 1 shows a recent synthesis of an 18-crown-6 hexaacid from three units of (+)tartaric acid [13]. This route illustrates all the key features in the syntheses of polycarboxylate crown ethers.

Tartaramide (1) in DMF was treated with one equivalent of NaH and then with excess benzylbromide to give the monobenzyl ether 2. Compound 2, like many other synthetic intermediates encountered, possesses a convenient hydrophilic/

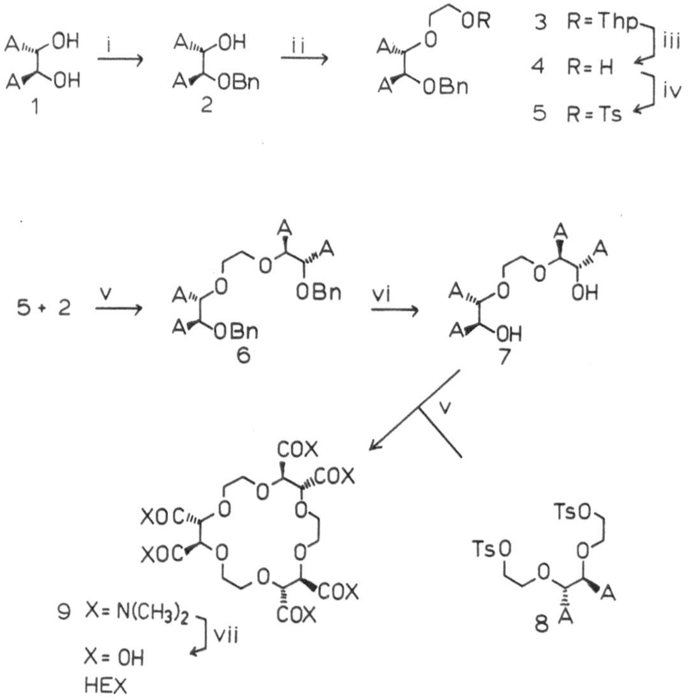

Scheme 1. Synthesis of an 18-crown-6 Hexaacid.
Reagents: (i) (1) NaH (1 equiv.)/DMF, (2) BnBr (excess); (ii) (1) TlOEt/DMF, (2) BrCH$_2$ CH$_2$OThp; (iii) H$_3$O$^+$/MeOH; (iv) TsCl/Et$_3$N/CH$_2$Cl$_2$; (v) NaH/DMF; (vi) H$_2$Pd–C; (vii) H$_3$O$^+$/reflux. Abbreviations: A = Me$_2$NC—, Bn = benzyl, Ts = tosyl, Thp = tetrahydropyranyl.
$$\underset{O}{\overset{\|}{}}$$

hydrophobic balance. Thus **2** can be purified by a simple sequence of solvent extractions, initially from hydrocarbon solvents into water and then into chlorocarbon solvents. Coupling of **2** with the tetrahydropyranyl ether of 2-bromoethanol was achieved using TlOEt in DMF to give **3**, purified by chromatography. Hydrolysis gave the alcohol **4** which was directly converted to a tosylate **5**. A second equivalent of the sodium salt of **2** (NaH/DMF) was then treated with **5** to yield the half crown **6**, which was purified by chromatography and then deprotected to give the diol **7**. A directly comparable sequence from **1** without monoprotection gave the ditosylate **8** [14]. The final macrocyclization again utilized NaH to yield the hexaamide **9** which was finally deprotected in refluxing aqueous HCl to give the hexaacid (HEX).

Although the hexaacid crystallized readily from water, it was soon apparent that it was not pure, but contained alkali metal and alkaline earth cation impurities. Even repeated crystallization from acid failed to remove these impurities. Finally, ion exchange using Dowex resin (extensively washed with high purity acid) and doubly distilled water as eluent, gave a metal-free sample of HEX [13]. The crystal structures of HEX and its Na$^+$, K$^+$, Tl$^+$ and Cs$^+$ complexes readily explain the strong propensity to resist purification by crystallization [15]. The free ligand HEX crystallizes as the tetrahydrate in which a hydrogen bonded dimer of water molecules is bound with the ligand cavity. Cation binding results in loss of a proton from the ligand and loss of one water molecule from the cavity. The water position is occupied by the cation, well above the plane of the macrocycle. The close oxygen contacts with the cation involve the ligand carboxylate, the remaining water and, at longer distances, the ether oxygens [15]. All complexes investigated are approximately isostructural with the free ligand thus the crystallization readily proceeds with inclusion of metal impurities.

3. Cation Complexation by Polycarboxylate Crown Ethers

The cation complexation behavior of the carboxylate crown ethers illustrated in Figure 1 was investigated by potentiometric titration. The principles have been discussed previously [10, 12]; the primary constants determined are cumulative

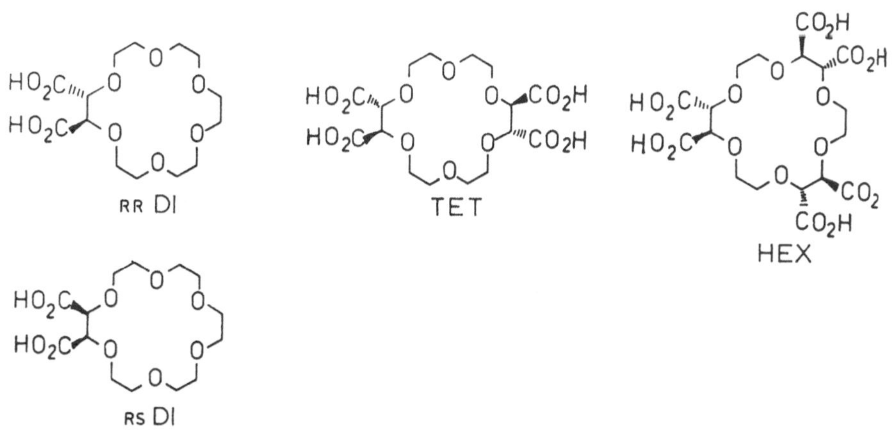

Fig. 1. Structures of the ligands considered with abbreviated names.

association constants which may be used to calculate stepwise formation constants for the 1 : 1 association of a cation with the ligand at various levels of protonation. The experimental procedure involves the pH-metric titration of the ligand as the free acid, with base (Me$_4$NOH). The resulting titration curve is analysed [16] to yield the ligand pK_as. Metal ion is then added to the system and a titration is repeated on the mixture. The titration curve, together with the previously determined pK_as, is then analysed [16] to yield the cumulative formation constants of the cation complexes.

Results for the four ligands of Figure 1, at various levels of protonation, are given in Table I [13]. The data clearly reveal the dominant role that electrostatic interaction plays in the complexation process: (i) There is a general trend to greater stability constant for a given cation as ligand charge increases (down a column), (ii) there is a clear trend which favors divalent over monovalent ions at all levels of ligand protonation, and (iii) many ligand/cation combinations exhibit a regular decrease in stability constant as the ligand is protonated. In addition, however, there are some clear structural effects. Most notable is the expected 'hole-size' effect [13]; in the present case, the larger cations are favored relative to smaller cations. This occurs despite the greater surface charge density of the smaller cations which would enhance any electrostatic interaction with the ligand.

The most important irregularity in the data concerns the Tl$^+$ complexes of HEX^{6-} and TET^{4-} and the K$^+$ complex of HEX^{6-}. In each of these cases, ligand protonation results in substantial stabilization of the complex in direct opposition to the expectation of electrostatic stabilization. These cases are the interactions of the most highly charged ligands with the cations of the lowest surface charge density. On a purely electrostatic basis, Tl$^+$ and K$^+$ are least able to organize the carboxylate donors in the face of the strong internal repulsions between the carboxylates. The fully deprotonated ligands are thus too large to provide a suitable set of donor atoms, and too rigid to be distorted, thus the complexes are destabilized. Monoprotonation, however, relieves some intramolecular repulsions, increases ligand flexibility and results in enhanced complex stability. This is an example of the principle that the 'guest organizes the host' [17].

Table I. Logarithm of stepwise formation constants.[a]

Ligand	Na$^+$	K$^+$	Tl$^+$	Ca^{2+}	Sr^{2+}	Ba^{2+}
R,R-HDI$^-$	2.4	3.2	4.6	*	4.2	*
R,R-DI^{2-}	3.3	4.2	5.7	5.6	5.9	6.5
R,S-DI^{2-}	2.5	3.1	3.3	4.3	5.8	5.6
H$_2$TET^{2-}	1.9	3.4	*	*	*	*
HTET^{3-}	4.1	4.7	4.8	7.0	6.1	6.2
TET^{4-}	4.5	4.8	3.6	8.6	8.0	7.2
H$_2$HEX^{4-}	4.0	4.0	5.1	6.7	6.8	6.5
HHEX^{5-}	5.1	5.3	5.8	8.6	8.8	8.8
HEX^{6-}	5.4	4.1	4.4	9.8	10.4	9.5

[a] Determined by potentiometric titration at 25°, $I = 0.05$ M with Me$_4$NCl. Values are calculated from the cumulative formation constants for the complexes and the ligand pK_as. The symbol * indicates that the complex was not required to achieve a fit of the experimental data. Uncertainty in log $K \pm 0.2$.

Since monoprotonation relieves the irregularity, it is possible that one additional donor site, a carboxylate, is required by the cation and the ligands act as seven oxygen donors. In this view, complexes of moderately charged ligands, or of cations of high surface charge density, could easily achieve seven coordination. In the Tl^+ and K^+ cases noted above, the energetic 'cost' of seven-coordination involving ligand distortion is inadequately balanced by the metal-ligand interaction energy gained, hence the lower stability constant.

The question of carboxylate participation in cation binding has been extensively examined. Evidence from infrared and nuclear magnetic resonance [10], from electron spin resonance [9] and from potentiometric titration experiments [10, 18] suggests that direct, cooperative carboxylate-cation interactions occur. The method of Eyring and coworkers [18] can be applied to the data of Table I to estimate the extent of this cooperative interaction. The method separates the total binding energy into three contributions as illustrated in Figure 2: (i) a cavity term (10), involving only the cation association with the neutral crown ether, (ii) an electrostatic term (12), involving only coulombic interactions between the ligand and the cation, and (iii) a cooperative term (11) involving the two effects together as implied by a coordinative carboxylate–metal cation interaction. For the majority of the cases of Table I, the cooperative term accounts for approximately half of the total binding energy, with electrostatic and cavity effects making up the other half [13]. These latter depend closely on the ion size, thus large cations (K^+, Tl^+, Ba^{2+}) have large cavity contributions and small electrostatic contributions. The converse applies to the smaller cations. The principal exceptions are the complexes of Ca^{2+}. With a very high surface charge density, Ca^{2+} complexation is dominated by the electrostatic factor with only minor (10%) contribution by the cooperative factor [13].

The data of Table I provides other insights into the complexaton process. A comparison of the configurational isomers R,R-DI and R,S-DI reveals that the complexes of the latter are uniformly less stable. As noted in the introduction, the RR-system derived from (+)tartaric acid results in an axial disposition of the carboxylates. The RS-system can only involve an axial–equatorial disposition of carboxylates, probably equilibrating between two equivalent forms in the free ligand. If cation complexation results in coordination by a ligand carboxylate as argued above, then this would freeze one conformation of the ligand. Complexation

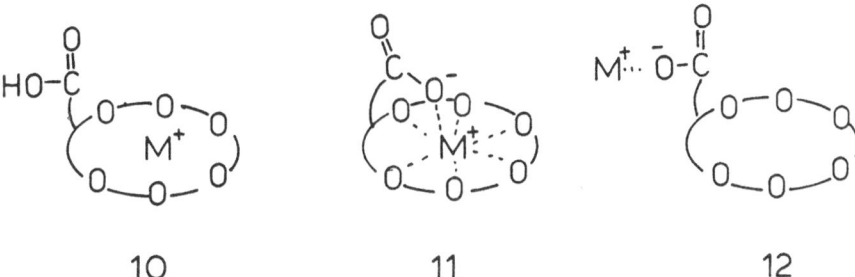

Fig. 2. Modes of cation-ligand association: 10 – cavity term; 11 – cooperative interaction; 12 – coulombic interaction.

would then be entropically less favored than in the RR-system. At the same time, the equatorial-CO_2^- is further away from the cation in the macrocycle and the electrostatic stabilization afforded would be less as well.

To summarize the overall picture of cation binding: these ligands normally bind cations with cooperative crown ether and carboxylate donor interactions. This results in a trend favoring larger cations. However, the complexes are also stabilized by a general electrostatic interaction of the ligand and the cation. This will be greatest for the smallest cations. Thus as ligand charge increases, a trend favoring large cations reverses to a trend favoring small cations. As well as increasing the charge, deprotonation also rigidifies the ligand. In some cases, the complexes of the low surface charge density cations Tl^+ and K^+ are destabilized, since the energetic advantage of carboxylate coordination is insufficient to achieve ligand distortion.

4. Application of Polycarboxylate Crown Ethers

The complexes of the polycarboxylate crown ethers of Figure 1 are substantially more stable than those of their parent, 18-crown-6. The stability constants of Table I, in fact, fall into the range usually associated with cryptands 2.2.1 and 2.2.2 [19] and EDTA [20]. However, relative to these ligands, the polycarboxylate crown ethers appear as relatively indiscriminant cation complexing agents. Very little molecular recognition based on ion size difference is occurring. Nonetheless, there are a number of features of the complexation of cations by polycarboxylate crown ethers which have lead to some simple but unique applications.

4.1. SIMULTANEOUS PROTON AND METAL ION BUFFERING

Polycarboxylate crown ethers bind cations over a wide range of pH. Although the protonated complexes tend to be less stable than the complexes of the fully deprotonated ligand, nonetheless cation binding does occur, even in acidic solution (pH 3). This is in sharp contrast to the cryptands and EDTA, in which protonated complexes are vastly less stable or do not form. Furthermore, the first pK_as of these ligands are much more basic [19, 20], thus cation complexation even at pH 7 cannot be easily achieved (some very stable EDTA complexes, with Ca^{2+} for example, can form into weakly acidic solution (pH 6)).

Furthermore, the polycarboxylate crown ethers have a series of protonation equilibria, both of the free ligand and of the complexes, which extend over a range of pH in which metal ions are complexed. This results in the possibility of simultaneous buffering both protons and metal ions with a single species [21]. One example of this potential is illustrated in Figure 3. Obviously the buffering capacity of such a dilute solution is limited. Even so, the calculated buffer capacity has significant values over part of the range where K^+ is bound (pH 3–7). As an example, consider the solution of 1 mM HEX containing 0.5 mM K^+ at pH 4.5. Addition of a further one quarter equivalent of KOH, to give a total K^+ concentration of 0.75 mM, results in a pH change to 4.68 and a change in free K^+ concentration to 2.7×10^{-5} M (> 95% complexed). This type of behavior is rare, although some pH buffer components will bind transition metal ions to achieve a similar effect [21].

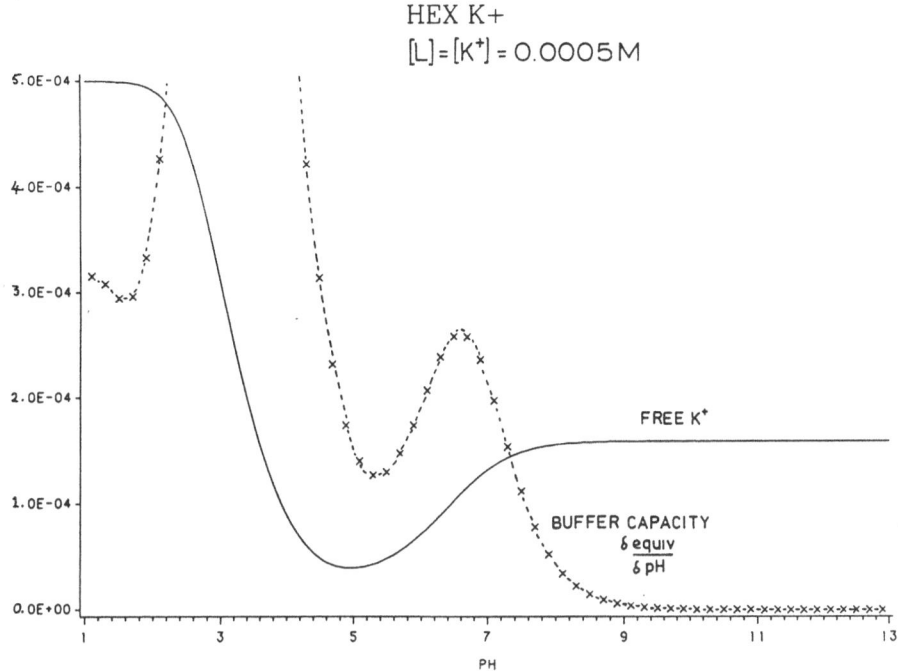

Fig. 3.   Calculated free potassium ion concentration and buffer capacity as a function of pH for a 1 : 1 mixture of HEX and K^+ at 5×10^{-4}M.

4.2. MODEL OF PRIMARY ACTIVE TRANSPORT

The cation complexes of polycarboxylate crown ethers are charged, thus resist extraction into organic solvent. This is in contrast to cryptands which can be used as organic soluble membrane transport carriers of cations. Our interests in membrane transport mechanisms [10, 12, 22] lead us to consider the implications of a theoretical framework of energy transduction in membranes proposed by Goddard [23]. Within the framework, strong parallels are drawn between simple gradient pumping systems, such as are widely known for crown ether and cryptand carriers [24, 26], and systems involving reaction pumping, or primary active transport. In order to convert a gradient pumping to a reaction pumping transport system, one of the translocated species must be intercepted in the receiving phase. It is the free energy of this association which energizes the transport cycle. We wished to explore the simple thermodynamic and kinetic predictions of the model proposed [23], hence we investigated the transport cycle illustrated in Figure 4 [27]. This system is formally an example of primary active transport in which the driving force for transport is provided by the association of the transported K^+ with the crown ether dicarboxylate in the right aqueous phase. The system exploits the property of strong complex formation in water and the low extractibility of the crown ether complex in direct contrast to the properties of the cryptand carrier in the same system.

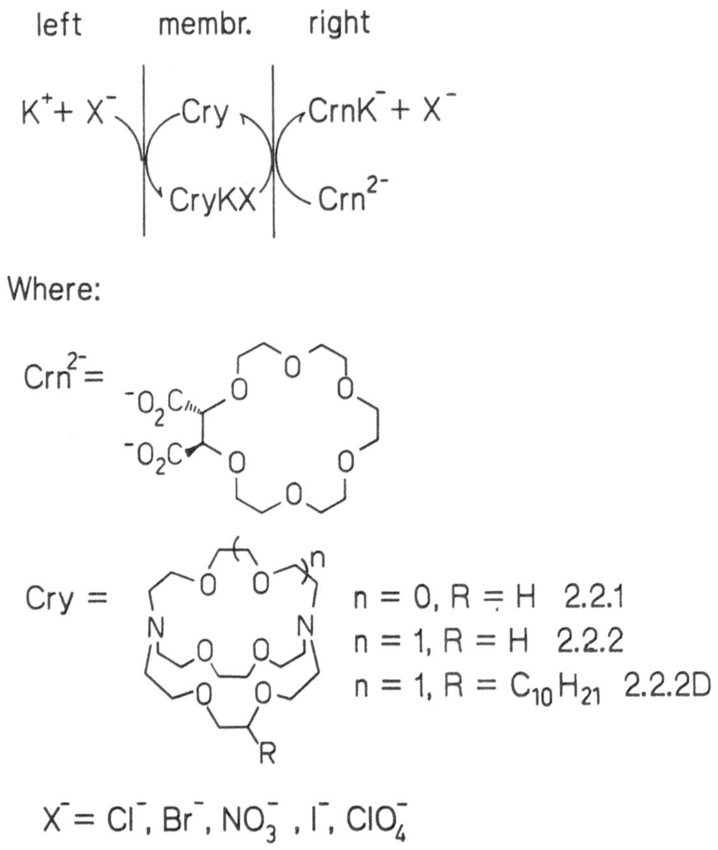

Fig. 4.   Schematic mechanism of a chemical model of primary active transport (reaction pumping).

In this system, as in all systems under diffusion control [22, 24], the flux is a bell-shaped function of the extraction constant, K_{ex}, of the carrier (Cry). As illustrated in Figure 5, at low values of log K_{ex}, the flux is low as only a small amount of KX is extracted into the membrane. As log K_{ex} increases, the flux increases to a maximum when the carrier is half saturated. A further increase in log K_{ex} results in a decrease in flux as the KX is held within the organic phase. The theoretical curve of Figure 5 may be calculated from a knowledge of the initial concentrations of the various species, and the value of the 1 : 1 association constant of the crown ether K^+ complex which energizes the transport [27]. The experimental points of Figure 5 were obtained from a series of experiments with the three carriers and five anions of Figure 4. In various combinations, they provided various values of log K_{ex}, and gave characteristic values of the flux in the transport experiment. The excellent agreement between theory and experiment serves to encourage our use of the theoretical framework of Goddard [23] as a design tool for development of new transport systems.

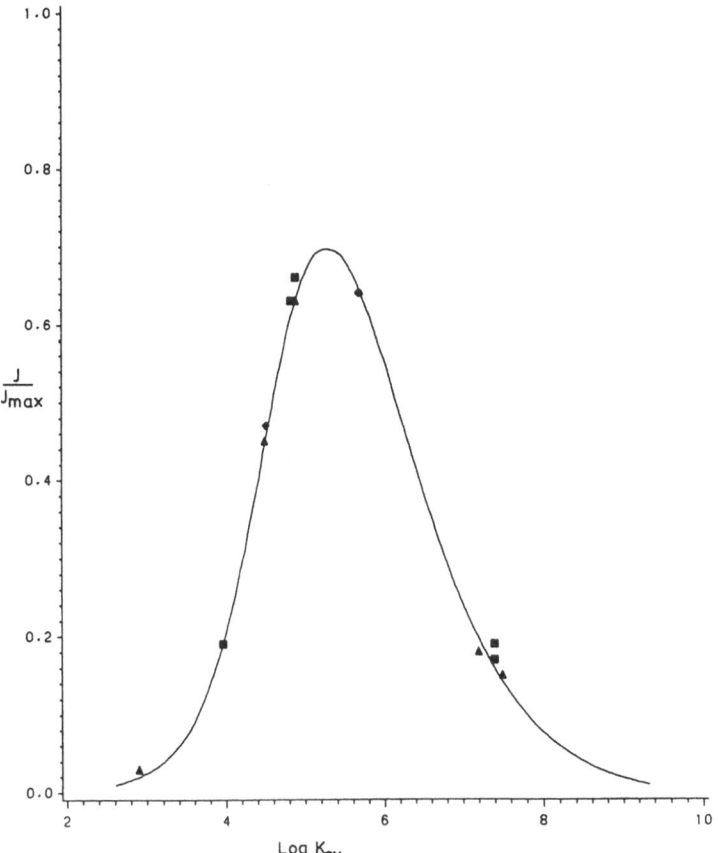

Fig. 5. Normalized membrane flux (J/J_{max}) as a function of extraction constant (log K_{ex}) for the transport system of Figure 4 [27]; 5.0×10^{-3} KX, pH 9.6; glycine/MTEAOH[11]1.0 $\times 10^{-3}$M Cry in CHCl$_3$ ∥ 5.0×10^3M Crn, 2.0×10^{-3}M KX, pH 9.6, glycine/MTEAOH, U-tube transport cell, stirring at 400 ± 5 rpm, 25°C. Theoretical curve calculated for 5% total transport. ■ = 2.2.2D as carrier, ▲ = 2.2.2 as carrier, ◆ = 2.2.1 as carrier.

Acknowledgement

The ongoing support of the Natural Sciences and Engineering Research Council of Canada, and of the University of Victoria is gratefully acknowledged.

References

1. C. J. Pedersen: *J. Am. Chem. Soc.* **89**, 7017 (1967).
2. J. M. Lehn: *Pure Appl. Chem.* **51**, 979 (1979).
3. D. J. Cram and V. N. Trueblood: *Top. Curr. Chem.* **98**, 43 (1981).
4. J. P. Behr, J. M. Girodeau, R. C. Hayward, J. M. Lehn, and J. P. Sauvage: *Helv. Chim. Acta,* **63**, 2096 (1980).
5. A. Anantanarayan, V. A. Carmicheal, P. J. Dutton, T. M. Fyles, and M. J. Pitre: *Synth. Commun.* **16**, 1771 (1986).
6. J. P. Behr, J. M. Lehn, and P. Vierling: *Helv. Chim. Acta,* **65**, 1853 (1982).

7. J. P. Behr, C. J. Burrows, R. Heng, and J. M. Lehn: *Tetrahedron Lett.* **26**, 215 (1985).
8. J. P. Behr, J. M. Lehn, D. Moras, and J. C. Theirry: *J. Am. Chem. Soc.* **103**, 701 (1981).
9. H. Dugas, P. Keroack, and M. Ptak: *Can. J. Chem.* **62**, 489 (1984).
10. T. M. Fyles and D. M. Whitfield: *Can. J. Chem.* **62**, 507 (1984).
11. H. O. Kalinowski, D. Seeback, and G. Crass: *Angew. Chem. Int. Ed. Engl.* **14**, 762 (1975).
12. L. A. Frederick, T. M. Fyles, N. P. Gurprasad, and D. M. Whitfield: *Can. J. Chem.* **59**, 1721 (1981).
13. P. J. Dutton, T. M. Fyles, and S. J. McDermid: *Can. J. Chem.* **66**, 1097 (1988).
14. A. Anantanarayan, P. J. Dutton, T. M. Fyles, and M. J. Pitre: *J. Org. Chem.* **51**, 757 (1986).
15. F. R. Fronczek, T. M. Fyles, and R. D. Gandour: unpublished observations.
16. I. G. Sayce: *Talanta* **15**, 1397 (1968); **18**, 653 (1971).
17. R. D. Gandour, F. R. Fronczkek, V. J. Gatto, C. Minganti, R. A. Schultz, B. D. White, K. A. Arnold, D. Mazzocchi, S. R. Miller, and G. W. Gokel: *J. Am. Chem. Soc.* **108**, 4078 (1986).
18. R. J. Adamic, E. M. Eyring, S. Petrucci, and R. A. Bartsch: *J. Phys. Chem.* **89**, 3752 (1985).
19. J. M. Lehn and J. P. Sauvage: *J. Am. Chem. Soc.* **97**, 6700 (1975).
20. L. G. Sillen and A. E. Martell: *Stability Constants of Metal-Ion Complexes.* Chem. Soc. Special Publication No. 17, (1964); No. 24, (1971).
21. D. D. Perrin and B. Dempsey: *Buffers for pH and Metal Ion Control.* Chapman and Hall, Chapter 7 (1974).
22. T. M. Fyles: *Can. J. Chem.* **65**, 884 (1987).
23. J. D. Goddard: *J. Phys. Chem.* **89**, 1825 (1985).
24. J. D. Behr, M. Kirch, and J. M. Lehn: *J. Am. Chem. Soc.* **107**, 241 (1985).
25. M. Okahara and Y. Nakatsuji: *Top. Curr. Chem.* **128**, 37 (1985).
26. R. M. Izatt, G. A. Clark, J. S. Bradshaw, J. D. Lamb, and J. J. Christensen: *Sep. Pur. Meth.* **15**, 21 (1986).
27. T. M. Fyles and S. P. Hansen: *Can. J. Chem.* **66**, 1445 (1988).

Journal of Inclusion Phenomena and Molecular Recognition in Chemistry 7 (1989), 183–191.
© 1989 *by Kluwer Academic Publishers.*

Chemistry and Functions of Recently Developed Macrocyclic Polyamines

EIICHI KIMURA
Department of Medicinal Chemistry, Hiroshima University School of Medicine, Kasumi 1-2-3, Minami-ku, Hiroshima 734, Japan

(Received: 1 February 1988)

Abstract. New, functionalized macrocyclic polyamines **1**, **9**, **17**, and **20** have been synthesized for investigation of their metal and molecular inclusion.

Key words. Macrocyclic polyamines, alkaline earth metals, cyclam, transition metal ion, cyclic voltammetry.

1. Introduction

Recently we have synthesized newly functionalized macrocyclic polyamines **1**, **9**, **17**, and **20** for the investigation of their metal and molecular inclusion.

2. A New Mg^{2+} Ion Receptor. Macrocyclic Polyamines Bearing an Intraannular Phenolic Group [1]

To explore a new potential of macrocyclic polyamines, we now have synthesized intraannular phenol-containing derivatives **1** and **2**, which were discovered to possess novel uptake features for alkaline earth metal ions. The homologous bifunctional host molecules **3**, **4** [2, 3] and **5** [4] have recently been reported and comparison with those congeners sheds light on the unique properties of the present phenol azamacrocycles.

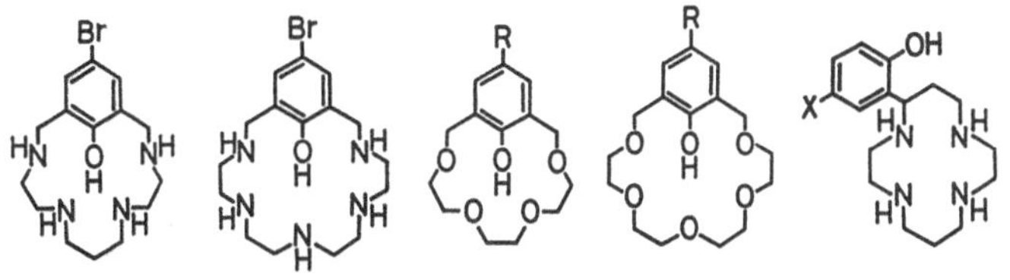

3a ; R= H   4a,b   5a ; X= H
b ; R= N=N-⟨⟩-NO2   b ; X= NO2
              NO2

The azacrown rings here are efficient acceptors of the phenol protons. Both the neutral phenol (λ_{max} 294 nm) and ionic phenoxide absorptions (λ_{max} 301 and 250 nm) are observed in the electronic spectra of **1** and **2** in EtOH and $CHCl_3$ solutions. The ratios for the neutral phenol form **2** to ionic phenolate form **6** with the pentaamine are estimated to be $1:1$ in anhydrous EtOH and $1:0.75$ in $CHCl_3$, on the basis of the UV absorptions for the sole phenol form (generated with CCl_3CO_2H, ε 2200 at 283 nm) and phenolate form (with NaOEt, ε 4400 at 301 nm and ε 8600 at 250 nm). Apparently, self dissociation of the phenol protons should be negligible with **3** and **4** in EtOH, and with **5a** in $CHCl_3$ and EtOH.

The phenol ionization in EtOH is further promoted by addition of *neutral alkaline earth metal salts* $MgCl_2$, $CaCl_2$, $SrCl_2$, *and* $Ba(SCN)_2$. In varying the [ligand]/[M^{2+}] ratio ([ligand] + [Mg^{2+}] = 1 mM), the total phenolate concentration reaches a maximum at [ligand]/[M^{2+}] = 1 with all the metals. Monovalent alkali metal salts (LiCl, NaCl, KCl, CsCl), on the other hand, do not dispel the phenolic protons of **1** and **2**. With crown homologues **3**, **4** basic conditions are needed to displace the phenol protons for M^+ and M^{2+}. The phenol-pendant N_4 homologues **5a**, **5b** do not interact with alkaline earth metal salts in EtOH solution.

The apparent constants for the $1:1$ complexation between **1** (or **2**) and M^{2+} in EtOH are calculable in terms of $K = [M^{2+}\text{-complex}]/[M^{2+}][H_{-1}L](M^{-1})$ (where $H_{-1}L$ is free ligand in phenolate anion form, and [M-complex] is [total phenolate anion]$-[H_{-1}L]$ using the UV spectral data. The results are summarized in Table I. As the M^{2+} size increases, complexation becomes more favorable with the larger macro ligand **2**. These size effects suggest that the metal inclusion into the size-fitted polyamine hole is a determining factor for the selective uptake. It is to be emphasized that the macrocyclic polyamine ligands are most appropriate for the smallest, Mg^{2+}, while crown ethers favor larger M^{2+} cations.

Furthermore, in EtOH the complexation of Mg^{2+} and Ca^{2+} with **1** or **2** is proved to be stronger than with 15-crown-5 or 18-crown-6, while that of Sr^{2+} and Ba^{2+} with **1** or **2** is weaker than with 18-crown-6. This is revealed by the change of the phenolate UV absorptions upon addition of crown ethers (1 eq.) to a $1:1$ mixture of **1** (or **2**) and MX_2 in EtOH.

Table I. 1:1 Metal complexation constants ($\log K$)

Metal Ion (Ionic Diameter, Å)	1[a] Cavity Size (1.4–2.0 Å)	2[a] (1.8–2.2 Å)	15-Crown-5[b] (1.7–2.2 Å)	18-Crown-6[b] (2.6–3.2 Å)
Mg^{2+} (1.30)	3.3	3.1	noncomplexation	unreported
Ca^{2+} (1.98)	2.9	2.9	2.1	3.9
Sr^{2+} (2.26)	2.3	2.6	2.6	> 5.5
Ba^{2+} (2.70)	1.6	2.4	unreported	7.0

[a] $K = [M^{2+}\text{-complex}]/[M^{2+}][H_{-1}L](M^{-1})$ in EtOH at 25°C. Standard deviation is ± 0.1.
[b] $K = [M^{2+}\text{-}L]/[M^{2+}][L](M^{-1})$ in MeOH at 25°C (R. M. Izatt, J. S. Bradshaw, S. A. Nielsen, J. D. Lamb, and J. J. Christensen: *Chem. Rev.* **85**, 271–339 (1985)).

Finally, the 1:1 complexation of **1** with Mg^{2+} was proved by isolation of its monoperchlorate salt, $Mg^{2+}\text{-}H_{-1}L\cdot ClO_4\cdot H_2O$ (as light yellow powder) from an aqueous EtOH solution.

The present results have thus revealed new potentials of macrocyclic polyamines for selective receptors of hard metal ions.

3. The *o*-Methoxyphenol-Pendent Cyclam Complexes 11. A Novel Molecule Designed for Intramolecular Redox-Coupling between Monodentate Catecholate and Metal Ions in a N_4 Macrocycle [5]

The coordination chemistry of bidentate catechol ligands (cat^{2-}) has long been a subject of chemical [6] as well as biochemical interest [7]. By contrast, the chemistry of monodentate catecholate ($catH^-$) complexes is virtually unknown. Recently, however, $Fe^{3+}\text{-}catH^-$ coordination was proposed as an active intermediate in catechol-cleaving dioxygenases, (e.g. protocatechuate 3,4-dioxygenase), whereupon the catechol becomes susceptible to O_2 attack [8].

With the intention of exploring the redox coupling between the monodentate catecholate and metal ions, we have designed a new cyclam ligand **9** that strategically places the N_4 macrocycle so as to hold metal ions during the course of the redox process close to the *o*-methoxyphenol, an equivalent of catechol. Earlier, we have reported the X-ray structure of the axial phenolate coordinating complexes (structure **10**), along with the mutually affected redox behavior of the phenolate ion and metal ions [4].

The new ligand **9** was synthesized as follows:

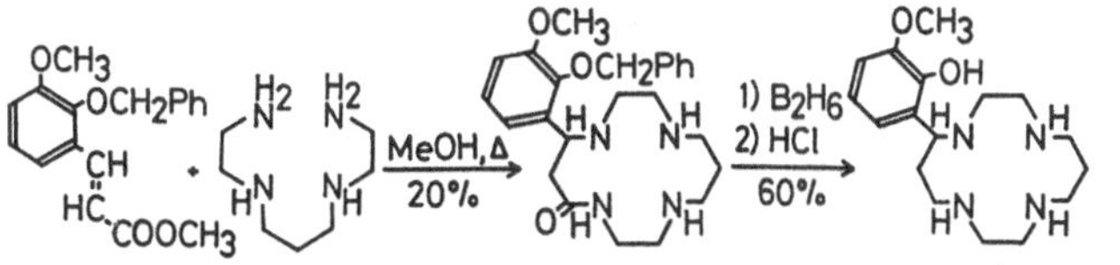

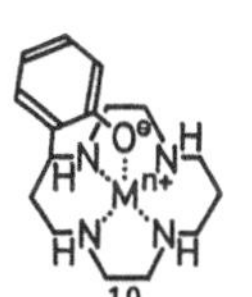

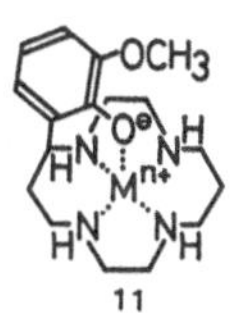

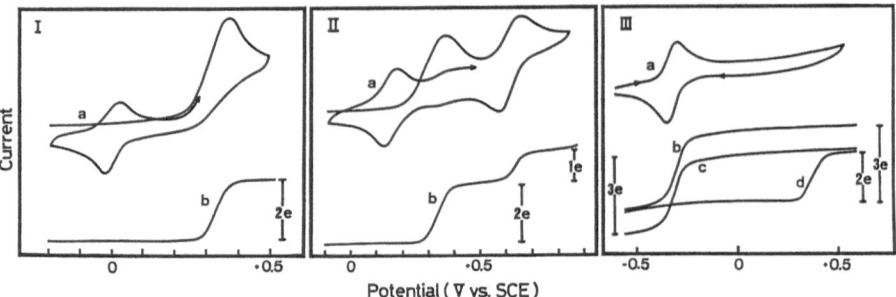

Fig. 1. Cyclic voltammograms (I-a, II-a, III-a) at a scan rate of 100 mV s^{-1} and RDE voltammograms (I-b, II-b, III-b, c, d) at an electrode rotation rate of 1000 rpm, and a scan rate of 10 mV s^{-1} on a glassy carbon disk electrode with 0.2 M Na$_2$SO$_4$ at 25°C. I for 1 mM Cu^{2+}-complex **3** at pH 10.0. II for 1 mM Ni^{2+}-complex at pH 7.3 (Tris buffer); curve *a* and *b* for Fe^{2+}-complex **3**, curve *c* for Fe^{3+}-complex **7** (aeration product of Fe^{2+}-complex **3**), curve *d* for free ligand **1**; no further oxidation wave was seen up to +0.5 V vs. SCE.

In an argon atmosphere **9** forms 1 : 1 complexes *in situ* having structure **11** with Ni^{2+} (pH > 7), Cu^{2+} (pH > 9) and Fe^{2+} (pH > 6), as established by pH-metric titration. The UV absorptions [λ_{max} 293 nm (ε 3600) and 247 nm (ε 8600) for Ni^{2+} (pH 8.2), 288 nm (ε 5700, *sh*) and 246 nm (ε 12 000) for Cu^{2+} (pH 10.0), and 287 nm (ε 3700) and 243 nm (ε 7600) for Fe^{2+} (pH 8.3)], being similar to each one of **10**, support the phenolate coordination in **11**. In electrochemical behavior, the Cu^{2+} complex **11** displays an identical CV (at pH 10.0, Figure 1-I) with that of the uncoordinated ligand, indicating little influence of Cu^{2+} on oxidation of the axial *o*-methoxyphenolate. Cu^{2+} is not oxidized in the measured potential range.

The CV and rotating disk electrode (RDE) voltammogram of the Ni^{2+} complex **11** (Figure 1-II) indicate the 2*e* oxidation (to **12**) at +0.33 V (pH > 7), followed by 1*e* oxidation (to **13**) at +0.62 V; in the subsequent CV sweep the reversible *o*-quinone/catechol (**12**⇌**14**) wave appears at +0.14 V, as was seen with the free ligand.

The most unusual synergistic oxidation behavior was revealed by the definite 3e oxidation of Fe^{2+} complex **11** *simultaneously at* −0.30 V (*pH 7.3 Tris buffer*) *on RDE* (see Figure 1-IIIb). The CV of **11** in Figure 1-IIIa shows no other redox wave up to +0.5 V. The potential of −0.30 V is too low for the 2*e* oxidation of free or mere polycation (such as Cu^{2+}, Ni^{2+})-binding *o*-methoxyphenolate. The 1*e* oxidation potential for Fe^{2+}/Fe^{3+} in **10** was −0.16 V [4]. We are thus tempted to conclude that Fe^{2+} is initially oxidized to Fe^{3+} (**15**) at the lower potential of −0.30 V under the influence of the stronger σ-donor, *o*-methoxyphenolate, and thereupon Fe^{3+} catalytically drains 2*e* out of this ligand to a possible quinone pendant **16**. All the attempts to prepare **16** in a large enough quantity for further identification by electrochemical oxidation at −0.10 V resulted in failure, mostly due to the immediate halt of the electric current. Mild aeration (20 min) of **11** initially oxidizes Fe^{2+} to Fe^{3+} [**15**, deep violet, λ_{max} 278 nm (ε 5500), 518 nm (ε 2150) at pH 7.0, in analogy to Fe^{3+} complex of **10** [4], which undergoes further 2*e* oxidation to **16** at −0.30 V (curve *c*).

Although the final product structure **16** remains open to question, the present Fe macrocyclic complexes **11** and **15** have offered the first prototype for synergistic intramolecular redox coupling between monodentate catecholate and metal ions to render the catechol unusually vulnerable to oxidation. Further modification of metal ions or the macrocyclic structure with a catechol pendant would find a novel redox system. Moreover, the reactivity of the remaining 6th axial position would be extremely interesting as a catalytic site.

4. New Cyclam with an Appended Imidazole. The First Biomimetic Ligation of Imidazole for Axial π-Interaction with Metal Ions [9]

We now have succeeded in synthesizing the first cyclam (17) bearing an appended imidazole that acts as an ideal axial donor for effective imidazole → metal π-interaction. Our synthetic procedure is simple and versatile, serving also for pyridyl [10] and phenol-appended cyclams [4].

Purple and pink crystals were precipitated when equimolar amounts of $NiSO_4$ and **17** were treated at 50°C for 10 min in an aqueous solution (pH = 8) in the presence of $NaClO_4$. The purple product is a kinetic one, which by longer treatment in warmer (>70°C) aqueous solution is completely converted to the pink as the ultimate thermodynamic product. The purple **18** and pink **19** crystals were subjected to X-ray structure analysis.

In six-coordinate **18** ($X = OClO_3^-$), the cyclam moiety is in a folded configuration with the imidazole nitrogen N_{16} and a perchlorate oxygen O_1 occupying the remaining two *cis* sites. The metal-macrocyclic nitrogen bond Ni—N_1, —N_4, —N_8, and —N_{11} lengths are 2.090(4), 2.087(5), 2.108(4), and 2.091(5) Å, respectively, which are in the normal range 2.05–2.10 Å for high-spin Ni^{II}—N bonds. The Ni—N(imidazole) bond length is extremely short at 2.067(5) Å. The Ni—O(perchlorate) bond length is 2.219(5) Å, to complete the octahedron.

Pink crystals of **19** ($X = CH_3CN$) were grown in CH_3CN—H_2O solution. The coordinate structure (Figure 2) reveals high-spin Ni^{II} in a planar cyclam ('chair form') with imidazole N_{16} and acetonitrile N_{20} at the axial positions. The bond lengths for Ni—cyclam N_1 2.060(11), —N_4 2.058(12), —N_8 2.121(12), and —N_{11} 2.046(12) of the 'metal-in' complex **19** are a little shorter than the corresponding ones of the 'metal-out' complex **18**. The imidazole ring stands vertically to the cyclam N_4 plane to become an axial donor. Moreover, the imidazole plane bisects (by 20°) the $N_4(N_1)$—Ni—$N_8(N_{11})$ angles, which, coupled with short Ni—N_{16}(Im) distance of 2.098(9) Å, may facilitate the imidazole → metal π-donation. Similar bond lengths (2.06–2.1 Å) [11] and orientations of the imidazole ring of the proximal histidine onto the porphyrin plane (20–22.5°) [11] are recognized in hemes or their models to serve Im–Fe^{II} π-interaction for the *trans* O_2 binding. Our new cyclam (**17**) is thus proved to be equipped with an ideal imidazole ligand for axial coordination.

A unique axial imidazole coordination in **19** affects the $Ni^{III/II}$ redox potential. Cyclic voltammetry (CV) shows a quasi-reversible voltammogram at $E_{1/2}$ of +0.54 V vs. saturated calomel electrode (SCE) (0.5 M Na_2SO_4, pH = 7.0, 25°C), which goes between +0.61 V of the pyridyl coordinate homologue [10] and +0.35 V of the phenol coordinate (**10**, M = Ni) [4], but is near to +0.50 V of the cyclam (without pendant) complex.

Another remarkable effect of the destined axial coordination of the imidazole is shown in the O_2 binding of its Co^{II} complex at room temperature in aqueous solution. The brown solid 1 : 1 Co^{II}–O_2 complex precipitated as the diperchlorate salt. Its ESR parameters in a frozen aqueous solution at 77 K are $g_\perp = 2.01$, $g_\parallel = 2.08$, $A_\perp^{Co} = 13.3$ G, $A_\parallel^{Co} = 20.0$ G, being identical to those reported for paramagnetic 1 : 1 Co^{II}–O_2 adducts [12]. Without the pendant imidazole, Co^{II}(cyclam) yields only diamagnetic 2 : 1 O_2 adducts (μ-peroxo complex). In the Fe complex of **17**, the $Fe^{III/II}$ redox potential (pH = 7, $I = 0.1$ M, $NaClO_4$, 25°C) is +0.00 V vs. SCE, to be compared with −0.16 V of (**10**, M = Ni) and +0.12 V of (pyridyl homologue). The Fe^{II}–**17** complex also forms an O_2 adduct with appearance of the O_2 → Fe CT band at 344 nm, which is currently under investigation.

5. A New Ditopic Receptor Molecule for Ionic Guest Molecules [13]

Macrocyclic polyethers (crown ethers) bind with cationic guests (e.g. primary ammonium cations) [14]. Meanwhile, anionic substrates (e.g. carboxylates) or

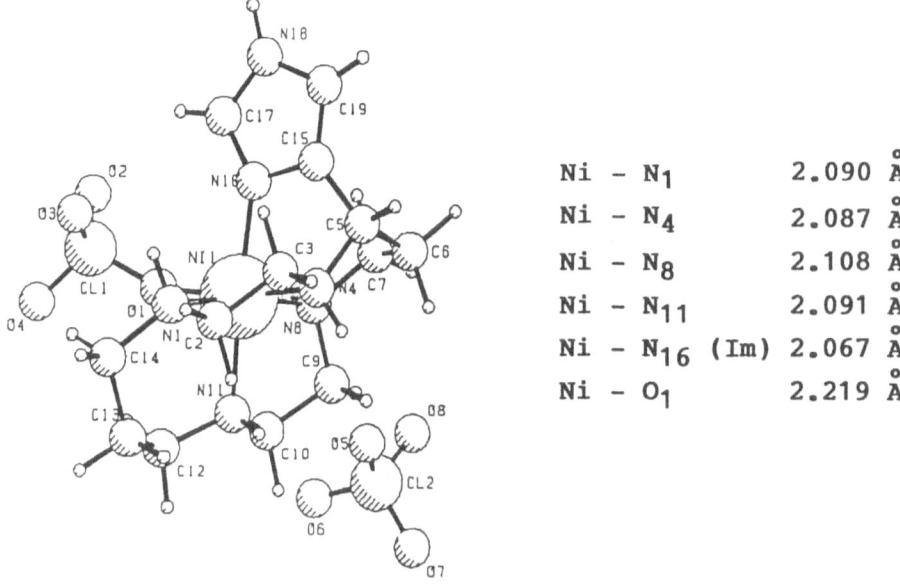

$$Ni - N_1 \quad\quad 2.090 \text{ Å}$$
$$Ni - N_4 \quad\quad 2.087 \text{ Å}$$
$$Ni - N_8 \quad\quad 2.108 \text{ Å}$$
$$Ni - N_{11} \quad\quad 2.091 \text{ Å}$$
$$Ni - N_{16} \text{ (Im)} \quad 2.067 \text{ Å}$$
$$Ni - O_1 \quad\quad 2.219 \text{ Å}$$

Fig. 2.   Crystal structure of imidazole-pendant cyclam.

electron donor substrates (e.g. catechols) are recognized by macrocyclic polyamine cations [15]. However, receptor molecules that can simultaneously recognize both cations and anions are very rare. Such ditopic hosts would offer efficient and selective recognition sites for ionic molecules by concerted binding actions.

Herein we report the first ditopic receptor molecule 20 composed of a macromonocyclic polyamine and crown ether, which indeed forms 1:1 complexes with ionic substrates such as amino acids 22–25, peptides 26, or catecholamine 27 in neutral aqueous solutions.

A measurement of host–guest interaction has been made with an anodic wave polarography in the same manner as those previously applied to the complexation of polycarboxylate [15] and catechols with [18]aneN$_6$ [16]. The final results along with the K_i values used for calculation are summarized in Table II.

The strong interaction of 20 with dopamine 27 is noteworthy. [18]aneN$_6$·3 H$^+$ alone binds with the catechol moiety of dopamine 27 with a β_L value of 1.1×10^3 M^{-1} [16]. Addition of 10 eq. of benzo-15-crown-5 has not affected its polarographic behavior at all. However, when covalently attached as in 20, the crown ether moiety interacts complementarily with the primary ammonium cation part of dopamine 27, as depicted in 21, resulting in a β_L value that is greater by a factor of ten. Since the

Table II. 1:1 Association constants β_L for (20)[a] with ionic substrates at 25°C and $I = 0.20$ M (NaClO$_4$)

Ionic Substrate	β_L, M^{-1}	Measured pH in Tris buffer
glycine (22)	1.50×10^2	6.5–8
β-alanine (23)	1.10×10^2	7–8
GABA (24)	1.02×10^2	7–8
6-amino-hexanoic acid (25)	1.05×10^2	7–8
diglycine (26)	6.87×10^1	7–8.5
dopamine (27)	2.92×10^4	7–8
catechol (28)	1.50×10^2	7–8

[a] $pK_a = 9.66, 9.13, 7.75, 4, \sim 2, \sim 1.$

experiment indicates involvement of $(i + j) = 5$ protons in this complexation, we assign $2\,H^+$ to the [18]aneN$_6$ part and $3\,H^+$ to the dopamine part. On the other hand, catechol 28, a monotopic guest of [18]aneN$_6 \cdot 3\,H^+$, does not enjoy affinity enhancement by the attachment of the crown ether moiety: the β_L value of 1.5×10^2 with 20 is almost the same, 1.6×10^2, as with [18]aneN$_6$.

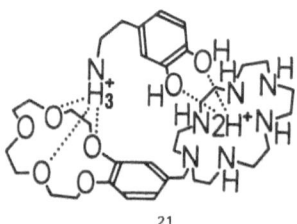

21

In view of the versatility and simplicity of the present synthetic method, the macromonocyclic polyamine linked with a crown ether forms a promising prototype for the design of a variety of polytopic recognition receptors that should find a number of applications.

References

1. E. Kimura, Y. Kimura, T. Yatsunami, M. Shionoya, and T. Koike: *J. Am. Chem. Soc.* **109**, 6212 (1987).
2. G. M. Browne, G. Fergeson, M. A. McKervey, D. L. Mulholland, T. O'Connov, and M. Parvez: *J. Am. Chem. Soc.* **107**, 2703 (1985).
3. K. Nakashima, S. Nakatsuji, S. Akiyama, T. Kaneda, and S. Misumi: *Chem. Pharm. Bull.* **34**, 168 (1986), and references therein.
4. E. Kimura, T. Koike, K. Uenishi, M. Hediger, M. Kuramoto, S. Joko, Y. Arai, M. Kodama, and Y. Iitaka: *Inorg. Chem.* **26**, 2975 (1987), and references therein.
5. E. Kimura, S. Joko, T. Koike, and M. Kodama: *J. Am. Chem. Soc.* **109**, 5528 (1987).
6. L. A. deLearie and C. G. Pierpont: *J. Am. Chem. Soc.* **108**, 6393 (1986), and references therein.
7. C.-W. Lee, D. J. Ecker, and K. N. Raymond: *J. Am. Chem. Soc.* **107**, 6920 (1985), and references therein.

8. T. Funabiki, A. Mizoguchi, T. Sugimoto, S. Tada, M. Tsuji, H. Sakamoto, and S. Yoshida: *J. Am. Chem. Soc.* **108**, 2921 (1986), and references therein.

9. E. Kimura, M. Shionoya, T. Mita, and Y. Iitaka: *J. Chem. Soc., Chem. Commun.*, 1712 (1987).

10. E. Kimura, T. Koike, H. Nada, and Y. Iitaka: *J. Chem. Soc., Chem. Commun.*, 1322 (1986).

11. G. B. Jameson, F. S. Molinaro, J. A. Ibers, J. P. Collman, J. I. Brauman, E. Rose, and K. S. Suslick: *J. Am. Chem. Soc.* **102**, 3224 (1980).

12. E. Kimura, M. Kodama, R. Machida, and K. Ishizu: *Inorg. Chem.* **21**, 595 (1982), and references cited therein.

13. E. Kimura, H. Fujioka, and M. Kodama: *J. Chem. Soc., Chem. Commun.*, 1158 (1986).

14. D. J. Cram and J. M. Cram: *Science* **183**, 803 (1974).

15. E. Kimura: *Top. in Curr. Chem.* **128**, 113 (1985).

16. E. Kimura, A. Watanabe, and M. Kodama: *J. Am. Chem. Soc.* **105**, 2063 (1983).

Journal of Inclusion Phenomena and Molecular Recognition in Chemistry 7 (1989), 193–201.

Molecular Recognition of Calixarene-Based Host Molecules

SEIJI SHINKAI
Department of Organic Synthesis, Faculty of Engineering, Kyushu University, Higashi-ku, Fukuoka 812, Japan

(Received: 1 February 1988)

Abstract. Water-soluble calixarenes with various functional groups were synthesized. It was found that they are capable of molecular recognition and serve as a new class of catalysts, ligands, and host molecules.

Key words. Calixarene, chiral, catalysis, uranium.

1. Introduction

'Calixarenes' are cyclic oligomers made up of benzene units in a similar way to that in which cyclodextrins are made up of glucose units and thus have been expected to be useful as a new class of host molecules. Although calixarenes can include several small molecules in the solid state, there exist only a few examples for the inclusion properties of calixarenes in solution [1–3]. This is in sharp contrast to cyclodextrins which can form a variety of host–guest-type solution complexes. The difference stems, we believe, from the poor solubility of calixarenes: that is, they are only sparingly soluble in several organic solvents, and insoluble in water. We considered, therefore, that experimental efforts should be directed primarily toward solubilization of calixarenes, which would lead to the exploitation of new host molecules and eventually to the development of refined calixarene-based enzyme mimics. With these objectives in mind, we synthesized several water-soluble anionic and cationic calixarene derivatives [3]. We have found that these water-soluble calixarenes act as excellent host molecules for the selective binding of metal ions and small organic molecules.

2. Synthesis of Water-Soluble Calixarenes

The synthetic route to water-soluble calixarenes is illustrated in Scheme 1. In order to introduce functional groups into each benzene ring one must choose the reaction having a quantitative yield, because the isolation of a fully-substituted product from lower-substituted by-products is fairly difficult. In Scheme 1 the key step to synthesize water-soluble anionic calixarenes is the sulfonation which proceeds quantitatively under the optimized conditions. The sulfonate group was converted to the nitro group to afford 2_m [4]. Finally, we obtained water-soluble cationic calixarenes, $4_m C_{n+1}$ from 3_m via O-alkyl-p-aminocalixarenes.

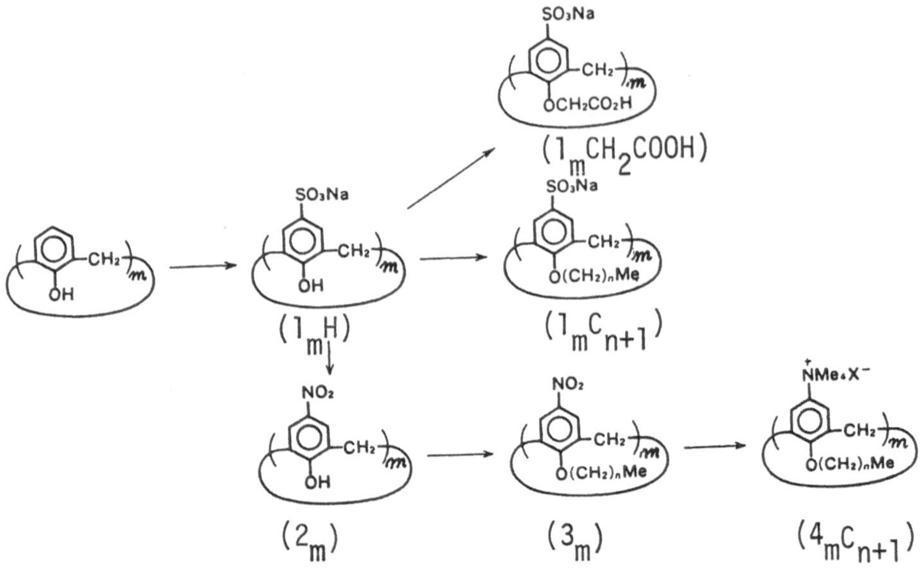

Scheme 1.

Another method to obtain *p*-substituted calixarenes is the rearrangement from the OH group. Gutsche *et al.* [5, 6] introduced the Claisen rearrangement but the products, *p*-allylcalixarenes, are fairly useless as intermediates for water-soluble calixarenes. We have found that 2_m is soluble in basic aqueous solution because of the lowered pK_a of the OH groups [7]. This suggests that introduction of electron-withdrawing *para* substituents may make the calixarene water soluble. We found that the Fries rearrangement of *O*-acylcalixarenes to *p*-acylcalixarenes occurs in 29–34% yield [8]. As expected, *p*-acylcalixarenes are soluble in water at pH > 10.

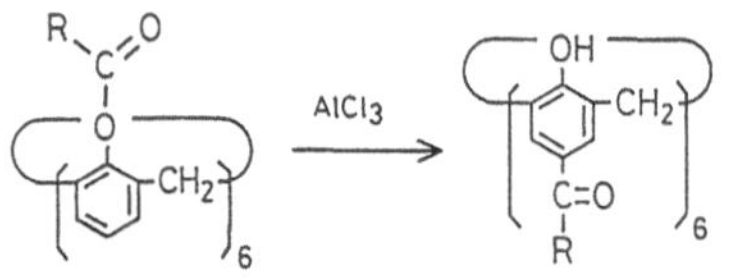

3. Stabilization of Arenediazonium Salts

In 1973, Gokel and Cram [9] found that crown ethers of the proper dimensions can solubilize several arenediazonium salts in nonpolar media. Subsequent spectroscopic studies established that the solubilization is caused by complexation, like the complexation between crown and metal cations, the linear Ar—$N^+\equiv N$ inserting into the hole of the crown ring with its oxygen atoms turned inward toward the positive charge as shown in **5**. It was already found that hexasulfonated calix[6]arenes 1_6C_{n+1} strongly bind cationic guest molecules [10]. Also, they can include amines through protonation and the apparent pK_a values are significantly lowered [11]. We thus considered that 1_6C_{n+1} may form stable complexes with the arenediazonium salt **6R**.

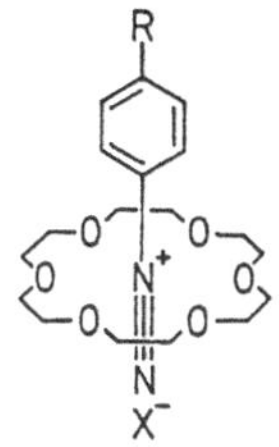

5 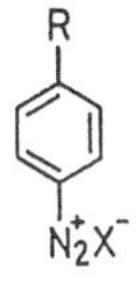   **6R** 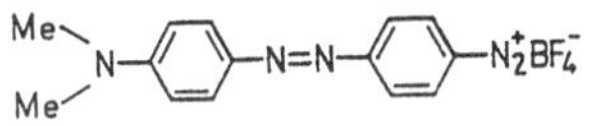   **7**

The first-order rate constants (k_d) for the dediazonation of **6H** and **6C$_6$** to the corresponding phenols are summarized in Table I. Examination of Table I reveals that in an aqueous system, neither 18-crown-6 nor anionic micelle suppresses the thermal decomposition of **6R** to a significant extent, but the k_d values decrease with increasing 1_6C_{n+1} concentration and in particular 1_6C_{12} can reduce the k_d to 20–23% of those observed in the absence of 1_6C_{n+1} [11]. To obtain insight into the stabilization mechanism, we examined the solvent effect on the dediazonation rate because the hydrophobic effect, which is partly reproduced by the solvent effect, may significantly contribute to the rate suppression [11]. We used 4-(4-dimethyl-aminophenylazo)benzenediazonium tetrafluoroborate **7** as a spectroscopic probe. Through the correlation of λ_{max} in a water–dioxane mixture we estimated the hydrophobicity of 1_6C_6 and 1_6C_{12} in water to be 25 and 85 vol% of dioxane, respectively. However, the k_d values for 1_6C_6 and 1_6C_{12} are incomparably smaller than those obtained in these mixed solvents. We believe, therefore, that the specific stabilization effect is due to the strong anionic field brought about by six sulfonate groups arranged on the edge of the calixarene cavity.

Table I. First-order rate constants (k_d) for thermal decomposition of **3R** at 30°C

Additive (conc./mM)	3H		3C$_6$	
	$10^5 \cdot k_d/s^{-1}$	k_d/k_0	$10^5 \cdot k_d/s^{-1}$	k_d/k_0
None	9.52 $(=k_0)$	1.00	1.60 $(=k_0)$	1.00
1_6C_1 (5.00)	8.62	0.91	1.27	0.79
1_6C_6 (5.00)	2.92	0.31	0.37	0.23
1_6C_{12} (2.00)	1.91	0.20	0.37	0.23
18-crown-6 (24.4)	9.25	0.96	–	–
SDS[a] (20.0)	8.84	0.93	–	–

[a] Sodium dodecylsulfate.

4. 'Hole-Size Selectivity' in Calixarene Complexes

Several groups have so far reported the metal ion selectivity of calixarenes which have ether or ester groups on the edge of the cylindrical architecture [12–14]. In contrast, almost nothing is known with certainty as to the selectivity of the calixarene cavity for neutral guest molecules. Does the calixarene cavity recognize the size of guest molecules? To answer this question we examined the host-guest behavior of water-soluble calixarenes with the different cavity size, 1_4C_4, 1_6C_4, and 1_8C_5 [15]. As guest molecules we selected Phenol Blue (PB) and Anthrol Blue (AB) with different molecular size.

PB : Me$_2$N—⟨⟩—N=⟨⟩=O

AB : Me$_2$N—⟨⟩—N=⟨⟩=O

When PB (or AB) is bound to 1_mC_4, the distinctive color change from blue to pink is observed. This occurs because of the stabilization of included protonated species (PBH$^+$, ABH$^+$) by the sulfonate groups [10, 11]. The total scheme for the complex formation (e.g., with PB) is illustrated as below.

$$PB + (1_mC_4) \xrightleftharpoons{K^0} PB \cdot (1_mC_4)$$

$$\Big\updownarrow K_a^0 \qquad\qquad \Big\updownarrow K_a^+$$

$$PBH^+ + (1_mC_4) \xrightleftharpoons{K^+} PBH^+ \cdot (1_mC_4)$$

First, we confirm that PB and AB form 1 : 1 complexes with 1_mC_4 on the basis of a continuous variation method. Based on the computer-assisted curve-fitting of the photo-titration data, we could determine the pK_a^0, pK_a^+, and K^+ values independently (Table II) [15]. Clearly, the K^+ values change sensitively in response to the cavity size: PB (small guest molecule) shows the selectivity order of $1_6C_4 > 1_4C_4 > 1_8C_4$ while AB (large guest molecule) shows the selectivity order of $1_8C_4 > 1_6C_4 > 1_4C_4$. The result suggests that PB can fit the cavity of 1_4C_4 and 1_6C_4 while AB is too large to be included in the cavity of these calixarenes. The largest difference in K^+ (31-fold) was attained between 1_6C_4–PB (largest K^+, 5.57×10^4 M^{-1}) and 1_4C_4–AB (smallest K^+, 1.82×10^3 M^{-1}). Thus, one may predict the guest selectivity of calixarenes (to some extent) on the basis of the 'hole-size selectivity'.

Table II. pK_a shifts and binding constants (K^+) for the calixarene complexes (30°C)

Guest		Calixarene		
		1_4C_4	1_6C_4	1_8C_4
PB[a]	ΔpK_a[c]	1.68	1.75	0.73
	$10^{-3} \cdot K^+ (M^{-1})$	47.2	55.7	13.5
AB[b]	ΔpK_a[c]	1.02	0.54	2.84
	$10^{-3} \cdot K^+ (M^{-1})$	1.82	9.30	15.0

[a] pK_a^0 (in the absence of 1_mC_4) = 4.60.
[b] pK_a^0 (in the absence of 1_mC_4) = 4.34.
[c] $\Delta pK_a = pK_a^+ - pK_a^0$.

The binding constants (K) for nonchromophoric guest molecules can be determined by a ^{1}H-NMR method. We estimated the K (25°C, D_2O) for the complexation of 1_mH and trimethylanilinium chloride **8**. In 1_4H and 1_6H, the resonance peaks of **8** shifted monotonously to higher magnetic field with increasing 1_mH concentration. The resultant plots of δ vs. $[1_mH]/[8]$ could be analyzed assuming the formation of 1:1 complexes: $K = 5610\,M^{-1}$ for 1_4H and $550\,M^{-1}$ for 1_6H. In contrast, the plot for 1_8H was biphasic with a break point at $[1_8H]/[8] = 0.5$. The finding supports, we believe, the proposed that 1_8H binds two **8** molecules in a stepwise manner.

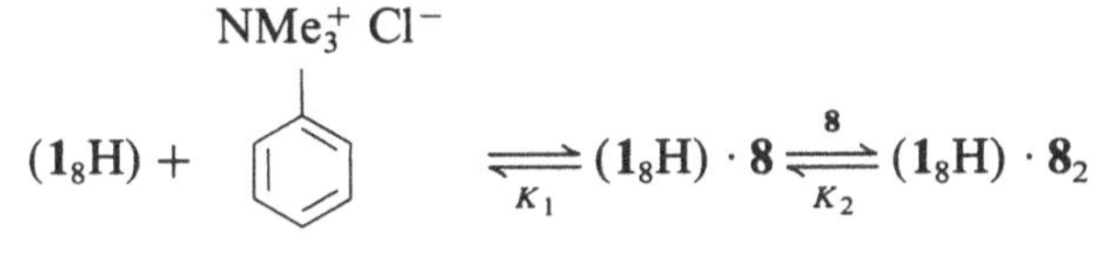

This unexpected result for 1_8H is the second example to prove the 'hole-size selectivity' in calixarene complexes.

5. Chiral Calixarenes

As described above, calixarenes are capable of including small molecules in solution. It thus occurred to us that introduction of chiral substituents into calixarenes would be of great value for development of a new class of chiral host molecules.

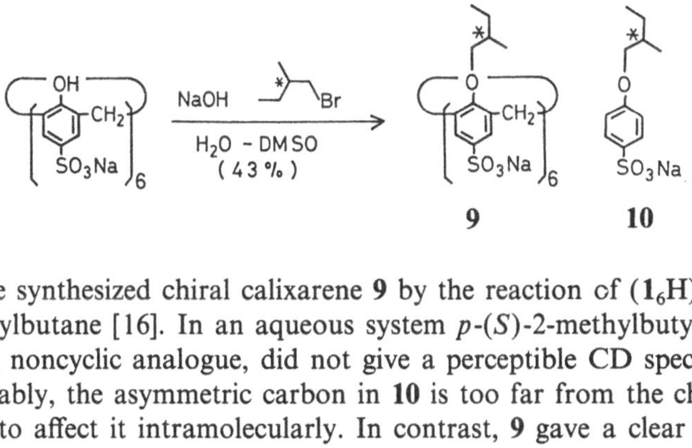

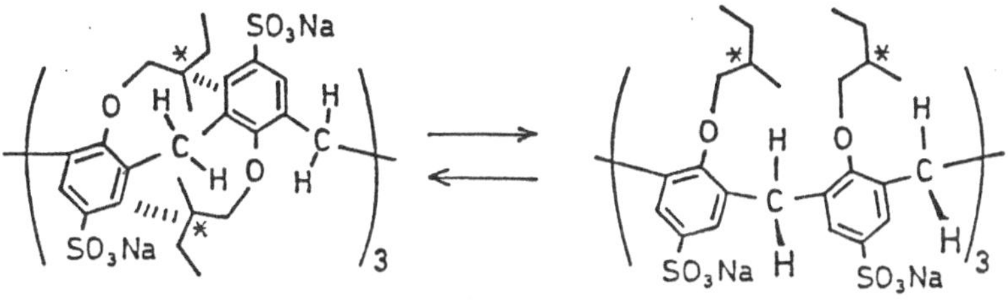

We synthesized chiral calixarene **9** by the reaction of (1_6H) and (S)-1-bromo-2-methylbutane [16]. In an aqueous system p-(S)-2-methylbutyloxybenzenesulfonate **10**, a noncyclic analogue, did not give a perceptible CD spectrum at 220–300 nm. Probably, the asymmetric carbon in **10** is too far from the chromophoric benzene ring to affect it intramolecularly. In contrast, **9** gave a clear CD spectrum in this wavelength region [λ_{max}236 ([θ] $-$ 14 100) and 269 nm ([θ] 3300)] [16]. These results suggest that the CD band of the benzene chromophore is induced by the (S)-2-methylbutyl groups in the neighboring benzene units but not by that in the same benzene unit. This situation is commensurate with the 'alternate' conformation in which (S)-2-methylbutyl group and the benzene ring are arranged alternately on the same side of the calixarene cavity.

Interestingly, we found that the CD band of **9** is weakened when the guest molecule is included in the cavity [16]. From the plots of [θ] vs. guest concentration we estimated the binding constants: $K = 1.4 \times 10^2\,\mathrm{M}^{-1}$ for hexan-1-ol, $1.2 \times 10^3\,\mathrm{M}^{-1}$ for heptan-1-ol, and $7.8 \times 10^3\,\mathrm{M}^{-1}$ for octan-1-ol. The result is rationalized in terms of the conformational change from 'alternate' to 'cone' upon inclusion of guest molecules, because the 'cone' conformation can arrange the (S)-2-methylbutyl groups and the benzene rings on the different sides of the calixarene cavity. Conceivably, the calixarene with the 'cone' conformation can provide a cavity more suitable to substrate-binding than that with the 'alternate' conformation.

The combination of chromophoric calixarene and chiral guest, which may give the ICD spectrum, is also interesting. As 2_m is soluble in basic aqueous solution and gives a λ_{max} at around 400 nm [7], we measured the CD spectrum of 2_m in the presence of chiral guest molecules (20°C, pH 12.8) [17]. Among them **11** ((R)-1,1'-binaphthyl-2,2'-phosphoric acid) afforded a distinct ICD band, and the greatest [θ] values were observed for 2_6 ([θ] $= -19\,000$ (392 nm), 2400 (450 nm)) and the next for 2_8 ([θ] $= -5900$ (410 nm)). In contrast, the ICD band was not detected at all for 2_4. A continuous variation plot showed that **11** and 2_6 form a 1:1 complex. One may conclude, therefore, that **11** fits the cavity of 2_6 (and probably of 2_8) but is too large to be included in the cavity of 2_4. On the basis of X-ray studies [2] and CPK molecular models the upper-rim diameters of calix[4]arene and calix[6]arene are estimated to be 3.8 Å and 5.0 Å, respectively. Thus, the molecular recognition pattern of calix[6]arene, found for the first time for **11**, is comparable with that of β-cyclodextrin (diameter 5.5–5.9 Å).

11

6. Molecular Design of Calixarene-Based Uranophiles

The selective extraction of uranium from sea water has attracted extensive attention from chemists because of its importance in relation to energy problems. In order to design a ligand that can selectively extract uranyl ion (UO_2^{2+}), one has to overcome a difficult problem: i.e., the ligand must strictly discriminate UO_2^{2+} from other metal ions present in great excess in sea water. A possibly unique solution to this difficult problem is provided by the unusual coordination structure of UO_2^{2+} complexes. X-ray crystallographic studies have established that UO_2^{2+} complexes adopt either a pseudoplanar pentacoordinate or hexacoordinate structure, which is quite different from the coordination structures of other metal ions. This suggests that a macrocyclic host molecule having a nearly coplanar arrangement of either five or six ligand groups would serve as a specific ligand for UO_2^{2+} (i.e., as a uranophile). This approach has been investigated by several groups [18–21]. For example, Tabushi et al. [19] synthesized a macrocyclic host molecule 12 having six carboxylate groups in the ring. Although the stability constant for 12 and UO_2^+ is pretty high ($\log K_{uranyl} = 16.4$ at pH 10.4 and 25°C), the selectivity for UO_2^{2+} is not satisfactory (e.g., $K_{uranyl}/K_{M^{n+}} = 80$–210 for Ni^{2+} and Zn^{2+}) and the synthesis is not easy [19].

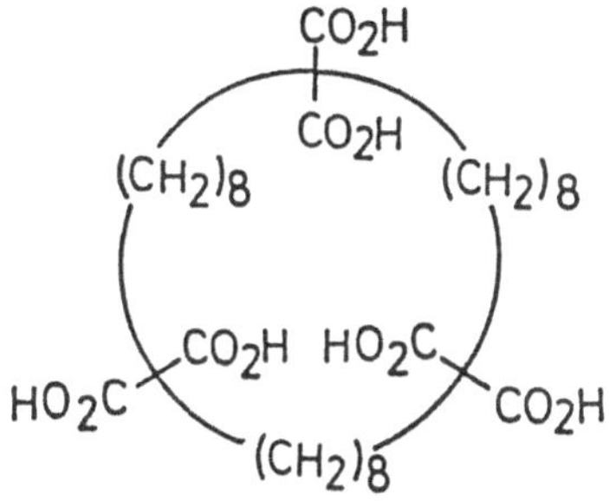

12

In the course of our studies on calixarenes, we noticed that calix[5]arene and calix[6]arene have an ideal architecture for the design of uranophiles, because the introduction of ligand groups into each benzene unit of these calixarenes provides exactly the required pseudoplanar penta- and hexacoordinate structures [22, 23]. We thus applied $1_m H$ and $1_m CH_2COOH$ as uranophiles. We found that, as shown in Tables III and IV $1_5 H$, $1_5 CH_2COOH$, $1_6 H$, and $1_6 CH_2COOH$ not only have high stability constants ($\log K_{uranyl} = 18.4$–19.2) but also an unusually high selectivity for UO_2^{2+} ($K_{uranyl}/K_{M^{n+}} = 10^{12}$–$10^{17}$) [23]. In contrast, the K_{uranyl} for $1_4 H$ and $1_4 CH_2COOH$ were dramatically decreased: they were smaller by about 16 log units than those for the pentamers and the hexamers [23]. The high affinity is rationalized in terms of the 'coordination-geometry selectivity': that is, the pentamers and the

Table III. Stability constants (K_{uranyl}) for calixarene derivatives and UO_2^{2+} (25°C)

Calixarene	pH	$\log K_{uranyl}$
$1_4 H$	6.5	3.2
$1_4 CH_2COOH$	6.5	3.1
$1_5 H$	10.4	18.9
$1_5 CH_2COOH$	10.4	18.4
$1_6 H$	10.4	19.2
$1_6 CH_2COOH$	10.4	18.7
$1_6 C_1$	6.5	3.2
12	10.4	16.4

Table IV. Selectivity factors for UO_2^{2+} ($K_{uranyl}/K_{M^{n+}}$)

Calixarene	Metal (M^{n+})	log $K_{M^{n+}}$	$K_{uranyl}/K_{M^{n+}}$
1_6H	UO_2^{2+}	(19.2)	1.0
1_6H	Mg^{2+}	a	$>10^{17}$
1_6H	Ni^{2+}	2.2	$10^{17.0}$
1_6H	Zn^{2+}	5.5	$10^{13.7}$
1_6H	Cu^{2+}	8.6	$10^{10.6}$
1_6CH_2COOH	UO_2^{2+}	(18.7)	1.0
1_6CH_2COOH	Mg^{2+}	a	$>10^{17}$
1_6CH_2COOH	Ni^{2+}	3.2	$10^{15.3}$
1_6CH_2COOH	Zn^{2+}	5.6	$10^{13.1}$
1_6CH_2COOH	Cu^{2+}	6.7	$10^{12.0}$

a The $K_{M^{n+}}$ is too small to determine by the polarographic method.

hexamers can provide the ligand groups arranged in a suitable way required for pseudoplanar penta- or hexacoordination on the edge of the calixarenes but the tetramers cannot. Similarly, the high selectivity is rationalized in terms of the moderate rigidity of the calixarene skeleton: that is, 1_6H and 1_6CH_2COOH firmly maintain the pseudoplanar hexacoordination geometry and cannot accommodate either to tetrahedral or to octahedral coordination geometry (Scheme 2).

Flexible hexadentate ligand

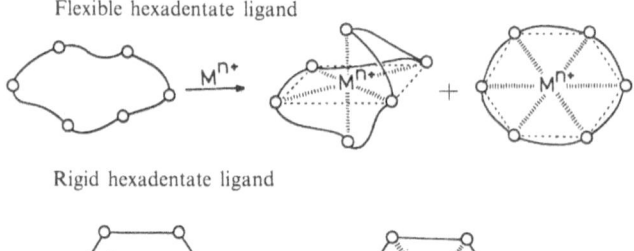

Rigid hexadentate ligand

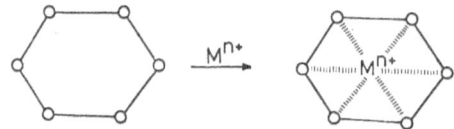

In general, there are two possible strategies for improving the metal selectivity of macrocyclic ligands: the first one is to enhance the stability constant for the target metal cation and the second one is to lower the stability constants for competing metal cations. The present study shows that calixarenes provide an ideal basic skeleton for the second strategy: they are moderately rigid, allowing the high metal selectivity to be realized although their conformational freedom still remains.

7. Conclusions

The foregoing results indicate that calixarenes are very useful as a basic skeleton to design a new class of catalysts, ligands, and host molecules.

References

1. C. G. Gutsche: *Acc. Chem. Res.* **16**, 161 (1983).
2. C. D. Gutsche: *Host Guest Complex Chemistry/Macrocycles*, Springer Verlag, Berlin (1985), p. 375.
3. S. Shinkai: *Pure Appl. Chem.* **58**, 1523 (1986).
4. S. Shinkai, K. Araki, T. Tsubaki, T. Arimura, and O. Manabe: *J. Chem. Soc., Perkin Trans. 1*, 2297 (1987).
5. C. D. Gutsche and J. A. Levine: *J. Am. Chem. Soc.* **104**, 2652 (1982).
6. C. D. Gutsche and L.-G. Lin: *Tetrahedron* **42**, 1633 (1986).
7. S. Shinkai, K. Araki, H. Koreishi, T. Tsubaki, and O. Manabe: *Chem. Lett.* 1351 (1986).
8. T. Arimura, S. Shinkai, T. Matsuda, Y. Hirata, H. Satoh, and O. Manabe: *Bull. Chem. Soc. Jpn.* **61**, 3733 (1988).
9. G. W. Gokel and D. J. Cram: *J. Chem. Soc., Chem. Commun.*, 482 (1973).
10. S. Shinkai, S. Mori, H. Koreishi, T. Tsubaki, and O. Manabe: *J. Am. Chem. Soc.* **108**, 2409 (1986).
11. S. Shinkai, S. Mori, K. Araki, and O. Manabe: *Bull. Chem. Soc. Jpn.* **60**, 3679 (1987).
12. S.-K. Chang: *Chem. Lett.*, 477 (1984).
13. R. Ungaro, A. Pochini, and G. D. Andreetti: *J. Incl. Phenom.* **2**, 199 (1984).
14. M. A. McKervey, E. M. Seward, G. Ferguson, B. Ruhl, and S. J. Harris: *J. Chem. Soc., Chem. Commun.*, 388 (1985).
15. S. Shinkai, K. Araki, and O. Manabe: *J. Chem. Soc., Chem. Commun.* 187 (1988).
16. S. Shinkai, T. Arimura, H. Satoh, and O. Manabe; *J. Chem. Soc., Chem. Commun.* 1495 (1987).
17. T. Arimura, S. Edamitsu, S. Shinkai, O. Manabe, T. Muramatsu, and M. Tashiro: *Chem. Lett.*, 2269 (1987).
18. A. H. Alberts and D. J. Cram: *J. Am. Chem. Soc.* **99**, 3380 (1977).
19. I. Tabushi, Y. Kobuke, K. Ando, M. Kishimoto, and E. Ohara: *J. Am. Chem. Soc.* **102**, 5948 (1980).
20. I. Tabushi, Y. Kobuke, and A. Yoshizawa: *J. Am. Chem. Soc.* **106**, 2481 (1984).
21. P. Fux, J. Lagrange, and P. Lagrange: *J. Am. Chem. Soc.* **107** 5927 (1985).
22. S. Shinkai, H. Koreishi, K. Ueda, and O. Manabe: *J. Chem. Soc., Chem. Commun.*, 233 (1986).
23. S. Shinkai, H. Koreishi, K. Ueda, T. Arimura, and O. Manabe: *J. Am. Chem. Soc.* **109**, 6371 (1987).

Journal of Inclusion Phenomena and Molecular Recognition in Chemistry 7 (1989), 203–211.

Organic Clays. Synthesis and Structure of Na5[calix[4]arene sulfonate]·12 H2O, K5[calix[4]arene sulfonate]·8 H2O, Rb5[calix[4]arene sulfonate]·5 H2O, and Cs5[calix[4]arene sulfonate]·4 H2O

JERRY L. ATWOOD*, ANTHONY W. COLEMAN, HONGMING ZHANG, and SIMON G. BOTT
Department of Chemistry, University of Alabama, Tuscaloosa, AL 35487, U.S.A.

(Received: 1 February 1988)

Abstract. The title calixarenes all exist in the solid state as bilayers of anionic calixarenes in the cone configuration. These layers alternate with inorganic regions which contain the cations and the water molecules. The overall structures bear a close resemblance to those found for clay minerals. The sodium salt crystallizes in the triclinic space group $P\bar{1}$ with $a = 10.998(6)$, $b = 13.582(5)$, $c = 14.472(5)$ Å, $\alpha = 74.01(3)$, $\beta = 89.09(4)$, $\gamma = 86.50(4)°$, and $Z = 2$ for $D_{calc} = 1.72$ g cm^{-3}. Refinement based on 4727 observed reflections led to a conventional $R = 0.050$. The potassium salt crystallizes in the triclinic space group $P\bar{1}$ with $a = 11.815(9)$, $b = 13.636(6)$, $c = 14.040(9)$ Å, $\alpha = 100.24(5)$, $\beta = 111.86(9)$, $\gamma = 95.14(9)°$, and $Z = 2$ for $D_{calc} = 1.77$ g cm^{-3}. Refinement based on 2977 observed reflections led to $R = 0.15$. The rubidium and cesium salts are isostructural and crystallize in the monoclinic space group $P2_1/n$ with parameters for Rb[Cs] $a = 11.603(5)$ [11.704(3)], $b = 28.607(8)$ [29.747(9)], $c = 12.512(5)$ [12.604(4)] Å, $\beta = 91.70(4)$ [91.63(2)°], and $Z = 4$ for $D_{calc} = 2.01$ [2.24] g cm^{-3}. Refinement based on 1750 [4257] observed reflections led to $R = 0.108$ [0.075]. Disorder of the cations was observed for the rubidium and cesium salts.

Key words. Calix[4]arene, crystal structure, clay mineral, alkali metal ion, layered structure, water soluble, sulfonate.

Supplementary Data relating to this article are deposited with the British Library as Supplementary Publication No. SUP 82074 (95 pages).

1. Introduction

There is currently significant interest in the chemistry of calixarenes, **1**. Synthesis of modified calixarenes and the formation of alkali metal, as well as main group,

1

---

*Author for correspondence.

transition, and inner transition metal complexes was reported during 1987 [1–18]. In the course of the preparation and investigation of water soluble derivatives, R = —SO_3H, —NO_2, we isolated crystals of Na_5[calixarene sulfonate]·12 H_2O and Na_5[calixarene sulfonate]·8 H_2O·acetone [19]. The bilayer structure found in both cases bore a striking resemblance to that of a clay of 2 : 1 layer type. If one considers the —SO_3^- groups as part of the hydrated layer, even the thickness of the layers of the calixarene sulfonates is similar to the related ones in clays. The similarities go further than metrical parameters: preparation of the ammonium salt led to a compound with much less water than for the sodium salt, and the bilayer structure was maintained [20]. We report herein the synthesis and structures of the sodium, potassium, rubidium, and cesium salts which further demonstrate the similarity of the [calixarene sulfonate]$^{5-}$ complexes to clay minerals.

2. Experimental

The starting compounds for the calixarene syntheses were purchased from Aldrich Chemicals and were used without further purification.

2.1. PREPARATION OF Na_5[CALIX[4]ARENE SULFONATE]·12 H_2O

In a typical synthesis, 10.0 g (0.024 mol) of the [calix[4]arene, R = —H and R′ = —H, were reacted with an excess of H_2SO_4 at 60°C for 3 h. This was followed by neutralization with $BaCO_3$ and the addition of Na_2CO_3 to pH 9. The solvent is removed and the compound is recrystallized from water. The overall yield is 15.2 g Na_5[calix[4]arene sulfonate]·12 H_2O, 60%. Large, colorless crystals of the compound may be obtained by slow evaporation.

2.2. PREPARATION OF K_5[CALIX[4]ARENE SULFONATE]·8 H_2O, Rb_5[CALIX[4]ARENE SULFONATE]·5 H_2O, AND Cs_5[CALIX[4]ARENE SULFONATE]·4 H_2O

The compounds were prepared as for the sodium analogue, except that K_2CO_3 (or Rb_2CO_3, Cs_2CO_3) was added to pH 9.

2.3. X-RAY DATA COLLECTION AND STRUCTURE DETERMINATION FOR Na_5[CALIX[4]ARENE SULFONATE]·12 H_2O

The crystals are stable for several hours in the absence of an atmosphere of water vapor, but were sealed in thin-walled capillaries to avoid decomposition during the X-ray experiment. Final lattice parameters as determined from 25 high-angle reflections ($2\theta > 40°$) carefully centered on a Enraf Nonius CAD-4 diffractometer are as follows: space group = $P\bar{1}$, $a = 10.998(6)$, $b = 12.582(5)$, $c = 14.472(5)$ Å, $\alpha = 74.01(3)$, $\beta = 89.09(4)$, $\gamma = 86.50(4)°$, and $Z = 2$ for $D_{calc} = 1.72$ g cm^{-3}. Parameters of data collection are as given in Reference [4]. Structure solution was accomplished by means of direct methods and the subsequent calculation of difference Fourier maps followed by partial cycles of least-squares refinement allowed location of all non-hydrogen atoms. The hydrogen bonding is so extensive that all the hydrogen atoms could be located on a final difference Fourier map.

Table I.   Final fractional coordinates for Na_5[calix[4]arene sulfonate]·12 H_2O

Atom	x/a	y/b	z/c	U(eqv)
S(1)	0.3871(1)	0.4671(1)	0.78667(9)	0.020
S(2)	0.9722(1)	0.3171(1)	0.76466(9)	0.023
S(3)	1.0847(1)	0.8013(1)	0.87652(9)	0.022
S(4)	0.4860(1)	1.0304(1)	0.82685(9)	0.021
Na(1)	0.5279(2)	0.2510(2)	0.8385(2)	0.033
Na(2)	0.7464(2)	0.0590(2)	0.9948(1)	0.030
Na(3)	0.1572(2)	0.5311(2)	0.9497(2)	0.035
Na(4)	0.3782(2)	0.3344(2)	0.0288(2)	0.039
Na(5)	0.2664(2)	1.2472(2)	0.6940(1)	0.030
O(1)	0.5929(3)	0.7464(3)	0.4415(2)	0.025
O(2)	0.8593(3)	0.6892(3)	0.4501(2)	0.024
O(3)	0.9067(3)	0.8570(3)	0.4852(2)	0.023
O(4)	0.6810(3)	0.8914(3)	0.4996(2)	0.026
O(5)	0.2816(4)	0.5244(3)	0.8099(3)	0.046
O(6)	0.4723(3)	0.4325(3)	0.8662(3)	0.036
O(7)	0.3596(3)	0.3798(3)	0.7534(3)	0.035
0(8)	0.9147(3)	0.3391(3)	0.8486(3)	0.036
O(9)	0.9158(3)	0.2351(3)	0.7376(3)	0.037
O(10)	1.1042(3)	0.2981(3)	0.7795(3)	0.031
O(11)	0.9677(3)	0.7867(3)	0.9273(3)	0.038
O(12)	1.1663(3)	0.7103(3)	0.9046(3)	0.034
O(13)	1.1369(3)	0.8934(3)	0.8853(3)	0.036
O(14)	0.4172(3)	1.1266(3)	0.7853(3)	0.030
O(15)	0.4117(3)	0.9505(3)	0.8833(3)	0.036
O.16	0.5905(3)	0.0457(3)	0.8808(3)	0.038
C(1)A	0.5548(4)	0.6831(4)	0.5264(3)	0.018
C(2)A	0.4754(4)	0.7239(4)	0.5858(3)	0.019
C(3)A	0.4284(4)	0.6568(4)	0.6671(3)	0.020
C(4)A	0.4585(4)	0.5531(4)	0.6890(3)	0.019
C(5)A	0.5410(4)	0.5148(4)	0.6321(3)	0.018
C(6)A	0.5907(4)	0.5792(4)	0.5514(3)	0.017
C(7)A	0.6875(4)	0.5350(4)	0.4945(3)	0.018
C(1)B	0.8884(4)	0.6060(4)	0.5239(3)	0.017
C(2)B	0.8093(4)	0.5254(4)	0.5451(3)	0.016
C(3)B	0.8402(4)	0.4381(4)	0.6168(3)	0.019
C(4)B	0.9467(4)	0.4281(4)	0.6693(3)	0.020
C(5)B	1.0241(4)	0.5087(4)	0.6498(3)	0.021
C(6)B	0.9966(4)	0.5975(4)	0.5778(3)	0.019
C(7)B	1.0794(4)	0.6863(4)	0.5573(3)	0.021
C(1)C	0.9579(4)	0.8471(4)	0.5711(3)	0.018
C(2)C	1.0422(4)	0.7646(4)	0.6112(3)	0.019
C(3)C	1.0847(4)	0.7524(4)	0.7306(3)	0.020
C(4)C	1.0489(4)	0.8214(4)	0.7542(3)	0.021
C(5)C	0.9706(4)	0.9055(4)	0.7124(3)	0.021
C(6)C	0.9266(4)	0.9197(4)	0.6217(3)	0.019
C(7)C	0.8338(4)	1.0081(4)	0.5806(3)	0.019
C(1)D	0.6377(4)	0.9220(4)	0.5748(3)	0.019
C(2)D	0.7070(4)	0.9801(4)	0.6184(3)	0.019
C(3)D	0.6618(4)	1.0093(4)	0.6966(3)	0.020
C(4)D	0.5450(4)	0.9856(4)	0.7305(3)	0.020
C(5)D	0.4734(5)	0.9311(4)	0.6855(3)	0.021
C(6)D	0.5181(4)	0.8976(4)	0.6089(3)	0.020

Table I. *Continued.*

Atom	x/a	y/b	z/c	U(eqv)
C(7)D	0.4403(4)	0.8379(4)	0.5610(3)	0.020
W(1)	0.4114(3)	0.1869(3)	0.9786(3)	0.035
W(2)	0.1793(3)	0.3551(3)	0.9549(3)	0.032
W(3)	0.8215(3)	−0.1153(3)	1.0555(3)	0.038
W(4)	0.0420(3)	0.4786(3)	1.1073(3)	0.048
W(5)	0.2675(4)	0.2255(4)	1.1651(3)	0.057
W(6)	0.3233(4)	0.4880(3)	1.0721(3)	0.051
W(7)	0.1695(4)	0.0908(4)	0.7186(4)	0.060
W(8)	0.2206(5)	0.3806(5)	1.2514(4)	0.082
W(9)	0.7054(3)	0.2421(3)	0.9253(3)	0.035
W(10)	0.9049(4)	0.0504(3)	0.8838(3)	0.043
W(11)	0.0310(4)	0.2600(3)	0.6905(3)	0.057
W(12)	0.5702(4)	0.2770(5)	1.1037(4)	0.074

Final refinement with the non-hydrogen atoms treated with anisotropic thermal parameters gave $R = 0.050$ ($R_w = 0.050$). The final values of the positional parameters are given in Table I. Tables of bond lengths and angles, anisotropic thermal parameters, and structure factors are available as supplementary material.

2.4. X-RAY DATA COLLECTION AND STRUCTURE DETERMINATION FOR K₅[CALIX[4]ARENE SULFONATE]·8 H₂O

Data were collected and manipulated in the same manner as for the sodium salt. The final lattice parameters are as follows: space group = $P\bar{1}$, $a = 11.815(9)$, $b = 13.636(6)$, $c = 14.040(9)$ Å, $\alpha = 100.24(5)$, $\beta = 111.86(9)$, $\gamma = 95.14(9)°$, and $Z = 2$ for $D_{calc} = 1.77$ g cm^{-3}. The water molecules were clearly resolved and no disorder was apparent. However, the hydrogen atoms could not be located and the final R value was high (0.15). This was presumably caused by an absorption problem and a crystal which scattered poorly (only 2977 observed reflections were obtained out of 5529 measured out to $2\theta = 44°$). The final values of the positional parameters are given in Table II. Tables of bond lengths and angles and structure factors are available as supplementary material.

2.5. X-RAY DATA COLLECTION AND STRUCTURE DETERMINATION FOR Rb₅[CALIX[4]ARENE SULFONATE]·5 H₂O

Data were collected and manipulated as for the sodium salt. The space group is $P2_1/n$ and the final lattice parameters are as follows: $a = 11.603(5)$, $b = 28.607(8)$, $c = 12.512(5)$ Å, $\beta = 91.70(4)°$, and $Z = 4$ for $D_{calc} = 2.01$ g cm^{-3}. Refinement based on 1750 observed reflections led to $R = 0.108$. The paucity of data together with the disorder of one of the rubidium atoms over three sites led to the high R value. The value of five water molecules should be viewed in the light of the overall structure. The final positional parameters and the observed and calculated structure factors are available as supplementary material.

Table II. Final fractional coordinates for K_5[calix[4]arene sulfonate]·8 H_2O.

Atom	x/a	y/b	z/c	U(eqv)
K(1)	0.2698(5)	0.0707(5)	0.5848(6)	0.056(19)
K(2)	0.5639(5)	0.4425(5)	0.3876(5)	0.052(28)
K(3)	0.8573(5)	0.4124(6)	0.2550(6)	0.062(28)
K(4)	0.7556(9)	0.7357(7)	0.5860(7)	0.087(16)
K(5)	0.9463(6)	0.2141(7)	0.4907(8)	0.088(42)
S(1)	0.2304(6)	0.3558(5)	0.6820(5)	0.027(1)
S(2)	0.3864(6)	−0.1819(5)	0.6523(5)	0.033(1)
S(3)	0.7766(6)	0.4959(5)	0.6518(5)	0.034(1)
S(4)	0.9945(6)	1.0075(5)	0.6580(5)	0.030(1)
O(1)	0.634(1)	0.308(1)	1.066(1)	0.035(4)
O(2)	0.701(1)	0.134(1)	1.029(1)	0.030(3)
O(3)	0.912(1)	0.194(1)	1.035(1)	0.029(3)
O(4)	0.878(1)	0.382(1)	1.057(1)	0.034(4)
O(5)	0.269(2)	0.449(1)	0.661(1)	0.041(4)
O(6)	0.207(2)	0.268(1)	0.597(1)	0.044(4)
O(7)	0.124(2)	0.363(1)	0.716(1)	0.043(4)
O(8)	0.307(2)	−0.133(2)	0.577(2)	0.065(5)
O(9)	0.321(2)	−0.256(2)	0.684(2)	0.069(6)
O(10)	0.471(1)	−0.230(1)	0.613(1)	0.037(4)
O(11)	0.739(2)	0.405(2)	0.570(2)	0.072(6)
O(12)	0.889(2)	0.559(1)	0.664(1)	0.052(5)
O(13)	0.678(2)	0.555(1)	0.634(1)	0.051(5)
O(14)	1.130(2)	1.014(2)	0.693(2)	0.058(5)
O(15)	0.931(2)	0.906(2)	0.621(2)	0.058(5)
O(16)	0.940(2)	1.064(2)	0.584(2)	0.071(6)
C(1)A	0.545(2)	0.314(2)	0.972(2)	0.028(5)
C(2)A	0.554(2)	0.402(2)	0.937(2)	0.028(5)
C(3)A	0.456(2)	0.416(2)	0.845(2)	0.035(6)
C(4)A	0.355(2)	0.337(2)	0.795(2)	0.026(5)
C(5)A	0.348(2)	0.247(2)	0.827(2)	0.025(5)
C(6)A	0.446(2)	0.234(2)	0.919(2)	0.025(5)
C(7)A	0.438(2)	0.136(2)	0.951(2)	0.029(5)
C(1)B	0.629(2)	0.061(2)	0.945(2)	0.022(5)
C(2)B	0.497(2)	0.057(2)	0.899(2)	0.021(5)
C(3)B	0.426(2)	−0.019(2)	0.813(2)	0.026(5)
C(4)B	0.474(2)	−0.087(2)	0.763(2)	0.023(5)
C(5)B	0.606(2)	−0.085(2)	0.812(2)	0.027(5)
C(6)B	0.679(2)	−0.013(2)	0.899(2)	0.022(5)
C(7)B	0.820(2)	−0.016(2)	0.952(2)	0.029(5)
C(1)C	0.931(2)	0.152(2)	0.947(2)	0.024(5)
C(2)C	0.890(2)	0.048(2)	0.905(2)	0.023(5)
C(3)C	0.910(2)	0.004(2)	0.815(2)	0.027(5)
C(4)C	0.972(2)	0.061(2)	0.774(2)	0.029(5)
C(5)C'	1.015(2)	0.164(2)	0.816(2)	0.025(5)
C(6)C	0.995(2)	0.211(2)	0.905(2)	0.021(5)
C(7)C	1.035(2)	0.325(2)	0.948(2)	0.029(5)
C(1)D	0.853(2)	0.407(2)	0.963(2)	0.019(5)
C(2)D	0.926(2)	0.380(2)	0.904(2)	0.023(5)
C(3)D	0.901(2)	0.408(2)	0.812(2)	0.026(5)
C(4)D	0.802(2)	0.459(2)	0.769(2)	0.028(5)
C(5)D	0.733(2)	0.487(2)	0.832(2)	0.029(5)
C(6)D	0.756(2)	0.457(2)	0.925(2)	0.023(5)

Table II. *Continued.*

Atom	x/a	y/b	z/c	U(eqv)
C(7)D	0.672(2)	0.483(2)	0.987(2)	0.024(5)
W(1)	0.845(2)	0.748(2)	0.803(2)	0.090(7)
W(2)	0.523(2)	0.655(2)	0.422(2)	0.064(5)
W(3)	0.484(2)	0.814(2)	0.303(2)	0.074(6)
W(4)	0.450(2)	0.337(2)	0.174(2)	0.077(6)
W(5)	0.633(2)	0.912(2)	0.564(2)	0.084(7)
W(6)	0.703(2)	0.133(2)	0.634(2)	0.086(7)
W(7)	1.105(2)	0.782(2)	0.743(2)	0.088(7)
W(8)	1.072(3)	0.591(2)	0.523(2)	0.107(9)

2.6. X-RAY DATA COLLECTION AND STRUCTURE DETERMINATION FOR
    Cs_5[CALIX[4]ARENE SULFONATE]·4 H_2O

Data were collected and manipulated as for the sodium salt. The compound is
isostructural with the rubidium analogue: $a = 11.704(3)$, $b = 29.747(9)$,
$c = 12.604(4)$ Å, $\beta = 91.63(2)°$, and $Z = 4$ for $D_{calc} = 2.24\ g\ cm^{-3}$. Refinement
based on 4257 observed reflections led to a final $R = 0.075$. One of the cesium
atoms is disordered over two sites and the four water molecules are disordered over
seven sites. The final positional parameters, bond lengths and angles, and structure
factors are available as supplementary material.

3. Results and Discussion

The calix[4]arene sulfonates of the alkali metals all exist as 5− anions at neutral
pH. The four protons of the sulfonate groups have been removed, as has one from
the hydroxyls [21]. The resulting 5− anion exists in the cone configuration which
is stabilized by the hydrogen bonding shown in Figure 1.

The cavity of the sodium salt of the calixarene contains two water molecules. In
this respect a similarity with α-cyclodextrin is noted [22]. Indeed, upon exposure to
an organic molecule with an appropriately sized hydrophobic region, the water
molecules are ejected in favor of the organic guest [19]. This was demonstrated by
the structure of Na_5[calix[4]arene sulfonate]·acetone·8 H_2O, in which the acetone
molecule is located in the cavity. One methyl group is embedded in the bottom of
the cone, the other is directed toward an —SO_3^- group, and the carbonyl oxygen
is oriented into the inorganic layer.

Figure 2 shows not only the cone configuration with the two guest water
molecules, but also the bilayer arrangement of the calix[4]arene anions. The
structure is divided into organic and inorganic layers. The former consists of the
bilayer of calixarene units, while the latter is comprised of the water molecules, the
sodium ions, and the —SO_3^- groups. This general structure is maintained throughout the series, as the representation of the rubidium analogue, Figure 3, illustrates.

The bilayer is stabilized in part by the electrostatic interaction of a sodium ion
from the inorganic region with a phenoxide oxygen atom and in part by hydrogen
bonding of one of the hydroxyl groups with a water molecule of the inorganic layer.

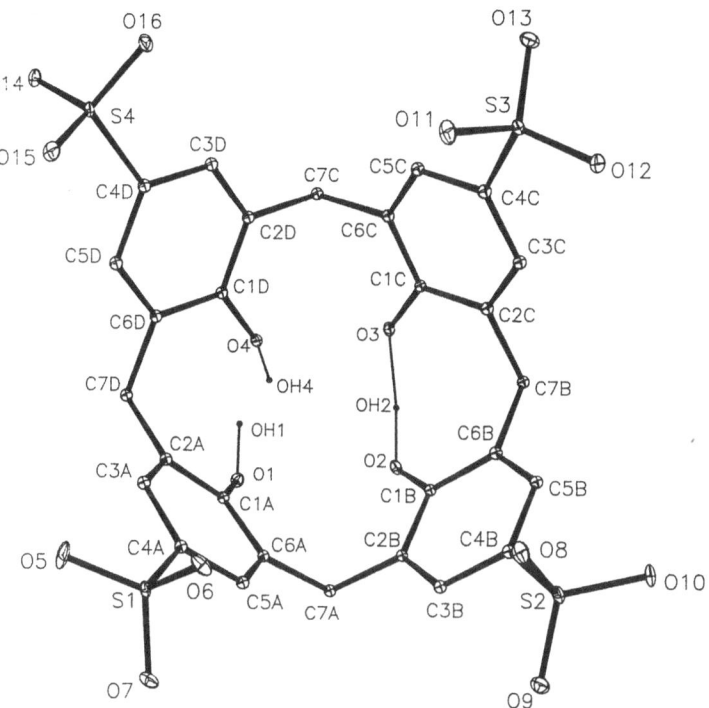

Fig. 1. Structure of the calixarene 5– anion viewed from the top of the cavity. Hydrogen atoms OH1 and OH2 stabilize the cone, while OH4 is used in hydrogen bonding to a water molecule of the inorganic layer.

It is also possible that the packing of the calixarenes such that the aromatic regions of one are positioned near those of its neighbors results in a favorable interaction.

While the structure bears a resemblance to bioorganic bilayer membranes, the overall similarity to the structures of clay minerals is striking. First, consider the repeat distance in the sodium salt of the calixarene, 13.7 Å, compared to

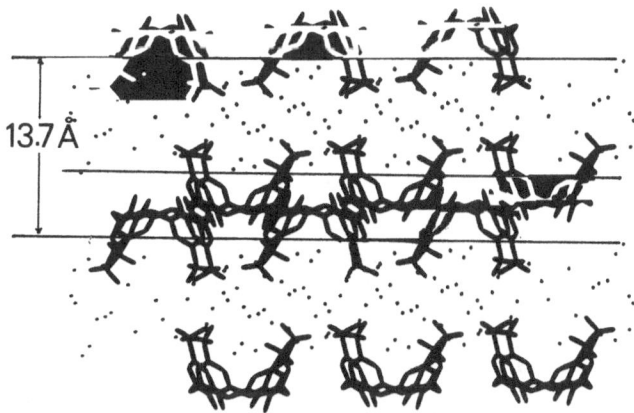

Fig. 2. The bilayer structure of Na_5[calix[4]arene sulfonate]·12 H_2O. The lines are least-squares best planes of the aromatic carbon atoms bonded to the —SO_3^- groups.

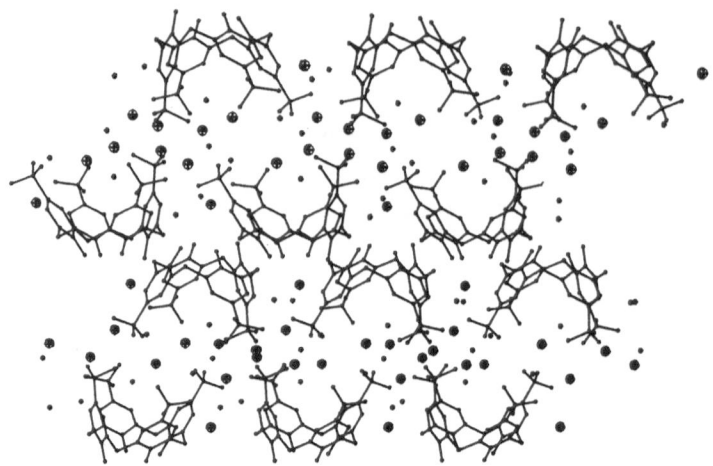

Fig. 3.   The bilayer structure for Rb$_5$[calix[4]arene sulfonate]·5 H$_2$O. The cross-hatched spheres are rubidium ions.

14.4–15.6 Å in smectite [23] and 15.0 Å in hydrated sodium vermiculite [24]. Second, the hydrated layer in sodium vermiculite is 9.4 Å, while the inorganic layer in Na$_5$[calix[4]arene sulfonate]·12 H$_2$O is 8.3 Å. A more complete comparison has been presented [19].

The chemical behavior of clays is also mimicked by the title complexes in two respects: the presence of cation-exchange capabilities and the reduced water content in the ammonium salt [20]. Currently, other similarities between clays and the water-soluble calixarenes are being probed. In particular, the size- and polarity-selective cavity of the calix[4]arene sulfonates is finding utility in the separation of organic substances, cations, anions, and neutral molecules, from aqueous feed streams.

Acknowledgement

We are grateful to the U.S. National Science Foundation for support of this research.

References

1. G. D. Andreetti, G. Calestani, F. Ugozzoli, A. Arduini, E. Ghidini, A. Pochini, and R. Ungaro: *J. Incl. Phenom.* **5**, 123 (1987).
2. P. D. Beer and A. D. Keefe: *J. Incl. Phenom.* **5**, 499 (1987).
3. A. W. Coleman, S. G. Bott, and J. L. Atwood: *J. Incl. Phenom.* **5**, 581 (1987).
4. S. G. Bott, A. W. Coleman, and J. L. Atwood: *J. Incl. Phenom.* **5**, 747 (1987).
5. C. D. Gutsche, M. Iqbal, and I. Alam: *J. Am. Chem. Soc.* **109**, 4314, (1987).
6. D. N. Reinhoudt, P. J. Dijkstra, P. J. A. Veld, K. E. Bugge, S. Harkema, R. Ungaro, and E. Ghidini: *J. Am. Chem. Soc.* **109**, 4761 (1987).
7. S. Shinkai, H. Koreishi, K. Ueda, T. Arimura, and O. Manaba: *J. Am. Chem. Soc.* **209**, 6371 (1987).
8. S. Shinkai, S. Mori, T. Arimura, and O. Manabe: *J. Chem. Soc., Chem. Commun.* 238 (1987).
9. G. Calestani, F. Ugozzoli, A. Arduini, E. Ghidini, and R. Ungaro: *J. Chem. Soc., Chem. Commun.* 344 (1987).

10. G. Ferguson, B. Kaitner, M. A. McKervey, and E. M. Seward: *J. Chem. Soc., Chem. Commun.* 584 (1987).
11. G. R. Newkome, Y. J. Joo, and F. R. Fronczek: *J. Chem. Soc., Chem. Commun.* 854 (1987).
12. V. Bohmer, L. Merkel, and U. Junz: *J. Chem. Soc., Chem. Commun.* 896 (1987).
13. V. Bohmer, H. Goldmann, R. Kaptein, and L. Zetta: *J. Chem. Soc., Chem. Commun.* 1358 (1987).
14. S. Shinkai, T. Arimura, H. Satch, and O. Manabe: *J. Chem. Soc., Chem. Commun.* 1495 (1987).
15. B. M. Furphy, J. M. Harrowfield, D. L. Kepert, B. W. Skelton, A. H. White, and F. R. Wilner: *Inorg. Chem.* **26**, 4231 (1987).
16. E. Paulus, V. Bohmer, H. Goldman, and W. Vogt: *J. Chem. Soc., Perkin Trans.* 2, 1609 (1987).
17. S. Shinkai, K. Araki, T. Tsubaki, T. Arimura, and O. Manabe: *J. Chem. Soc., Perkin Trans. I* 2297 (1987).
18. H. Casabianca, J. Royer, A. Satrallah, A. Taty-C, and J. Vicens: *Tetrahedron Lett.* **28**, 6595 (1987).
19. A. W. Coleman, S. G. Bott, S. D. Morley, C. M. Means, K. D. Robinson, H. Zhang, and J. L. Atwood: *Angew. Chem. Int. Ed. Engl.* **27**, 1361 (1988).
20. S. G. Bott, A. W. Coleman, and J. L. Atwood: *J. Am. Chem. Soc.* **110**, 610 (1988).
21. S. Shinkai, T. Mori, T. Tsubaki, T. Sone, and O. Manobe: *Chem. Lett.* 1351 (1986).
22. B. Klar, B. Hingerty, and W. Saenger: *Acta Crystallogr.* **B36**, 1154 (1980).
23. S. W. Bailey: *Crystal Structures of Clay Minerals and their Identifications*, Ed. G. W. Brindley and G. Brown, Monograph 5 of Mineralogical Society, London, 1980.
24. B. Mason and L. G. Berry: *Elements of Mineralogy*, Freeman, San Francisco, 1968.

Journal of Inclusion Phenomena and Molecular Recognition in Chemistry **7** (1989), 213–226.

Molecular Recognition by Macropolycyclic Hosts

IAN O. SUTHERLAND
Department of Organic Chemistry, Robert Robinson Laboratories, P.O. Box 147,
Liverpool L69 3BX, England

(Received: 1 February 1988)

Abstract. The formation of complexes between crown ethers and aklylammonium cations may, to some extent, be modelled using standard molecular mechanics methods and an appropriate charge distribution scheme. Monocyclic crown ethers may be developed to give chromoionophores suitable for use in optical fibre based ion sensors. The incorporation of two crown ether systems into polycyclic host molecules which show highly selective complexation of guest bis-alkylammonium cations is described. The scope of these ditopic receptors may be extended by using metalloporphyrins in place of one or both of the crown ether binding sites.

Key words. Aza crown ethers, ditopic hosts, metalloporphyrins, molecular modelling, chromoionophores.

The formation of molecular complexes in solution by protein host molecules (Equation 1) has long been recognised as an essential feature of many biological processes [1]. The current wide interest in synthetic host molecules dates from the discovery of the crown ethers by C. J. Pedersen in 1967 [2], their development by the pioneering work of D. J. Cram and his coworkers [3], and the invention of cryptand host molecules by J.-M. Lehn [4].[1]

$$\text{Host} + \text{Guest} \underset{\longleftarrow}{\overset{K_s}{\rightleftharpoons}} \text{Host} \cdot \text{Guest} \tag{1}$$

These early studies led to our initial interest in cyclophanes and macropolycyclic compounds of the aza-crown ether type [5] which has developed into the research described in this paper.

The relative importance of the different types of non-covalent binding forces between guest and host molecules that can lead to an appropriate value for the association constant for complexation (K_s of Equation 1) depends upon the solvent [6]. For most of the examples that will be discussed in this paper complexation takes place in organic solvents ($CHCl_3$, CH_2Cl_2 and their perdeuterated derivatives) and is based upon electrostatic attraction between the two components, in some cases this is manifested as hydrogen bonding.

Hydrophobic interactions, which are of major importance for complexation in aqueous solvents [7], are not available in organic solvents so that for most examples involving high binding energies (>6–10 kcal mol^{-1}) one of the components of the complex is either a cation, or less frequently, an anion [8]. For cationic guests the host molecule must contain electron rich atoms, usually N or O, as binding sites. In principle, the ideal host for any particular guest contains a cavity with size, shape and charge distribution complementary to those of the guest as shown diagrammatically in Figure 1 for guest D-ala–D-ala. The principles of host design are therefore simple but the translation of the curved lines of Figure 1 into a real host molecule for D-ala–D-ala is a formidable problem which to date has been solved only by nature [9] (vancomycim and related antibiotics).

Relatively simple synthetic host molecules, such as coronands [2], cryptands [4], and spherands [3], for example 1–3, contain circular or spherical cavities which are ideally suited for simple spherical guest metal cations or, in appropriate cases, for alkylammonium cations. Guest recognition for both types of guest can be achieved by careful design of the host molecule for all three structural types.

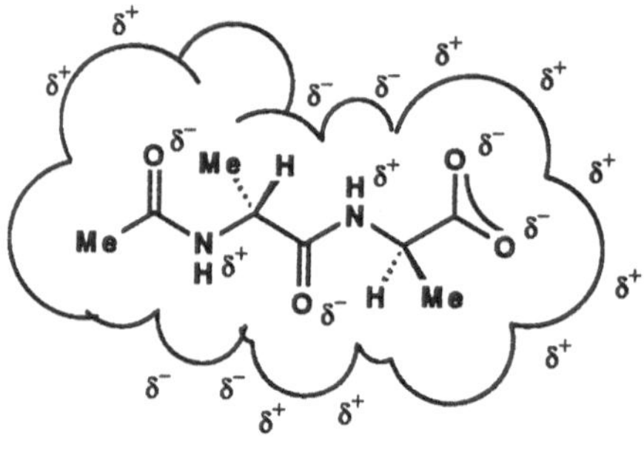

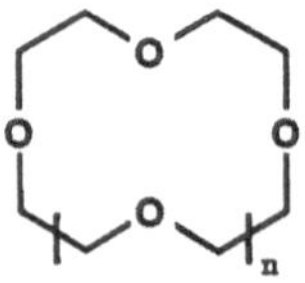

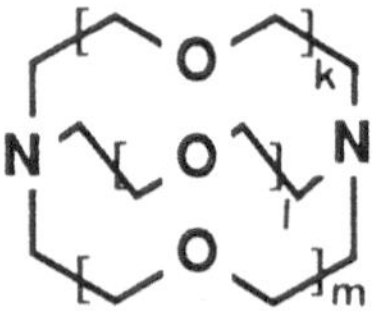

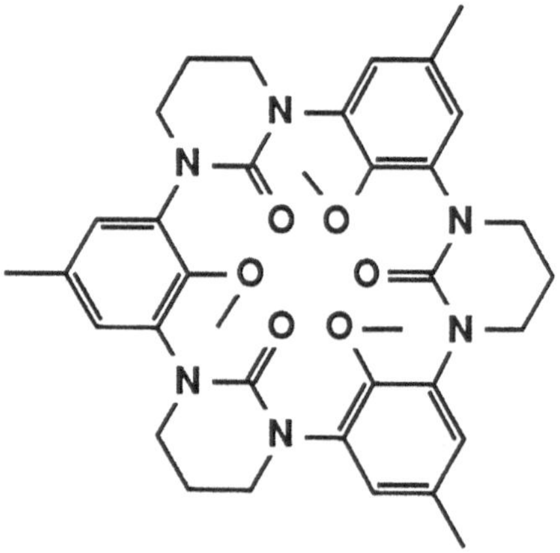

1   Coronands   2   Cryptands   3   Spherand

Fig. 1.

Host molecules 1–3 have usually been designed by studies of CPK molecular models [3] but, to some extent, these classical models are now being replaced by computer based modelling [10]. This requires a graphics terminal together with software to provide graphical input and output of structures and their three dimensional manipulation, in addition molecular mechanics programs are needed for calculating conformational energies and intermolecular interactions.[2] Molecular mechanics programs will usually locate energy minima with acceptable accuracy but the calculated steric energies and, to a lesser extent, molecular geometries depend upon the parameters of the force field. These parameters include the usual terms for bond lengths, bond angles, torsion angles, and non-bonded interactions, but for calculations involving complexation by electrostatic attraction additional parameterisation for charge distribution is of great importance. For our work we have used the charge scheme developed by R. J. Abraham and his coworkers [11]. This charge scheme, used with the force field [12] contained in the COSMIC package,[3] gives reasonable values for the energetics of inversion of the chair conformations of

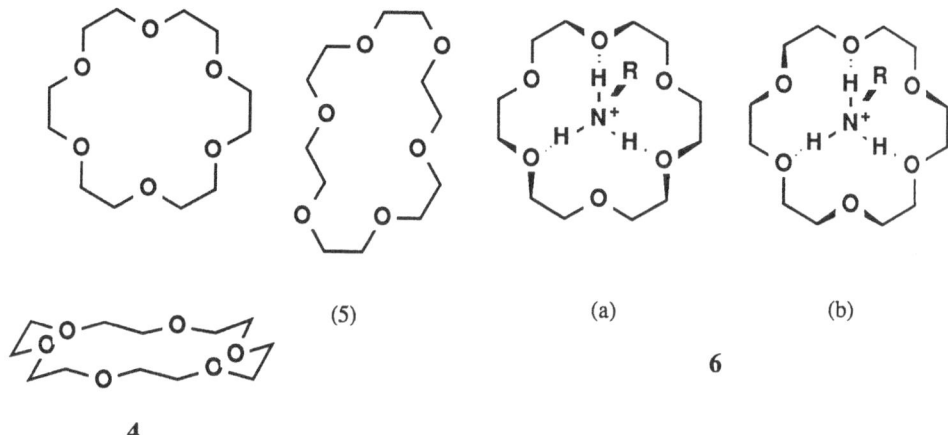

Fig. 2.

six-membered rings (cyclohexane, dioxan, tetrahydropyran and N-methylpipe-ridine) and the relative energies of different conformations of crown ethers. For example it correctly assigns a lower steric energy to the conformation **5** of 18-crown-6 which is found in the crystal structure of the free host as compared with the well known D_{3d} conformation **4** which is found in 18-crown-6 complexes. The difference in these steric energies (ΔE 8.4 kcal mol^{-1}) resides very largely in the different electrostatic energies (ΔE 7.3 kcal mol^{-1}) of the two conformations.

Calculated binding energies of alkylammonium cations by 18-crown-6, using the same charge scheme **8** and the COSMIC force field are much too large (ΔE 50–60 kcal mol^{-1}) but the calculations neglect the effects of the solvent, polarisability of the two interacting molecules, and counter-ion association. The results are in accord with the very limited guest recognition shown [13] in the formation of the complexes 18-crown-6·RN$^+$H$_3$ and they also indicate that, although the two possible minimum energy arrangements of guest and host in the complex show well defined N$^+$—H—O hydrogen bonds, there is little or no distinction in binding energy between hydrogen bonding to the three oxygen atoms on the upper face of the host **6a** as compared with the oxygen atoms on the lower face of the host **6b**.

The most favorable conformations of free crown ethers are those in which the unfavourable O—O interactions associated with *gauche* C—C bonds are avoided as far as possible by the inclusion of one or more *anti* C—C bonds in the ring system. In contrast, the most suitable conformations of crown ethers for complexation of cations are those in which all the C—C bonds are *gauche*. At the present time, computer based molecular modelling provides only a qualitative approach to the design of host molecules for specific guests and the rationalisation of experimental observations. Accurate molecular modelling will require: (i) correct assessment of the effects of solvent, molecular polarisability, and counter ion, (ii) the use of well based schemes for charge distribution and force fields that are appropriate for these charges, and (iii) the identification of global conformational energy minima for both the free host and the complex. Thus many different conformations of comparable steric energy can be found for most macrocycles (>12 members) and although only a limited number have been recognised in crystal structures relating to any particular ring system the situation in solution is not easily defined. For example, NMR spectroscopy will only define solution conformations within the limitations of signal averaging by molecular motion [14] and other physical data may also represent the average of two or more equilibrating conformations. None of these problems is trivial but, nevertheless, it would be surprising if computer based molecular modelling does not become a reliable procedure for host design and the investigation of guest/host interactions.

Although most of our work has been concerned with complexes of alkylammonium cations [15] we have recently been interested in the development of optical fibre sensors [16] for metal cations.[4] This involves the use of a chromoionophore [17] which can be placed at the end of an optical fibre, the chromoionophore responds to cation complexation with a colour change that can be detected using a suitable light source and detection system at the other end of the optical fibre. Chromoionophores are of two types, neutral and ionisable. The former contain a polarised chromophore, as shown diagrammatically in 7a and 7b, which responds to the capture of a cation with a shift in absorption spectrum to shorter wavelength 7a or longer wavelength 7b. A number of neutral chromoionophores 8 and 9 of both types were investigated but they were found to be unsuitable due to their limited spectroscopic response and their low sensitivity [18]. The ionisable chromoionophores 10 have proved to be much more suitable. These respond to metal ion complexation by forming the salts 11, which, for suitable chromophores, gives a very substantial change in the absorption spectrum.

A number of ionisable chromoionophores 12 and 13 have been tested and compound 12b, which changes colour from yellow to purple on cation complexation, proved suitable for the detection of K^+ (KCl in H_2O) in the pH range 7–9 at concentrations (1–100 mM) similar to those found in blood plasma (blood $K^+ \sim 4.5$ mM). The K^+/Na^+ selectivity ratio (6.4) was too low to determine K^+

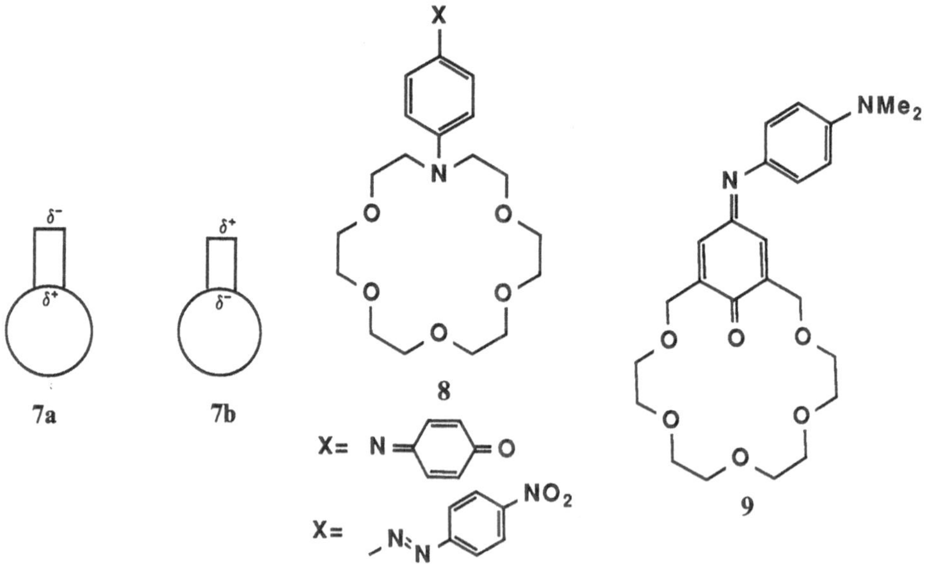

Fig. 3.

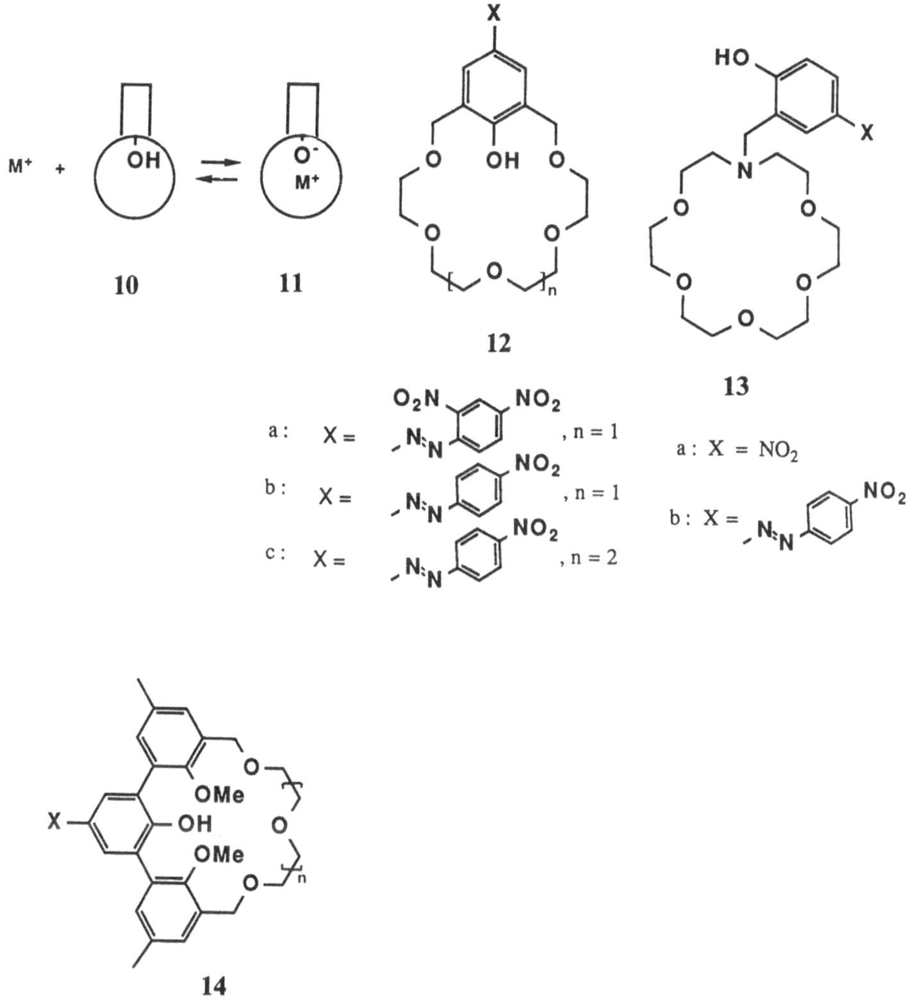

Fig. 4.

levels in blood (Na^+/K^+ ratio ~ 30) and current work is focused upon improving this selectivity by using a suitable modified hemispherand system [19], such as **14**, or a valinomycin mimic [20] such as a suitably modified dibenzo-30-crown-10.

A high level of recognition of guest alkylammonium cations by hosts of the crown ether type was identified as an important objective at the outset of our work. Because complexation of primary alkylammonium cations by simple monocyclic crown ethers, such as 18-crown-6, gives only a very low level of guest recognition [13] our initial studies were directed towards the inclusion of the guest cation in the cavity of a cryptand host [4], as in **15**. Such a molecular inclusion in solution would clearly require the use of a host molecule in which the binding forces are directed towards the centre of the molecular cavity, as indicated by the arrows in **15** to avoid the competitive formation of externally bound complexes such as **16**. Without this stereoselective inward binding exclusion complexes **16** will tend to be formed to

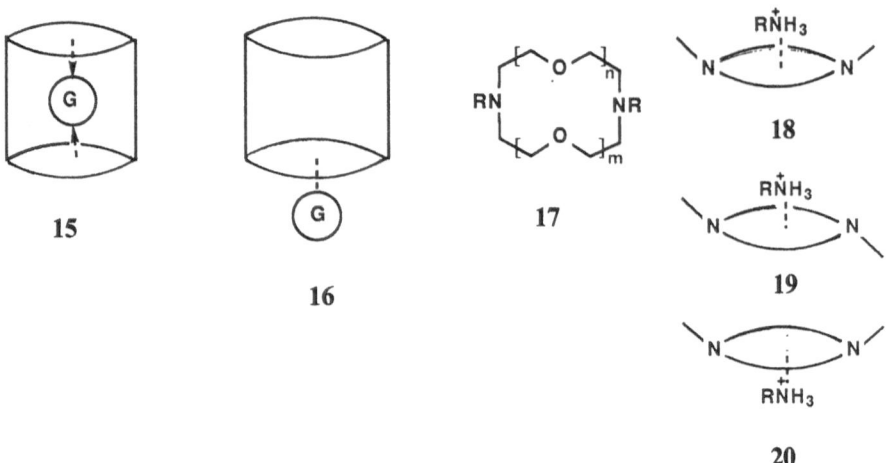

Fig. 5.

avoid unfavourable van der Waals interactions which inevitably result from imperfect design of the host cavity. We were fortunate to find [21] that diaza crown ethers with twelve-membered ($17, m = n = 1$) and fifteen-membered ($17, m = 1, n = 2$) rings formed only the *cis,cis*-complexes **18** with guest alkylammonium cations RNH_3^+ whereas the larger eighteen-membered diaza-crowns ($17, m = n = 2$ or $m = 1, n = 3$) formed mixtures of *cis,cis*-**18**, *cis,trans*-**19** and *trans,trans*-**20** complexes. These studies, based upon ^1H NMR spectroscopy, have been fully described elsewhere.

Following these preliminary studies, it was possible to design host molecules of the type shown in **15**, based upon diaza-12-crown-4 and diaza-15-crown-5 receptors, which would form inclusion complexes with either one bisalkylammonium cation $H_3N^+(CH_2)_nN^+H_3$ **21** or two simple primary alkylammonium cations RN^+H_3 **22**. The use of tricyclic compounds of this type has been described elsewhere [23]. The cryptands **23**, which are based upon two identical diaza-crown ether moieties, were synthesised by the simple one step procedure shown in Scheme 1, whereas it was necessary to use the stepwise procedure [24] summarised in Scheme 2 for the synthesis of analogous cryptands **24** in which the two diaza crown ether moieties are different.

It was found that, in general, the rather rigid tricyclic hosts **23** and **24** selected one or two members from the series of *bis*-cations $H_3N^+(CH_2)_nN^+H_3$ (Table I) and that guest selection could be assessed readily from the ^1H NMR spectra of a $1:1:1$ mixture of the host with a pair of competing guest cations [23]. A clear distinction between the spectra of free and complexed *bis*-cations results from the large induced high field shifts (2–4 ppm) of the guest CH_2 protons because they lie in the shielding zones of the two aromatic systems [Ar in **23** and **24**] of the bridges. The selectivity is summarised in the Table[5] and from the results it can be seen that guest selection depends upon the length of the CH_2ArCH_2 bridge and the size of the aza crown macrocycles. If it is assumed that for the optimum guest $H_3^+N(CH_2)_nN^+H_3$ the $N^+—N^+$ separation in the fully extended (all *anti*) conformation of the guest l Å is

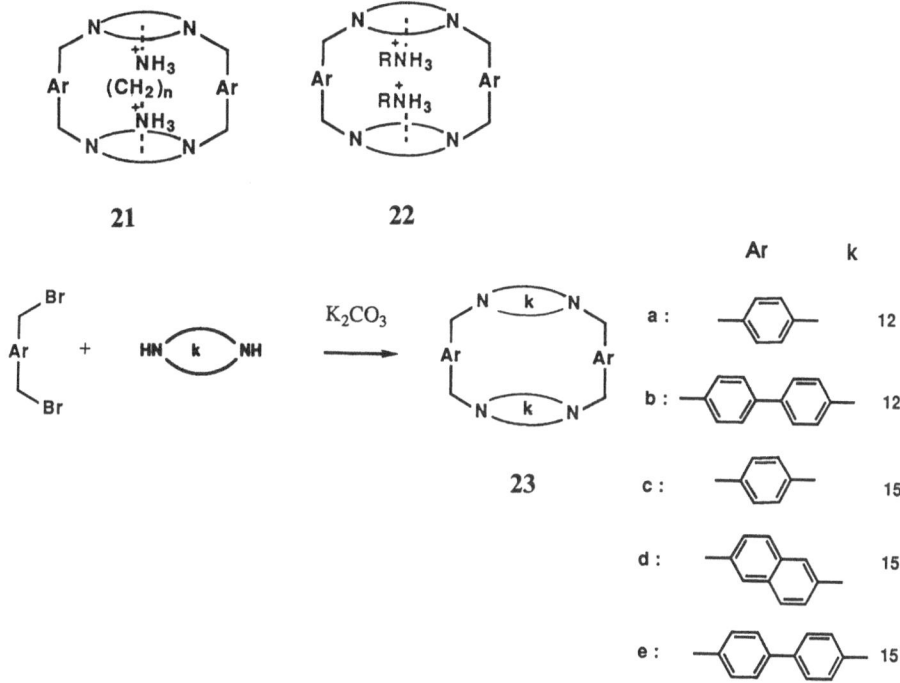

21   **22**

Scheme 1. Synthesis of symmetrical cryptands (**23**).

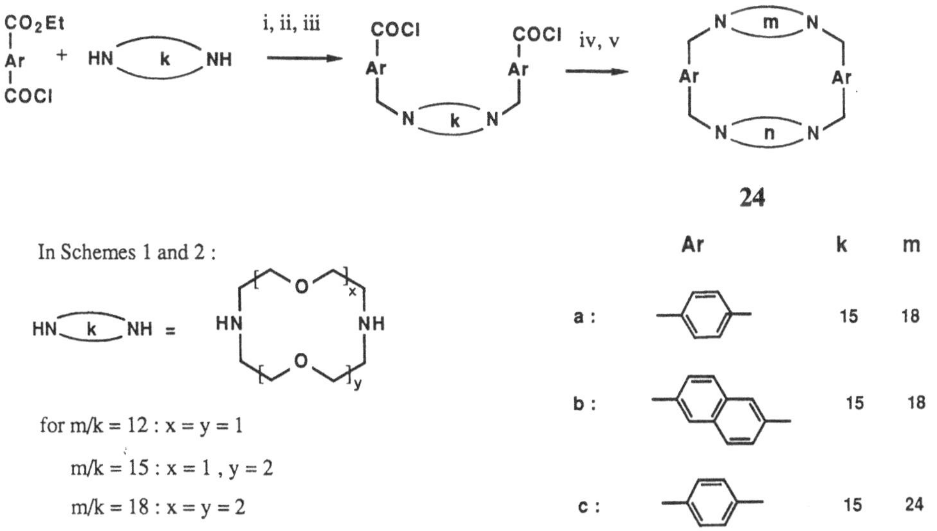

24

In Schemes 1 and 2 :

HN⌒k⌒NH =

for m/k = 12 : x = y = 1

m/k = 15 : x = 1 , y = 2

m/k = 18 : x = y = 2

Ar	k	m
a :	15	18
b :	15	18
c :	15	24

Reagents : i, Et₃N ; ii, NaOH / EtOH , then H⁺ ; iii, (COCl)₂ ; iv, HN⌒m⌒NH, Et₃N

v, BH₃ / THF, then H⁺ / EtOH

Scheme 2. Synthesis of asymmetrical cryptands.

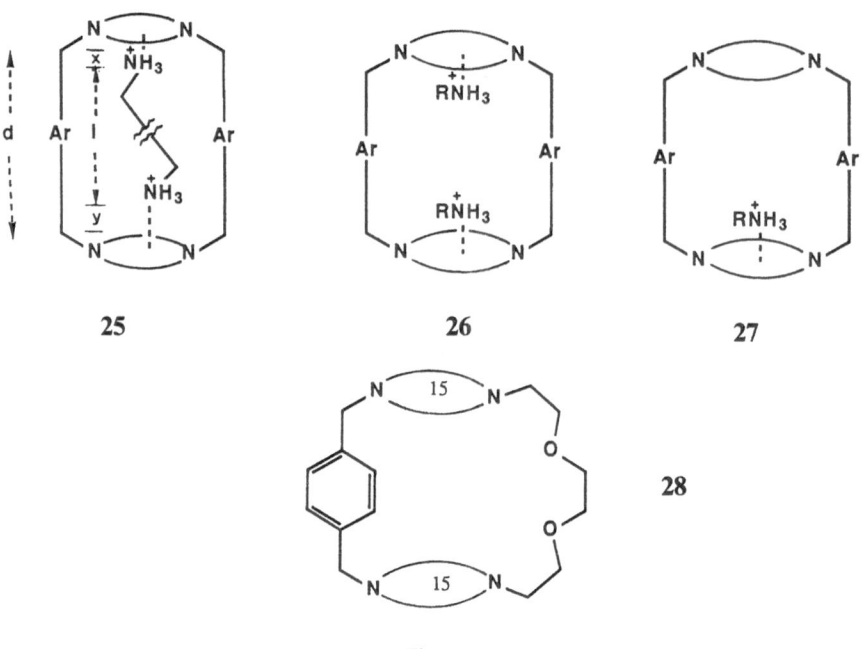

Fig. 6.

Table I. Selectivity in complexation of salts $H_3(N^+(CH_2)N^+H_3 \cdot 2 \, NCS^-$ by cryptands **23** and **24** in CD_2Cl_2.

Ar	Host			Guest Selectivity[a]		$d - x - y$[b]
	Ring[c]	Sizes[a]	$d/\text{Å}$[b]	n	$l/\text{Å}$	Å
	12	12	5.8	2	3.7	3.6
	15	15		2	3.7	4.2
B[e]	15	18		3	4.9	4.7
	18	18		3, 4	5.5	5.2[d]
	15	15	7.9	4	6.2	6.3
N	15	18		4, 5	6.7	6.8
	18	18		5	7.3	7.3[d]
	12	12	10.1	5, 6	8.0	7.9
BP	15	15		6	8.6	8.5
	18	18		7	9.8	9.5[d]

[a] The value of n refers to the optimum guest(s) in the series of bis-cations $H_3N^+(CH_2)_nNH_3^+$, the value of l refers to the $N^+—N^+$ distance in the extended conformation of this guest or the average distance for a pair of guests.

[b] For distance d see **25**; x and y are based upon best agreement between experimental values of l and calculated values of $d - x - y$: for $n = 12$, $x = 1.1$; $n = 15$, $x = 0.8$; $n = 18$, $x = 0.3$ Å.

[c] The numbers refer to k and m in formulae **23** and **24**.

[d] The value for n is taken from the work of J. M. Lehn ref. [26].

[e] B = ⟨benzene⟩ , N = ⟨naphthalene⟩ BP = ⟨biphenyl⟩

equal to the distance d (see formula **25**) minus the sum of x and y, the penetration of the $^+NH_3$ groups into the diaza-crown ether macrocycles (see formula **25**), then best values for x and y can be obtained from the data in the Table. Calculated values of $d - x - y$ were then compared with values of l for all 10 hosts for which data is available. Agreement between the calculated and observed data is excellent, confirming the general correctness of the structure of the complexes shown diagrammatically in **25** which is also supported by crystal structure data reported by J.-M. Lehn and co-workers [25].

The hosts **23**, based upon 12- and 15-membered diaza-crown ethers are potentially hosts for simple alkylammonium cations RN^+H_3 and they were expected to form 2:1 **26** and 1:1 **27** complexes of the inclusion type. This was found to be the case for the hosts **23a**, **23b** and **23d** with either $MeN^+H_3·NCS^-$ or $MeN^+H_3·ClO_4^-$ as the guest salt, but even for low guest to host ratios (1:1) mixtures of 2:1 **26** and 1:1 **27** complexes were formed. Complexation is detectable, and can be analysed for 2:1 and 1:1 complexes, using ^1H NMR spectroscopy, individual guest CH_3 signals for the two complexes are resolved at low temperatures ($-90°C$) where guest exchange is slow on the NMR time scale. At higher ratios of guest salts (2:1 and 4:1) the ^1H NMR spectra showed that only the 2:1 complex **26** was obtained. There is no difficulty in fitting two MeN^+H_3 cations into the cavity of hosts **23b** and **23d** but the small cavity of **23a** appears to be barely adequate in size on the basis of CPK molecular models. Subsequently it also proved impossible to construct a computer based model of the complex **23a** $\cdot$ 2 MeN^+H_3. Nevertheless the ^1H and ^{13}C NMR spectra of a 2:1 mixture of host **23a** and $MeN^+H_3·ClO_4^-$ showed two guest CH_3 signals at low temperatures ($< -70°C$) shifted to low field (δ 2.9 and 27.9 ppm) and high field (δ 0.2 and 25.5 ppm) relative to the spectrum of the free salt (δ 2.5 and 27.3 ppm) and the ^1H NMR spectrum of a 4:1 mixture showed CH_3 signals for the two complexed cations and the free cation in a 1:1:2 ratio. This host **23a** forms a 2:1 complex and, in the absence of other evidence, we conclude that it is an inclusion complex **26** in which the two guest CH_3 groups occupy very different environments because of the limited space within the cavity. Even in this case the formation of the 2:1 complex **26** appears to be more favorable than the formation of a 1:1 complex **27** and we conclude that the entry of the first MeN^+H_3 cation opens the cavity so that the second cation can enter more easily.

Diaza-12-crown-4 has been shown [27] to form 1:1 complexes with secondary alkylammonium cations $RR'N^+H_2$. The ^1H NMR spectrum of a 1:1 mixture of host **23a** and $Me_2N^+H_2·ClO_4^-$ showed a signal [28] for the guest CH_3 group (δ 1.4 ppm) almost 1.5 ppm to high field of the signal from the free cation. Addition of excess of the guest salt, up to a 4:1 guest to host ratio, gave a CH_3 signal for a 1:1 complex together with a second CH_3 signal corresponding to the excess of the guest salt. Evidently the small cavity of host **23a** can accommodate only one molecule of the larger $Me_2N^+H_2$ cation.

Hosts **23** and **24** have rigid CH_2ArCH_2 bridges and consequently show very high selectivity for the *bis*-cations $H_3N^+(CH_2)_nN^+H_3$. It was of interest to examine the results of introducing less rigid bridges into the tricyclic system, accordingly host **28** was examined [29]. This host selects the bis-cations with $n = 2$ and 3 almost equally readily on the basis of competition experiments of a type similar to those used for the other tricyclic hosts **23** and **24**. The bis-cations with $n = 4$ and 5 are also complexed equally readily but significantly less readily than the shorter pair of cations ($n = 2$ and 3). We conclude that the shorter cations are complexed by a conformation of the host which has a low energy conformation for the $CH_2(CH_2OCH_2)_2CH_2$ bridge and that the larger cations are complexed by a second conformation of the host which has a higher energy (longer) conformation of this bridge. Such a difference could result from a change in torsion angle about an OC—CO bond from *gauche* to *anti*. Thus hosts such as **28** with flexible bridges may use different conformations of the bridge to form complexes and show selectivity for guest molecules which reflects this flexibility.

Hosts **23**, **24** and **28** are ditopic and contain two electron rich receptor sites; they are therefore suitable hosts for bis-cations or two mono-cations. The development of synthetic receptors for anionic species has been rather slower than the development of synthetic receptors for cations and cryptand species of this type are rather rare, although some interesting examples have been reported. For ditopic hosts, electron deficient receptor sites can be incorporated to give the new cryptand species shown diagrammatically in Scheme 3 in which the ellipses represent electron rich (for example diaza-crown ether) receptor sites and the rectangles represent electron deficient receptor sites. This scheme also indicates possible guest species below each class of ditopic host and clearly the mixed system **29** is of particular interest because it can potentially complex both components of a guest salt.

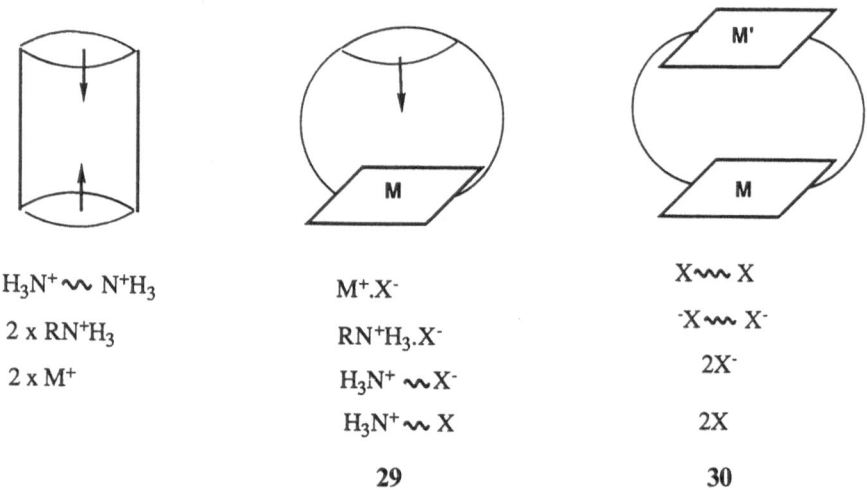

Scheme 3. Synthetic ditopic receptors and appropriate guest species.

The electron deficient receptors, indicated by rectangles in the new classes of ditopic host **29** and **30**, are conveniently based upon metalloporphyrins **31** because their coordination chemistry has been extensively investigated [30] and they are readily available synthetically with side chain functionality that permits their incorporation into macropolycyclic systems of types **29** and **30**. Whereas a suitable choice of ring size for the aza-crown ether receptor ensures guest binding only in the inwards direction, as indicated by the arrows in Scheme 3, a metalloporphyrin can

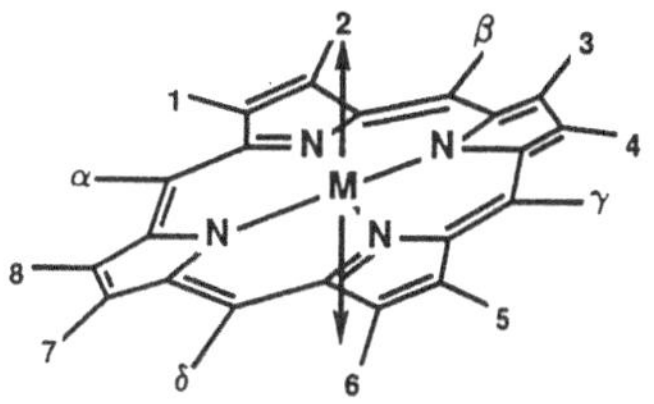

Fig. 7.

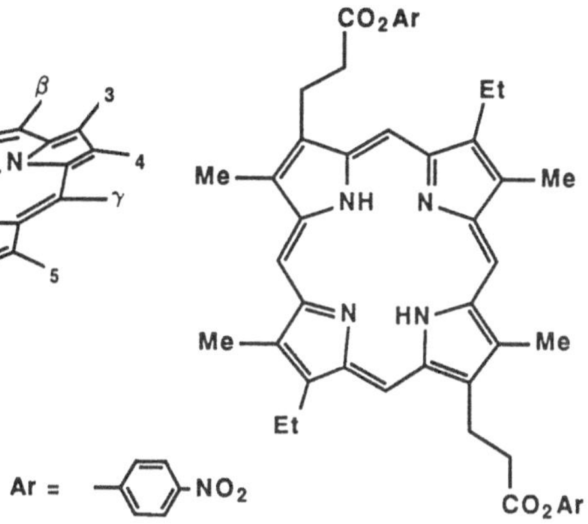

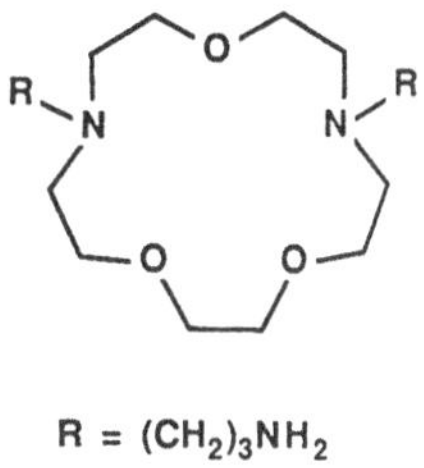

usually bind ligands at either face, as indicated by the arrows in **31**. However, for octahedral coordination at the central metal atom of the electron deficient receptor site in **29** and **30** the outer face of the receptor can be blocked by using a bulky ligand which is unable to enter the cavity. Bridges may be constructed using substituents at the β-positions of the pyrrole rings [positions 1–8 in **31**] or the meso-positions [α–δ in **31**] of the porphyrin system. The only ditopic receptors of type **29** that we have studied are derived from the crown capped porphyrin **34**, synthesized by reaction of the bis-*p*-nitrophenyl ester **32** with the diaza-15-crown-5 derivative **33** [31, 32]. The unmetallated capped porphyrin **34** can be metallated readily by reaction with an appropriate metal salt under the usual conditions [30] and a range of metallated systems **35** was prepared, the Zn(II) and Cu(II) derivatives were selected for most studies of complexation.

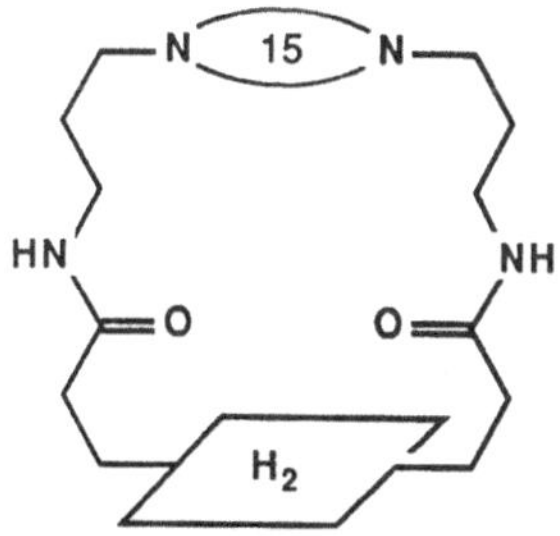

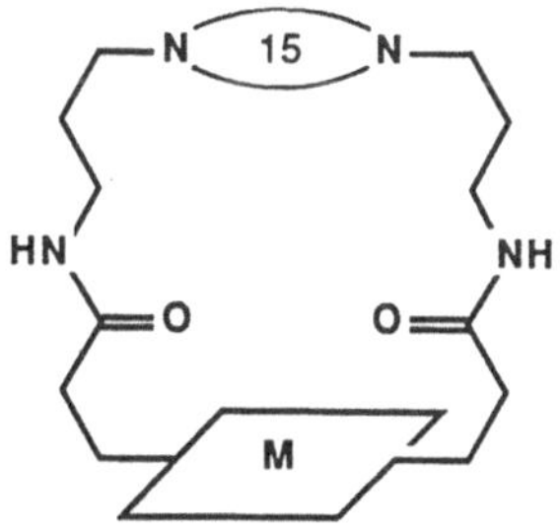

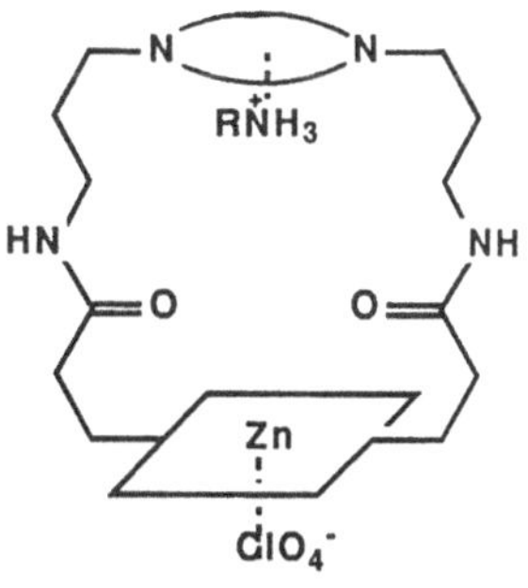

M = Zn(II), Cu(II), Co(II),

Mn(III), Fe(III)OH, Sn(IV)

Fig. 8.

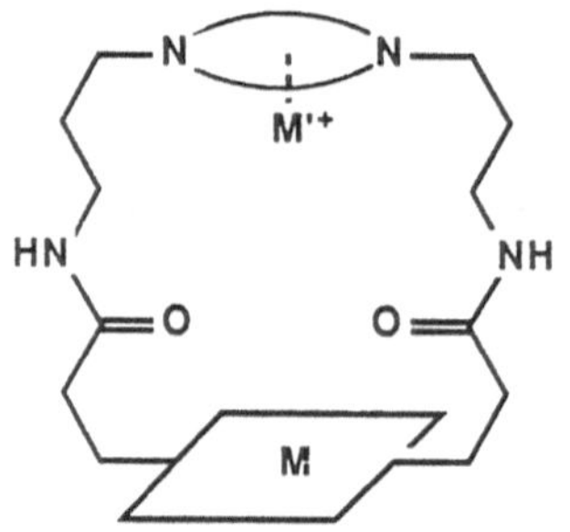

36 M = Zn ; M' = Cu^{2+}, Fe^{2+}, Fe^{3+}, Mn^{2+}, Co^{2+}, Ni^{2+}

37 M = Cu ; M' = Cu^{2+}, Fe^{3+}, Fe^{2+}, Mn^{2+}

Fig. 9.

The fluorescence spectrum of a methanolic solution of the Zn(II) derivative (**35**, M = Zn) was quenched in the presence of paramagnetic metal salts; from the quenching data it was possible to obtain association constants for the formation of the bimetal complexes **36**. Similar results could be obtained from the analogous Cu(II) derivative (**35**, M = Cu). One or more of the counter ions in these bimetal-lated complexes, **36** and **37**, are probably bound to the metalloporphyrin moiety but the spectroscopic data does not provide clear evidence for this. Similar treatment of the ditopic receptors (**35**, M = Zn and Cu) with salts of Na^+, Mg^{2+}, Ba^{2+} and Zn^{2+} did not lead to fluorescence quenching although at least some of these cations must be complexed by the diaza-crown ether system. It was con-cluded [32] that fluorescence quenching was brought about by the magnetic field of a complexed paramagnetic cation rather than through any other effect, but an electron transfer mechanism for fluorescence quenching could not be completely ruled out [33].

The crown capped porphyrin (**35**, M = Zn) proved to be a rather disappointing receptor for alkylammonium salts [22]. The absorption spectrum of the Zn(II) porphyrin was modified in the presence of the salts $RN^+H_3 \cdot ClO_4^-$ and from the spectroscopic data it was possible to obtain association constants for complexa-tion. The host showed little response to changes of R in the guest salt but the ${}^1H$ NMR data, together with the changes in the absorption spectrum, suggest that the complexes have the structures shown in **38**. The rather flexible links between the crown ether and porphyrin moieties are presumably one reason for the lack of guest recognition [compare with the semi-flexible ditopic receptor **28**].

Finally, examples of the third type of ditopic receptor **30**, containing two electron deficient metalloporphyrin receptor sites, can be synthesised by the route outlined in Scheme 4 [34]. The properties of these face-to-face porphyrins **39** as selective hosts have not yet been investigated although rather different face-to-face porphyrins have been investigated by Collmann and coworkers in a different context [35]. The diarylpophyrin system of **39** was selected to provide a well defined structural unit and it is hoped that this approach to a largely pre-or-ganised cavity will prove to be effective.

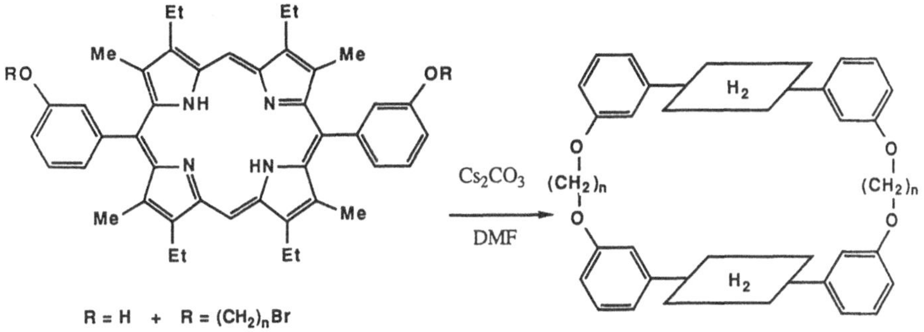

39

Scheme 4. Synthesis of face-to-face porphyrins.

Acknowledgements

The work that has been reported in this paper has been carried out by a number of enthusiastic collaborators; I am grateful to them for their skill, patience and tenacity in working with compounds of high molecular weight which have in many cases demanded a high level of experimental ability.

Notes

[1] It is a pleasure to note the award of the Nobel Prize to Dr. Charles Pedersen and Professors Don Cram and Jean-Marie Lehn for their major contributions to the area of host–guest chemistry.

[2] Many such software packages are available. We have programs from commercial (CHEMX, Molecular Design Ltd), industrial (COSMIC, Dr. J. G. Vinter of Smith, Kline and French) and academic (MACROMODEL, Dr. Clark Still, Columbia University) sources.

[3] We thank Professor R. J. Abraham for making the computer programme CHARGE available and Dr. J. G. Vinter of Smith, Kline and French for a generous gift of the COSMIC package.

[4] This work has been carried out in collaboration with Dr. J. F. Alder (Department of Analytical Science, UMIST). It is a pleasure to acknowledge the important contributions of Dr. Alder's group who have developed all of the instrumentation and optical fibre preparation.

[5] These data also include results obtained by J.-M. Lehn and coworkers reported in Ref. [26].

References

1. T. E. Creighton: *Proteins*, W.H. Freeman, New York, 1984; J. Darnell, H. Lodish, and D. Baltimore, *Molecular Cell Biology*, Scientific American Books, New York, 1986.

2. C. J. Pedersen: *J. Am. Chem. Soc.* **89**, 2495, 7017 (1967); C. J. Pedersen and H. K. Frensdorff: *Angew. Chem., Int. Edn. Eng.* **11**, 16 (1972).

3. D. J. Cram and K. N. Trueblood: *Top. Curr. Chem.* **98**, 43 (1981); D. J. Cram, R. A. Cormack, and R. C. Helgeson: *J. Am. Chem. Soc.* **110**, 571 (1988), and preceding papers in the series 'Host–Guest Complexation'.

4. M. W. Hosseini and J.-M. Lehn: *J. Am. Chem. Soc.* **109**, 7047 (1987); J.-M. Lehn, *Pure Appl. Chem.* **50**, 871 (1978); **51**, 979 (1979).

5. S. J. Leigh and I. O. Sutherland: *J. Chem. Soc., Chem. Commun.*, 414 (1975).

6. J. S. Bradshaw, S. L. Baxter, J. D. Lamb, R. M. Izatt, and J. J. Christensen: *J. Am. Chem. Soc.* **103**, 1821 (1981); Y. Inoue and T. Hakushi: *J. Chem. Soc., Perkin Trans.* 2, 935 (1985).

7. C. F. Lai, K. Odashima, and K. Koga: *Tetrahedron Lett.* **26**, 5179 (1985); K. Odashima and K. Koga: in *Cyclophanes*, Vol. 2, eds. P. M. Keehn and S. M. Rosenfeld, Academic Press, New York, 1983, Ch 11; F. Diederich, K. Dick, and D. Griebel: *J. Am. Chem. Soc.* **108**, 2273 (1986).

8. E. Kimura: *Top. Curr. Chem.* **128**, 113 (1985); F. Vögtle, H. Sieger, and W. M. Muller: *Top. Curr. Chem.* **98**, 107 (1981).

9. D. H. Williams: *Acc. Chem. Res.* **17**, 364 (1984).

10. G. Wipff, P. Weiner, and P. A. Kollman: *J. Am. Chem. Soc.* **104**, 3249 (1982); P. A. Kollmann, G. Wipff, and U. C. Singh: *J. Am. Chem. Soc.* **107**, 2212 (1985).

11. R. J. Abraham, L. Griffiths, and P. Loftus: *J. Comp. Chem.* **3**, 407 (1982); R. J. Abraham and B. Hudson: *J. Comp. Chem.* **5**, 562 (1984); **6**, 173 (1985).

12. J. G. Vinter, A. Davis, and M. R. Saunders: *J. Computer-Aided Mol. Design* **1**, 31 (1987).

13. F. de Jong and D. N. Reinhoudt: *Stability and Reactivity of Crown-Ether Complexes*, Academic Press, New York, 1981.

14. J. Sandström: *Dynamic NMR Spectroscopy*, Academic Press, London, 1982; I. O. Sutherland in *Applications of NMR Spectroscopy to Problems in Stereochemistry and Conformational Analysis*, ed. Y. Takeuchi and A. P. Marchand, VCH Publishers, Florida, 1986.

15. A. B. Kyte, K. A. Owens, I. O. Sutherland, and R. F. Newton: *J. Chem. Soc., Perkin Trans. 1*, 1921 (1987), and earlier papers in this series.

16. J. F. Alder, D. C. Ashworth, R. Narayanaswamy, R. E. Moss, and I. O. Sutherland: *Analyst* **112**, 1191 (1987).

17. M. Takagi and K. Ueno: *Top. Curr. Chem.* **121**, 39 (1984); K. Sugihara, T. Kaneda, and S. Misumi: *Heterocycles* **18**, 57 (1982); H.-G. Lohr and F. Vögtle: *Acc. Chem. Res.* **18**, 65 (1985); S. Misumi, Y. Kai, H. Morii, K. Miki, and N. Kasai: *J. Am. Chem. Soc.* **107**, 4802 (1985); I. Tanigawa, K. Tsuemoto, T. Kaneda, and S. Misumi: *Tetrahedron Lett.* **25**, 5327 (1984).

18. D. C. Ashworth and R. E. Moss: to be published.

19. D. J. Cram and S. P. Ho: *J. Am. Chem. Soc.* **108**, 2998 (1986).

20. P. D. J. Grootenhuis, P. D. Van der Wal, and D. N. Reinhoudt: *Tetrahedron* **43**, 397 (1987).

21. L. C. Hodgkinson, M. R. Johnson, S. J. Leigh, N. Spencer, I. O. Sutherland, and R. F. Newton: *J. Chem. Soc., Perkin Trans. I*, 2139 (1979).

22. I. O. Sutherland: *Chem. Soc. Rev.* **15**, 63 (1986).

23. S. Mageswaran and I. O. Sutherland: *J. Chem. Soc. Chem. Commun.*, 722 (1979).

24. R. K. Lewis: Ph.D. Thesis, Liverpool, 1985.

25. J. P. Kintzinger, F. Kotzyba-Hilbert, J.-M. Lehn, A. Pagelot, and K. Saigo: *J. Chem. Soc. Chem. Commun.*, 833 (1981).

26. F. Kotzyba-Hilbert, J.-M. Lehn and P. Vierling: *Tetrahedron Lett.*, 941 (1980).

27. J. C. Metcalfe and J. F. Stoddart: *J. Am. Chem. Soc.* **99**, 8317 (1977); J. Krane and O. Aune: *Acta Chem. Scand.* **34B**, 397 (1980).

28. S. Mageswaran and I. O. Sutherland: to be published.

29. A. Kumar, S. Mageswaran, and I. O. Sutherland: *Tetrahedron* **42**, 3291 (1986).

30. K. M. Smith: *Porphyrins and Metalloporphyrins*, Elsevier, Amsterdam, 1985.

31. C. K. Chang: *J. Am. Chem. Soc.* **99**, 2819 (1977).

32. N. M. Richardson, I. O. Sutherland, and P. Camilleri: *Tetrahedron Lett.* **26**, 3739 (1985).

33. V. Thanobal and V. Krishnan: *J. Am. Chem. Soc.* **104**, 3643 (1982); S. G. Schulman: *Fluorescence and Phosphorescence Spectroscopy: Physicochemical Principles and Practice*, Pergamon, Oxford, 1977.

34. T. Lane, I. O. Sutherland, and C. H. Yap: to be published.

35. J. P. Kollman, P. Denisevich, Y. Konai, M. Marrocco, C. Koval, and F. C. Anson: *J. Am. Chem. Soc.* **102**, 6027 (1980).

Journal of Inclusion Phenomena and Molecular Recognition in Chemistry 7 (1989), 227–245.

The Making of Molecular Belts and Collars[1]

J. FRASER STODDART
Department of Chemistry, The University, Sheffield S3 7HF, Great Britain

(Received: 1 February 1988)

Abstract. This article relates the first encouraging steps towards the fulfilment of a long-standing research goal aimed at turning the chemistry of laterally-fused six-membered rings through 90° ... or, more specifically, the making of (i) beltenes, in which 1,4-cyclohexadiene rings are linked in a polycyclic array by lateral fusion through their carbon–carbon double bonds, (ii) collarenes, in which alternating benzene and 1,4-cyclohexadiene rings are fused to form macropolycyclic hydrocarbons and (iii) cyclacenes, which may be considered as two annulenes joined to each other by carbon–carbon single bonds between every other atom around the annulene rings. The synthesis of the key macropolycyclic compound, which is a potential precursor of [12]beltene and [12]collarene, exploits the amazing stereoelectronic control that exists in the Diels–Alder reaction between a bisdiene and a bisdienophile with the appropriate structural features and reactivity characteristics.

Key words. Cavitands, Diels–Alder reactions, stereoelectronic control, kohnkene, clathrate formation, chemical sensor, [12]collarene, [12]beltene, [12]cyclacene, organic zeolites.

Over the years, the thrust of my research effort has been dictated largely by the challenge to design and synthesise molecular receptors to bind with a particular substrate or to promote a chosen reaction. However, we do try to invest a small amount of our time in chasing fantasies, rather than pursuing, what at least we think are realities. It is a flight of fancy that I am going to relate to you in this lecture. Just forget all about molecular recognition, as either a ground state or a transition state phenomenon, and let us take a trip into dreamland in search of new synthetic molecular materials and receptors that are rigid and have some of the topological features of belts and braces not to mention clips and collars.

Rigid molecules with belt- and collar-like structures, although relatively rare, are not exactly new: carbohydrates provide perhaps the best known examples in the shape of the cyclodextrins [1]: then chance has played its part in the spontaneous assembly of cucurbituril [2] from glyoxal, urea, and formaldehyde: and careful design and synthesis has led to the realisation of the so-called cavitands [3]. Although all three of these hosts have proved to be rich hunting grounds for both ground and transition state receptor chemistry, they have their limitations: cavitand manufacture often involves a lot of synthetic effort, cucurbituril is a one-off job, and cyclodextrins, despite yielding increasingly to reliable chemical modification [4], remain locked into something of a structural straight-jacket. Against this background, we have been seeking an approach to the design and synthesis of rigid doughnut-shaped molecules which would offer a lot of flexibility in terms of (i) structure, (ii) size, (iii) shape, (iv) electronic characteristics, (v) functionality, and (vi) properties. Topologically, the molecules I am going to describe in this lecture are similar to the cyclodextrins, cucurbituril, and the cavitands: structurally, however, they constitute a completely new class of potential receptor molecule with, I suspect, their own unique physical and rich material properties.

Kekulene (1)

Polyacenes (2)

Pentacene (3)

Since Kekulé proposed [5] the first satisfactory structure for benzene in 1865 to the synthesis of kekulene (1), for example, in 1978 by Diederich and Stabb [6], the cyclic homologation of the benzenoid nucleus has proceeded in the planes of the benzene rings. To any benzene ring, one can envisage *ab*-, *ac*-, and *ad*-fusion of another benzene ring: kekulene (1) contains twelve benzene rings, six fused *ac* and six fused *ad*. Pure *ad*-fusion leads, of course, to the polyacenes (2): they have been isolated and characterised up to heptacene [7] and very recently a much improved synthesis of pentacene (3) has been reported [8]: they are linear molecular strips. Imagine, however, the bending of a polyacene containing a repeating unit of twelve benzene rings into a molecular belt in which the plane of the macropolycycle is orthogonal to the mean plane of the benzene rings now forced to adopt boat-like conformations: we have just constructed (Figure 1) [12]cyclacene (4). Bent benzene rings are not uncommon and exist, for example [9], in the [*n*]paracyclophane series of compounds where $8 \geq n \geq 4$. It will be useful during this lecture for me to project this 3-dimensional structure on to a 2-dimensional surface as shown in Figure 1. This constitutional formula suggests that [12]cyclacene (4) – a constitutional isomer of kekulene (1) – might best be thought of as two antiaromatic [24]annulenes linked by twelve carbon–carbon bonds. The polyacenes (2) and [*n*]cyclacenes could become very important organic compounds if the predictions [10] that their con- densed phases might possibly show high temperature ferromagnetism and warm superconductivity are realised. Whatever their electronic character, and always assuming they are stable enough to be isolated, what are we going to do with them chemically?

Scheme 1 outlines some of the directions that could be pursued starting from the [*n*]cyclacenes. We could attempt to (i) effect electrophilic-like substitutions upon the unsaturated rings, (ii) reduce the benzene-like rings to 1,4-cyclohexadiene rings and obtain the so-called [*n*]beltenes [11], which could, in turn, become substrates for (iii) allylic substitutions, (iv) cleavage of the carbon–carbon double bonds with ozone to give cyclic [*n*]ketones, and (v) epoxidation of the carbon–carbon double bonds to afford [*n*]epoxides. In principle, this addition could occur on either the inside or the

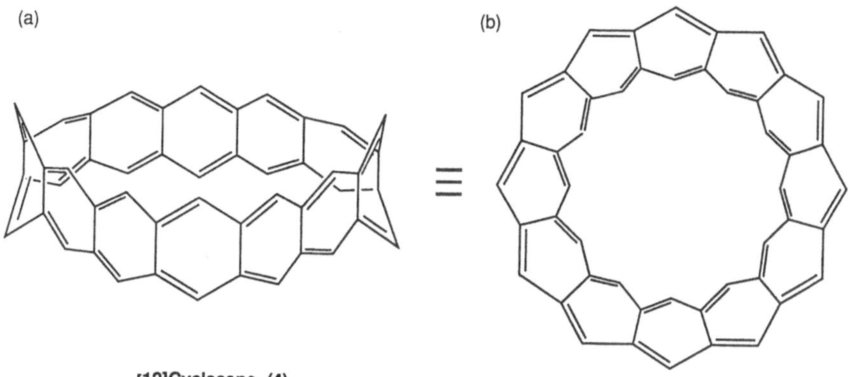

(a)

(b)

[12]Cyclacene (4)

Fig. 1.   (a) Conformational and (b) constitutional formulae for [12]cyclacene. They indicate how **4** can be considered to be composed of two [24]annulenes linked by twelve carbon–carbon bonds.

outside of the beltene – as could (vi) hydrogenation, (vii) hydroxylation, and (viii) halogenation, for example.

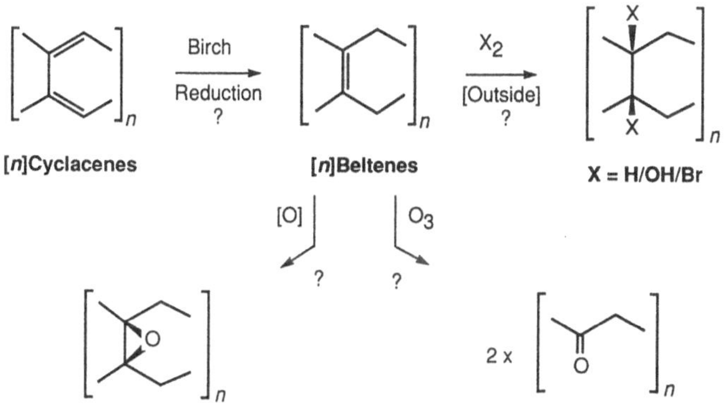

[n]Cyclacenes   [n]Beltenes   X = H/OH/Br

Scheme 1

How are we going to make macropolycycles built up of laterally-fused six-membered rings? We recommend a synthetic approach that relies upon the Diels–Alder reaction. I find it quite remarkable that so far, to my knowledge, no one has used this well-known reaction, where not one but, two bonds are formed simultaneously between a diene and a dienophile, to make a molecular receptor. Just imagine incorporating the diene and the dienophile into the same molecule: then, an intramolecular Diels–Alder reaction, always assuming it is stereochemically feasible, will lead to intramolecular macropolycyclisation. Or more realistically, by employing a bisdiene and a bisdienophile, the Diels–Alder reaction could be used (Scheme 2) sequentially to build laterally-fused six-membered rings into a macropolycycle.

Why appeal to the Diels–Alder reaction? I suggest two reasons. Firstly, its practice extends over half a century [12]: the reaction (i) can involve cheap and readily-available starting materials, (ii) gives six-membered rings, (iii) goes in high

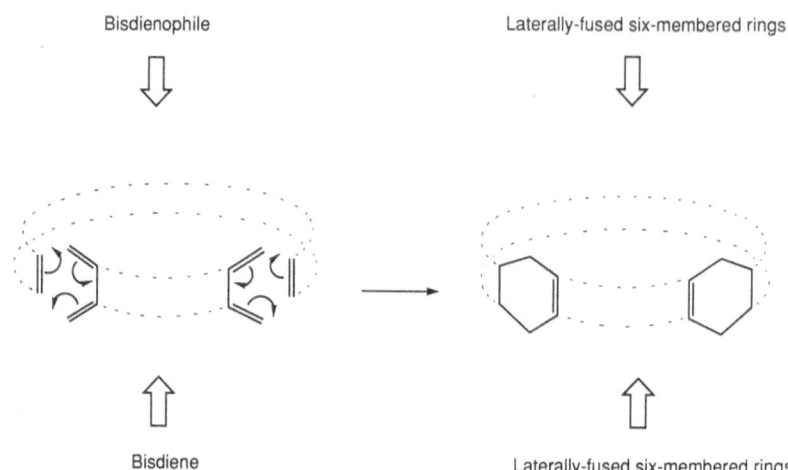

Scheme 2

yields very often, (iv) can be performed in a range of solvents including water [13], (v) can be carried out over a wide range of temperatures, (vi) can be promoted by very high pressures [14], and (vii) can be catalysed by Lewis acids [15]. Secondly, the reaction mechanism is well understood [16]: the Diels–Alder reaction exhibits (i) high regioselectivity, (ii) complete stereospecificity, i.e. *cis*-addition, and (iii) high stereoselectivity, *nota bene*, with respect to bicyclic systems, for 'close' stereochemistry, the descriptors *endo* and *exo* are usually used to define relative configurations, whereas for 'remote' stereochemistry, the terms *anti* and *syn* can be employed as relative configurational descriptors.

The bisdiene 5   The bisdienophile 6

With reference to Scheme 2, the main problem was identifying a bisdiene and a bisdienophile that worked: after 10 years of learning the hard way, we finally uncovered two building blocks that worked well [17, 18]: they are 2,3,5,6-tetramethylene-7-oxabicyclo[2.2.1]heptane (**5**), first described in the literature [19] by Professor Pierre Vogel from Lausanne in Switzerland in 1974 and the *syn*-isomer **6** of 1,4:5,8-diepoxy-1,4,5,8-tetrahydroanthracene, first isolated and characterised [20] by Professor Harold Hart at Michigan State University in 1983.

The choice of the bisdiene **5** as one of the starting materials was based on the following observations and features: (i) it can be prepared [21] in four steps from the maleic anhydride adduct of furan, 100 g of the latter yielding 10 g of the bisdiene **5**: (ii) it forms [19, 21–23] monoadducts *ca.* 100 times faster than bisadducts and so repetitive Diels–Alder reactions can be done in discrete steps by starting with mild conditions and making them progressively more forcing: (iii) in its cycloadditions with bisdienophiles such as benzoquinone, *one* of two possible diastereoisomers predominates overwhelmingly (>95 : <5), indicating that high stereoselectivity is operating at the 'remote' stereochemical level. The choice of the bisdienophile **6** as the other starting material was based on the following observations and features: (i) it can be prepared [20] from 1,2,4,5-tetrabromobenzene and furan, 100 g of the latter yielding 10 g of the bisdienophile **6** after chromatography to remove any by-products and the *anti*-isomer of 1,4 : 5,8-diepoxy-1,4,5,8-tetrahydroanthracene (X-ray crystallography [24] established the relative configurations of both isomers beyond any doubt): (ii) when **6** is reacted with anthracene only *one* of the three possible diastereoisomeric bisadducts is formed [20], indicating that high stereoselectivity is also operating at the 'close' stereochemical level. Both the bisdiene **5** and the bisdienophile **6** share another two important characteristics: (i) they are rigid molecules with concave and convex surfaces (Figure 2); bring them together in the correct way and they should be able to close a molecular loop and make a molecular belt; (ii) it should be possible to remove the oxygen atoms somehow (e.g. deoxygenation or dehydration) at the end of the sequence of Diels–Alder reactions.

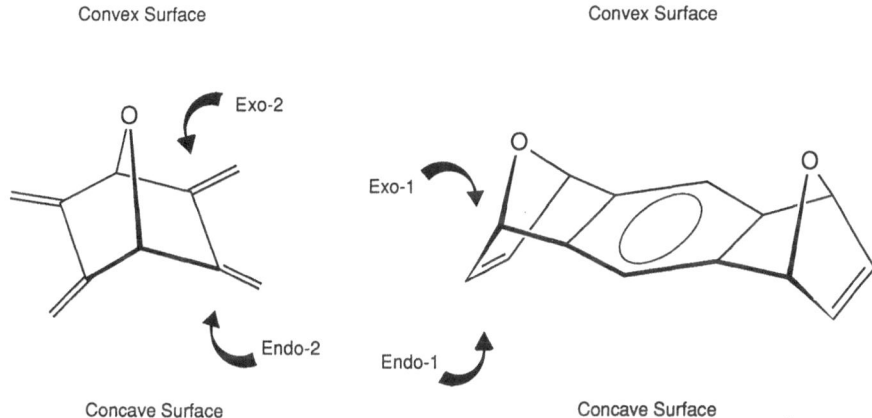

Fig. 2.   The shapes of the bisdiene **5** and the bisdienophile **6**, showing their convex and concave surfaces as well as their *exo* and *endo* faces ('2' represents a diene unit whereas '1' represents a dienophilic unit).

Given the fact that both **5** and **6** have diene and dienophilic components respectively with *endo* and *exo* faces, there are *eight* different ways of bringing the two components together to give *four* diastereoisomeric 1 : 1 adducts, i.e. there is a two-fold degeneracy in the reaction pathways with respect to the reaction products. In practice, one of the two diastereoisomeric transition states can be discounted for steric reasons in each case in the formation of the *syn/endo*-H, *syn/exo*-H, *anti/endo*-H, and *anti/exo*-H isomers, **7**, **8**, **9**, and **10**, respectively.

Examination of framework molecular models shows that, while the *anti* isomers **9** and **10** can only undergo polymerisation, the *syn* isomers **7** and **8** can, in principle at least, dimerise and cyclise: the *syn/exo*-H isomer **8** might conceivably be capable of cyclisation, i.e. it could undergo an intramolecular Diels–Alder reaction. At

Syn/Endo-H Isomer 7

Syn/Exo-H Isomer 8

Anti/Endo-H Isomer 9

Anti/Exo-H Isomer 10

worst, only **9** and/or **10** with the *anti* configuration might be obtained: at best, we might get only **7** and/or **8** with the *syn* configuration: in truth, we were prepared to accept a mixture of all four isomers and face some demanding chromatography: in reality, we were only able to isolate (Scheme 3) and characterise **7**, the *syn/endo*-H isomer. When you reflect on the fact that the reaction has the option to go down eight different pathways, then the exclusive choice of *one* is all the more remarkable. I alluded earlier to the high stereoelectronic control of 'close' and 'remote' stereo-chemistries *separately*: now, we see that when these two stereochemistries exist *in tandem*, the stereoelectronic control is just as tight. This realisation opens up the exciting prospect of the existence[2] of a land of molecular 'lego'.

Reaction of the bisdiene **5** with the bisdienophile **6** afforded (Scheme 3) the 1:1 adduct **7** and the 2:1 adduct **11** in yields of 24 and 61% respectively after chromatography. The 'close' stereochemistry in both these adducts could be estab-

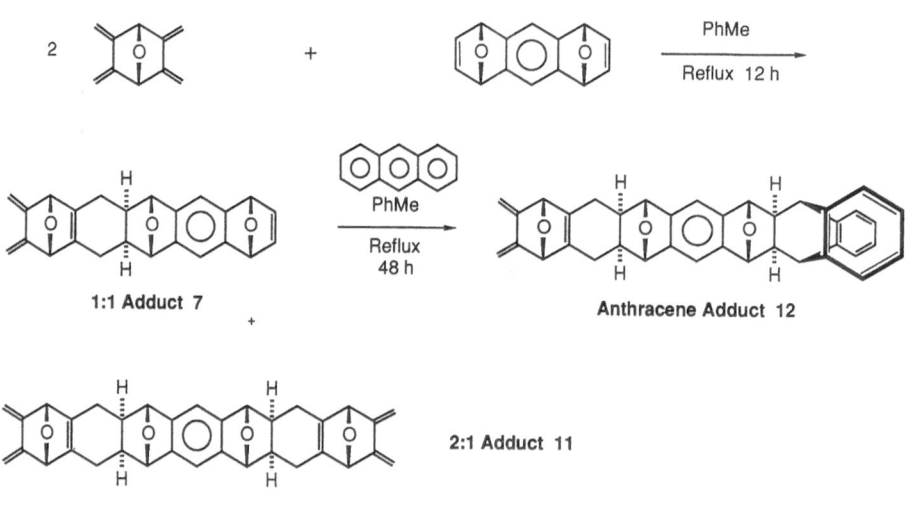

1:1 Adduct **7**

Anthracene Adduct **12**

2:1 Adduct **11**

Scheme 3

lished easily by ^{1}H NMR spectroscopy: the sharp singlets observed in each case for the methine protons at the newly formed ring junctions indicate [19, 21–23] that these hydrogen atoms have the *endo* configuration. The 'remote' stereochemistry (*nota bene*, when the oxygen atoms in the bisdiene and bisdienophilic components of the adducts are on the same side, the relative configuration is defined *syn* and when they are on opposite sides it is defined *anti*) was quite another matter and it soon became clear that this configurational information could only be secured with certainty by appealing to X-ray crystallography. Neither adduct could be persuaded to yield good quality single crystals and so we had to go in search of a derivative of the 1:1 adduct **7**; eventually, after several months and numerous disappointments, the anthracene adduct **12** of **7** (see Scheme 3) afforded good single crystals

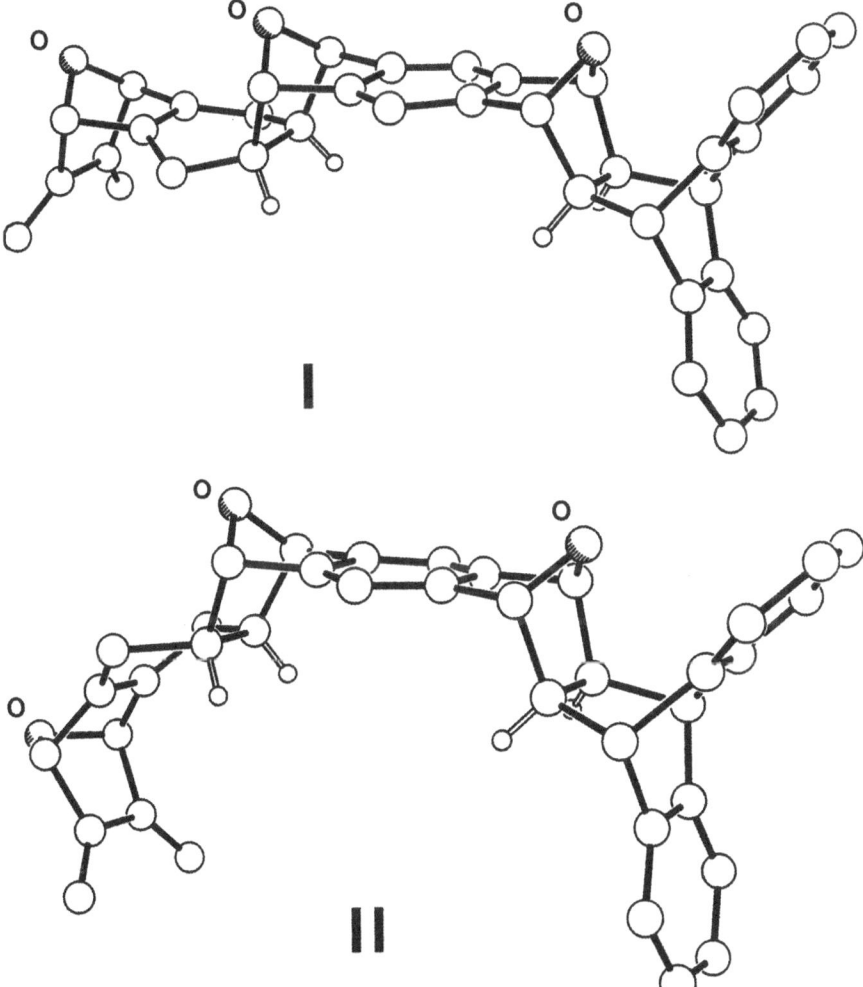

Fig. 3. Ball-and-stick representations of the solid state structures of the two crystallographically independent molecules (with the same configuration but differing in conformation) of the anthracene adduct **12**.

suitable for X-ray crystallography after employing chloroform-1,2-dichloroethane-light petroleum in a vapour diffusion method [25]. Although X-ray structural analysis revealed (Figure 3) the presence of two crystallographically-independent molecules, I and II, both have the same relative configuration consistent with the original 1:1 adduct having the *syn/endo*-H configuration **7**. Molecules I and II differ mainly in the conformations of their substituted cyclohexene rings: compounds of this type have pincer-like qualities to them such that, if they were adorned in the region of their claws with appropriate functional groups, then they could provide expandable and contractible molecular grooves and clefts, adaptable to a whole range of molecular recognition phenomena (cf. Ref. [26]). What happens stereochemically on 'the left of centre' during the formation of the 1:1 adduct **7** presumably also happens on 'the right of centre' during the formation of the 2:1 adduct **11**. This implies the production of a 2:1 adduct with a *syn/endo*-H//*syn/endo*-H configuration **11** having C_{2v} symmetry. This implication is supported strongly by the observation of (a) only *nine* signals in the ^{1}H-decoupled ^{13}C NMR spectrum for the *nine* heterotopic carbon atoms in **11**, and (b) only *eight* resonances in the ^{1}H NMR spectrum for the *eight* heterotopic hydrogen atoms in **11**. Also, the subsequent macropolycyclisation (*vide infra*) of **11** with **6** establishes the stereochemical assignment I have just made to **11** beyond question.

The amazing stereoselectivity which is exercised during the formation of the 1:1 adduct **7** – and indeed the 2:1 adduct **11** – is worthy of comment. The *syn/endo*-H stereochemistry of **7** could, in principle, ensue (Figure 4) from transition states exhibiting *syn/endo*-1/*exo*-2 or *syn/exo*-1/*endo*-2 geometries. The former can be ruled out on steric grounds: the latter is presumably favoured for stereoelectronic reasons. As far as the dienophilic component of **6** is concerned, there is evidence [23] of the pyramidalisation of the carbon–carbon double bond such that the electron density in the π-bonding orbital is much greater on the *exo*-1 face than on the *endo*-1 face: and so *exo*-1 attack is favoured. What controls the approach of the dienophilic component of **6** to the diene orbitals of **5**? We believe that stereoelectronic control could be dictated in this instance by secondary factors similar to those postulated by Paquette [27] to account for the favoured *endo*-2 attack of the 2,3-dimethylene-7-methanobicyclo[2.2.1]heptane constitution by dienophiles.

At last, we were in a position to attempt the cyclodimerisation (Scheme 4) of the 1:1 adduct **7**. After heating **7** under reflux at ca. 150°C in xylene for 2 days, we managed to isolate, in 3.5% yield, after painstaking chromatography, a compound

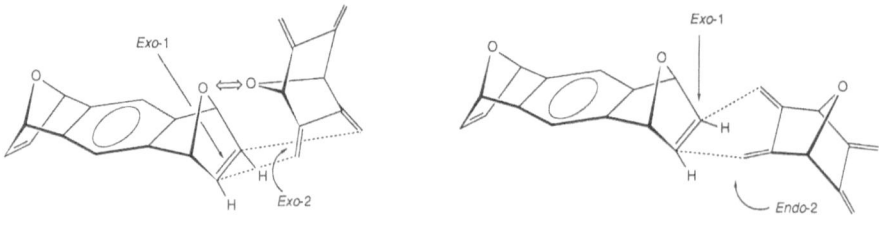

Syn/Exo-1/Exo-2 Transition State   *Syn/Exo-1/Endo-2* Transition State

Fig. 4.   The sterically disfavoured *syn/exo*-1/*exo*-2 and stereoelectronically favoured *syn/exo*-1/*endo*-2 transition states which afford a 1:1 adduct **7** with *syn/endo*-H stereochemistry.

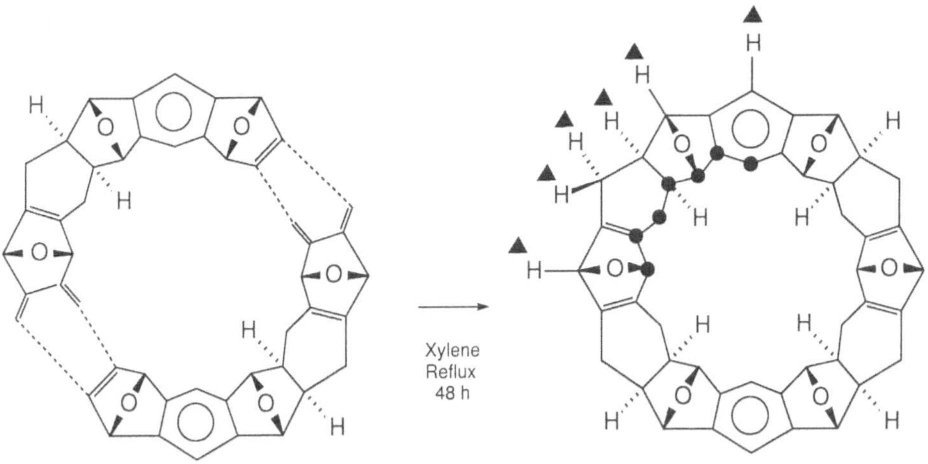

7 x 2

Xylene
Reflux
48 h

Kohnkene (13)

Scheme 4

which got us really excited when we examined its spectroscopic properties: (a) positive ion FABMS gave a very intense peak at 713 for $[M + H]^+$, (b) the ^{1}H-decoupled ^{13}C NMR spectrum revealed *seven* signals, and (c) the ^{1}H NMR spectrum, *six* resonances. The macropolycycle **13** formulated in Scheme 4 has D_{2h} symmetry, consistent with the molecule having *seven* heterotopic carbon atoms and *six* heterotopic hydrogen atoms. Now, I am sure you will all agree that the real advances in synthetic chemistry occur at the laboratory bench and not on the office desk: in this instance, it was Franz Kohnke from the University of Messina who brought **13** into the world. I believe it is the professional duty of more senior scientists to recognise openly and encourage vigorously the remarkable achievements of their younger colleagues: in recognition of the outstanding talent and considerable tenacity shown by Dr Kohnke in synthesising **13**, I propose to call this compound – the first macropolycyclic one of its kind to be isolated and characterised – *kohnkene*[3]. It should be added that this proposal has been advanced after considering and rejecting many alternatives: it should also be emphasised that it was made in the face of considerable opposition from the maker of **13**.

Has anyone been trying to draw kohnkene (**13**)? Without a plan, it is not easy. Here are some tips. Refer to Figure 5. Draw (**A**) a clock-face. Sketch (**B**) a 12-pointed star. Add spokes of equal length to the star emanating out from the points and drawn (**C**) collinear with the centre. Construct (**D**) another larger 12-pointed star. Equipped with this basic saturated carbon skeleton, it is a trivial matter (**E**) to introduce the unsaturation and add the oxygen atoms whilst also denoting the stereochemistry. We find the clock numbering system illustrated in **E** on kohnkene (**13**) in Figure 5 particularly useful when discussing its physical properties and chemical reactivity. While rings can be identified as occurring on the hour at 1, 2, 3, . . . o'clock, the ring junctions correspond to half-past the hour and so can be identified as 1:30, 2:30, 3:30, . . . etc.

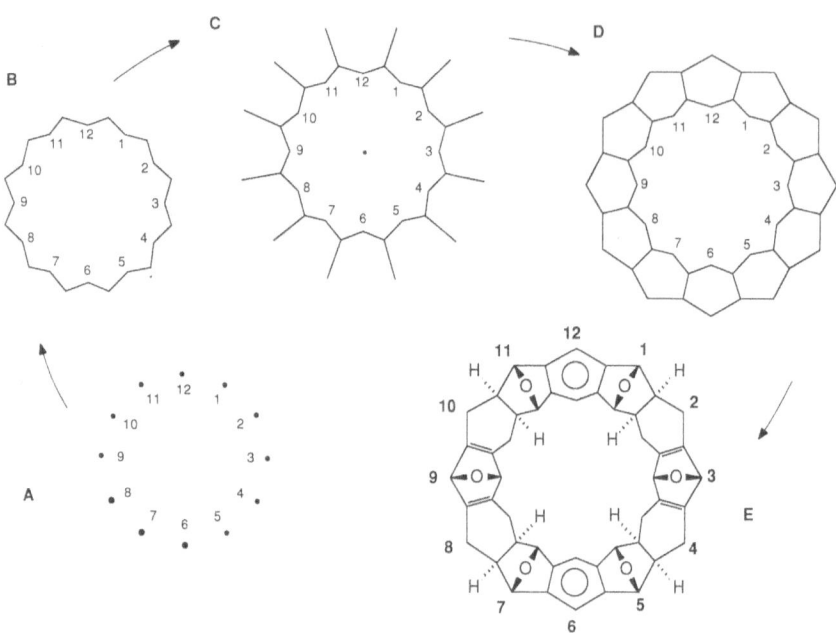

Fig. 5. The steps (A–E) in the recommended approach to drawing the configurational formula of kohnkene (13).

We have learnt to draw kohnkene (13) and to identify its parts but remember we have only obtained it in very low yield (3.5%) from the 1 : 1 adduct 7. This is not very encouraging if kohnkene (13) is ever going to be any more than an academic curiosity. Now, it is well known that Diels–Alder reactions have *negative* volumes of activation [14], i.e. the coming together at the transition state of a diene and a dienophile is accompanied by a reduction in the molar volume of the reactants: thus, Diels–Alder reactions are accelerated by very high pressures. Previously, I recounted how we had first made (Scheme 4) kohnkene (13) from the 1 : 1 adduct 7 by inducing it to cyclodimerise thermally. The problem with 7 is that it is not all that stable: it is prone to polymerisation on standing. By contrast, the 2 : 1 adduct 11 can be stored for months on end without any sign of decomposition. And so we decided in the high pressure experiment, carried out at Reading University under the expert guidance of Dr Neil Isaacs, to react (Scheme 5) 11 with the bisdienophile 6 at 10 kbar in dichloromethane for 8 days at 60°C: we obtained a welcome increase in the yield of kohnkene (13) – to 20%. Armed with half-gram quantities of 13, we could now do more chemistry on it. First of all, however, let me tell you about the difficult, but ultimately successful, solid state characterisation of kohnkene (13).

Good, single crystals of kohnkene (13) for X-ray crystallography were not at all easy to grow: finally, they emerged from slow evaporation of a chloroform solution of 13 during several weeks. They were not all that stable and Dr David Williams and Sandra Slawin at Imperial College London were only able to collect the required data by sealing the single crystal in a tube so that it was kept for the duration of the data collection in a vapour pressure of the solvent. Why are the

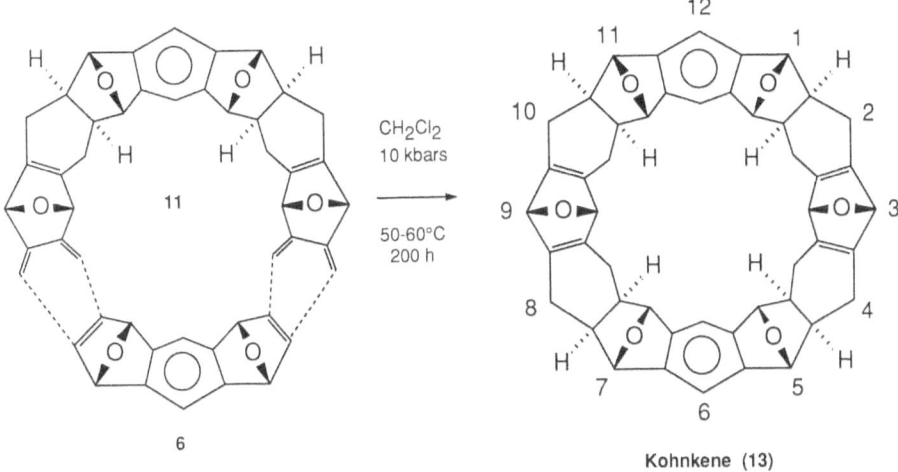

Scheme 5

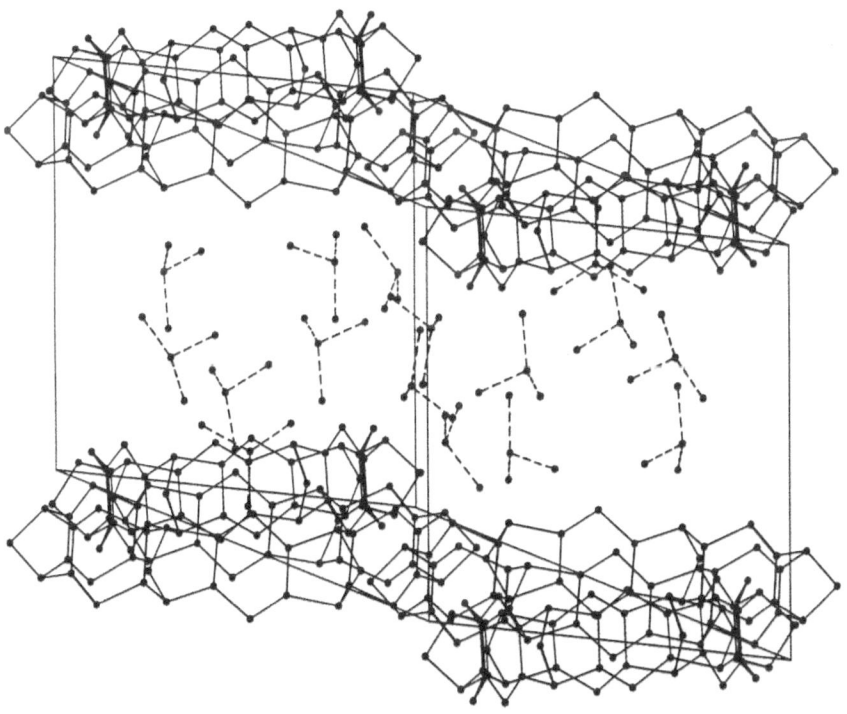

Fig. 6.   The unit cell in the solid state structure of kohnkene (**13**). Note that there are 12 disordered chloroform molecules included in the unit cell in 16 different orientations.

crystals so unstable? The explanation follows from how the molecules choose to pack in the unit cell (Figure 6). This picture reveals that the molecules are arranged in parallel approximate fall-to-face layers with 12 disordered chloroform molecules sandwiched between them in the unit cell in 16 discrete orientations. This clathration phenomenon explains the instantaneous loss of chloroform with concomitant collapse of the crystals on removal from the mother liquor. The X-ray crystal structure (Figure 7) of **13** discloses the elegance of the molecular structure: six

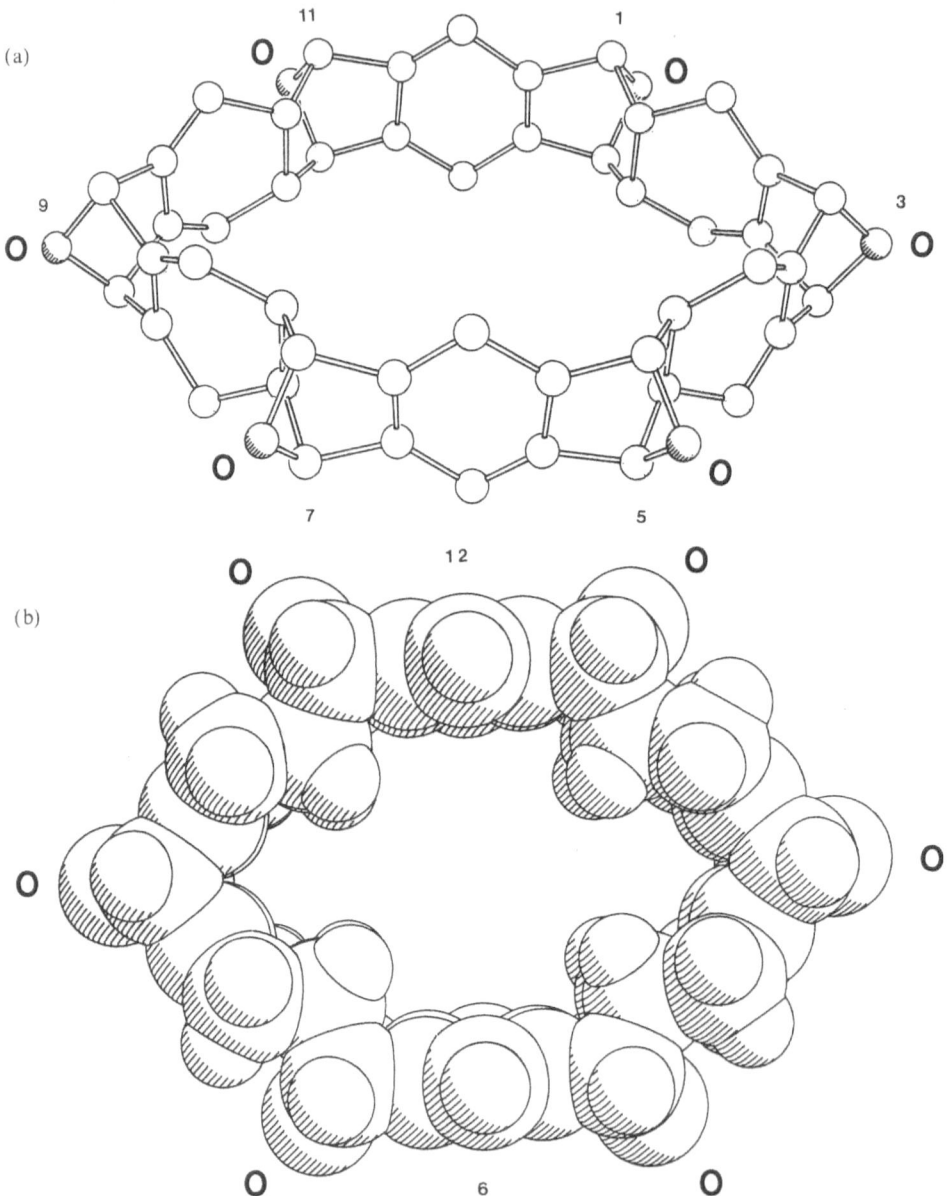

Fig. 7. (a) Ball-and-stick and (b) space-filling representations of the solid state structure of kohnkene (**13**).

oxygen atoms are almost evenly spaced around the outside of an elliptical molecular belt. The interplanar separation between the two benzene rings at 6 and 12 o'clock is 7.9 Å. This leaves enough room (ca. 5 Å between the van der Waals surfaces of the benzene rings) for a substituted benzene molecule to be accommodated orthogonally with respect to these two benzene rings. We argued that, particularly for a phenyl group (Ph) carrying an electron withdrawing substituent (X), the resulting edge-to-face interaction [28, 29] with 13 could be sufficiently stabilising electrostatically to allow PhX molecules to enter inside its rigid molecular cavity. Indeed, in collaboration with the UWIST Chemical Sensor Group, led by Dr Ron Thomas in Cardiff, we were able to demonstrate [30] the piezoelectric quartz crystal detection of nitrobenzene ($PhNO_2$) using kohnkene (13) as a detector coating.

Let us return to the synthetic trail with [12]cyclacene (4) firmly in our sights. Looking at kohnkene (13), one is reminded that there is more than one way to skin a rabbit. Why not try (a) deoxygenation at 3 and 9 o'clock, followed by (b) dehydration at 1, 5, 7, and 11 o'clock, and then finally, (c) dehydrogenation at 2, 4, 8, and 10 o'clock? Deoxygenation of 13 went [31] smoothly (Scheme 6) in 43%

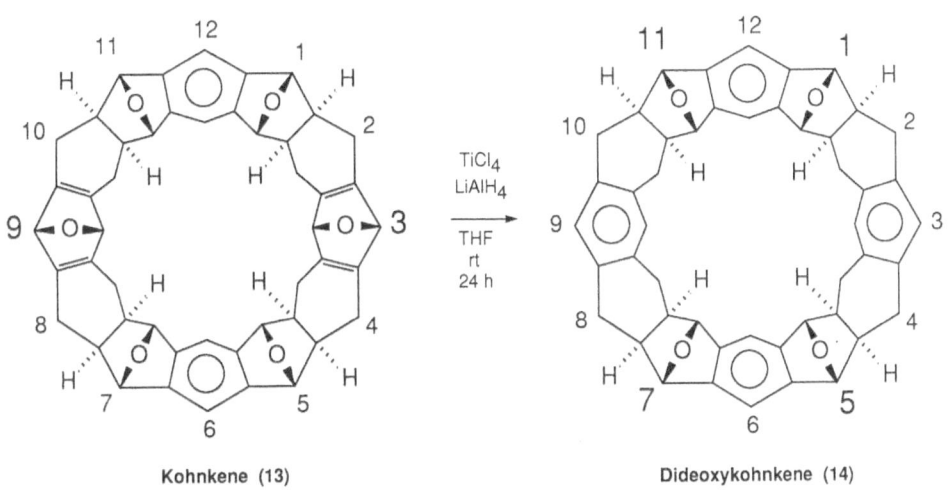

Kohnkene (13)   Dideoxykohnkene (14)

Scheme 6

yield to afford dideoxykohnkene (14). The positive ion FABMS had a peak of 690 for M^+ and NMR spectroscopic data were totally in agreement with the structural assignment given to 14. The vapour diffusion method [25], using chloroform–methanol, produced single crystals suitable for X-ray cyrstallography, which revealed (Figure 8) that the rigid cavity now has an approximately square cross-section with distances between the mean planes of the parallel-disposed aromatic rings of 8.9 Å (3 to 9 o'clock) and 9.6 Å (6 to 12 o'clock). The most striking feature of the structure is the Celtic cross-like hydrophobic cavity within which a water molecule is trapped like a gem. Yet, the water molecule does not enter into any hydrogen bonding with 14. The water oxygen atom is >3.5 Å removed from any of the four pairs of inward pointing methine hydrogen atoms in

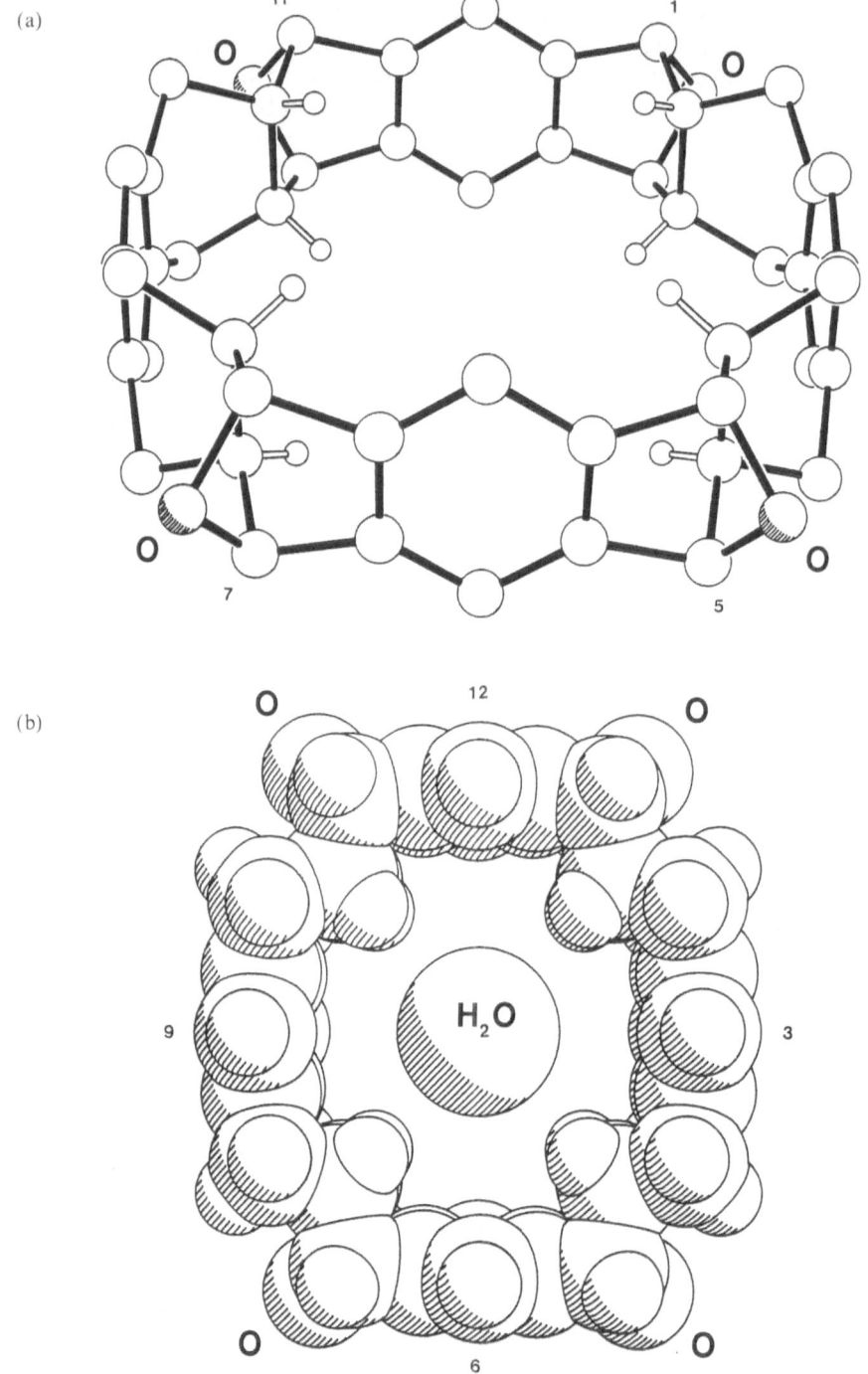

Fig. 8. (a) Ball-and-stick and (b) space-filling representations of the solid state structure of dideoxykohnkene (14). In view of the disorder in the positions of the hydrogen atoms, the included water molecule has been represented in (b) as a sphere with radius equivalent to the envelope of a water molecule.

14. The individual hydrogen atoms, which are >2.7 Å away from any potential interactive sites within the cavity, are directed principally towards the pairs of methine hydrogen atoms at 1:30 and 7:30.

Naïvely, we might have anticipated that dehydration (Figure 9) of dideoxykohnkene (**14**) would afford a hydrocarbon **15** containing two anthracene and two benzene units. However, in the knowledge [7, 8] that partially hydrogenated polyacenes reshuffle their aromatic rings to maximise their resonance

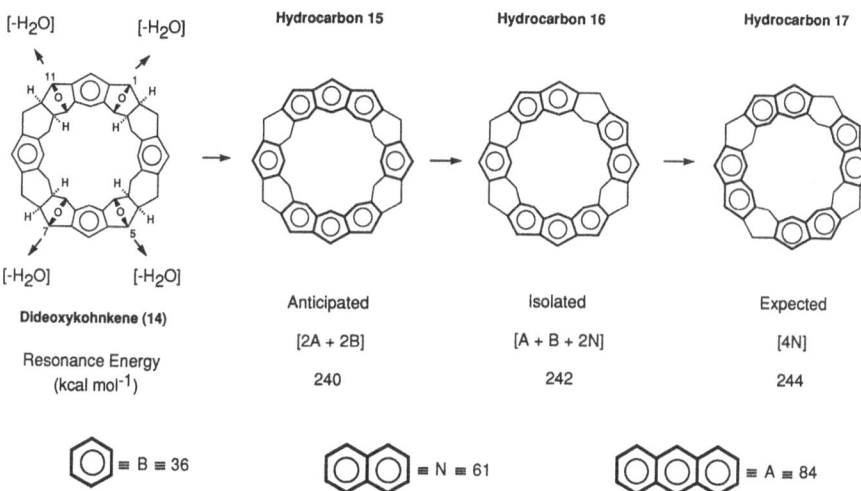

Fig. 9. The dehydration of dideoxykohnkene (**14**) showing three possible hydrocarbons, i.e., **15–17**, which could be formed. The resonance energies have been estimated, to a first approximation, on the basis of the literature values for benzene, naphthalene, and anthracene.

energies, we might have expected the hydrocarbons **17** with four naphthalene units to have emerged as the thermodynamically most stable product. In the event, dehydration gave (Scheme 7) a mixture of hydrocarbons, all displaying molecular

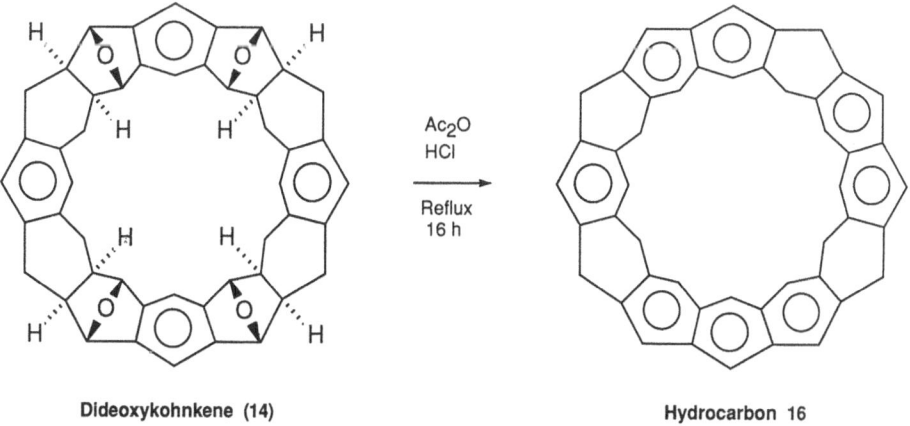

Scheme 7

ions with m/z 608 in their EIMS. After chromatography on silica gel using a lot of hot benzene, the hydrocarbon **16** of low solubility and high thermal stability was isolated and characterised by high field ^{1}H NMR spectroscopy. The presence of *four* anisochronous AB systems, along with chemical shift evidence for a benzene and an anthracene unit, clinches the constitution of this hydrocarbon as being **16** with two naphthalene units in addition to the benzene and anthracene units. Of the three hydrocarbons (i.e. **15**, **16**, and **17**) in Figure 9, only **16** has four constitutionally heterotopic cyclohexadiene units which are essential to explain the ^{1}H NMR spectroscopic data. In addition to the peak at m/z 608, the EIMS of **16** shows an intense M^{2+} ion at m/z 304 as well as the characteristic patterns for loss of hydrogen observed in the EIMS of compounds such as 1,2-dihydronaphthalene and 9,10-dihydrophenanthrene. The behaviour of **16** in the mass spectrometer augurs well for the eventual synthesis of [12]cyclacene (**4**).

The hydrocarbon **16** is a common and immediate precursor of [12]collarene[4] (**18**) and [12]beltene (**19**) as well as [12]cyclacene (**4**). With [12]beltene (**19**) singled out as our next synthetic objective, Birch reduction of **16** has revealed (Scheme 8) the presence in the desorption electron impact mass spectrum of a dodeca-hydro[12]cyclacene derivative (with m/z 612), which is probably [12]collarene (**18**), the most stable isomer with six alternating benzene rings along the route to [12]beltene (**19**). Although the road to **19** is an uphill one in energy terms, we feel confident that this synthetic objective – seen [11] not so long ago as requiring 'a major synthetic effort' – can now be reached: the incentives to prepare [12]beltene (**19**) are numerous, not least of all because of its high symmetry and molecular appeal.

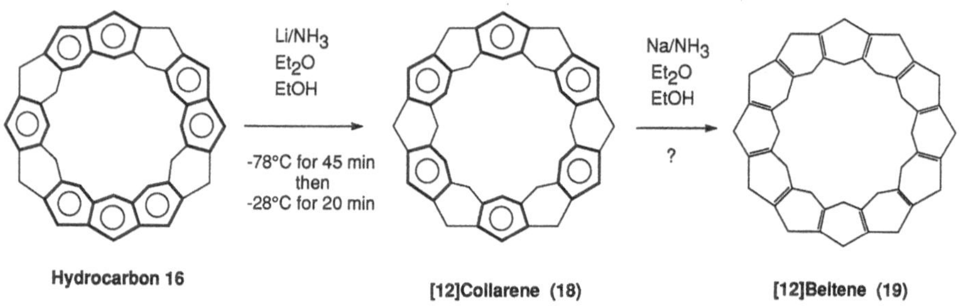

Hydrocarbon 16   [12]Collarene (18)   [12]Beltene (19)

Li/NH₃ Et₂O EtOH — -78°C for 45 min then -28°C for 20 min — Na/NH₃ Et₂O EtOH — ?

Scheme 8

And now, let us look into the future for a moment. What can one see amongst the galaxy of molecular stars? I suggest a veritable constellation of macropolycyclic compounds, given the supply of bisdienes and bisdienophiles that exists. And, not to be forgotten, some of the synthetic intermediates and by-products provide the additional opportunity to increase the availability of rod-like molecules containing linearly-annulated rings [32]. And then, what about the production of molecular columns and molecular nets, not to mention molecular barrels, such as the macrobipolycyclic compound **20**, that Dr Kohnke has designed and all but made. In the final analysis, organic zeolites could soon be within our grasp [33–35].

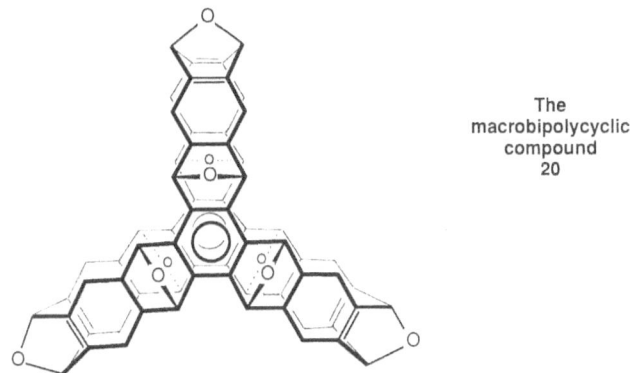

The
macrobipolycyclic
compound
20

As yet, we do not know much about the molecular recognition properties of this new family of compounds. But, there is no doubt that we have stumbled upon a simple way of creating new molecular materials that are rigid, ordered, and large. They can be assembled in a stepwise manner under a level of stereochemical control [36] that is quite breathtaking in its precision. The chemistry of laterally-fused six-membered rings has been turned through 90°. There is also a philosophical message to emerge from these experiences: it is a simple but a fundamental one – let us not try to plan everything in our research: let us leave something to chance: let us find time to dream a little and allow Nature to take over and unfold its wonders before us.

Acknowledgements

During the course of this lecture, I have given credit – where it was so handsomely earned – to the individuals involved directly in the making and in the assessment of these new molecular belts and collars. Behind the scenes, however, many other people have played crucial roles: they include Mr Peter Ashton (mass spectrometry) and Dr Catriona Spencer (NMR spectroscopy) who were always willing to respond to impossible requests without a moment's hesitation. Moral support, underlined by some financial backing, came freely from Prof. G. Stagno d'Alcontres (University of Messina) and Dr Bob Handscombe (Univeristy of Sheffield). Dr Peter Beadle, formerly of the BP Venture Research Unit, forced us to sharpen up our ideas and objectives during the writing of a research proposal which went through numerous drafts before being finally rejected. Now that the ideas have been shown to work, there is no shortage of generous financial sponsors, including the Science and Engineering Research Council in the United Kingdom.

Notes

[1]This article is dedicated to the memory of the late Professor Iwao Tabushi. The tale told here is based on a lecture delivered on 6 November 1987 at the US/Japan Joint Seminar on Molecular Recognition in Miami and is a highly personal account of the trapping at the end of a ten-year hunt for what became known as 'the crazy molecule' at Sheffield. The lecture was dedicated to the three pioneers of small molecule molecular recognition, the 1987 Nobel Laureates in Chemistry, Charles Pedersen, Jean-Marie Lehn, and Donald Cram – three men attracted as much by the art of chemistry as by its science.

[2]Reference was also made at the US/Japan Joint Seminar in Miami to the concept of molecular 'lego' by Professor Richard D. Gandour from Louisiana State Univeristy.
[3]With the help of Dr Alan McNaught at The Royal Society of Chemistry in London we have arrived at the following systematic name for kohnkene (13): rel(1R,4S,4aS,7aR,8R,10S,10aS,13aR,14R,17S,17aS, 20aR,21R,23S,23aS,26aR)1,4 : 6,25 : 8,23 : 10,21 : 12,19 : 14,17-hexaepoxy-1,4,4a,5,6,7,7a,8,10,10a,11,12, 13,13a,14,17,17a,18,19,20,20a,21,23,23a,24,25,26,26a-octacosahydro-2,16 : 3,15-dimethenoundecacene.
[4]We propose to refer to the macropolycyclic hydrocarbons which are comprised to alternating benzene and 1,4-cyclohexadiene rings as [n]collarenes.

References

1. M. L. Bender and M. Komiyama: *Cyclodextrin Chemistry*, Springer-Verlag, New York (1978); W. Saenger: *Angew. Chem. Int. Ed. Engl.* **19**, 344 (1980); J. Szejtli: *Cyclodextrins and their Inclusion Complexes*, Akademiai Kiado, Budapest, 1982; R. Brewlow: *Chem. Brit.* **19**, 126 (1983); W. Saenger: *Inclusion Compounds* (Volume 3, Ed. J. L. Atwood, J. E. D. Davies, and D. D. MacNicol), pp. 231, Academic Press, London (1984); J. Szejtli: *Inclusion Compounds* (Volume 3, Ed. J. L. Atwood, J. E. D. Davies, and D. D. MacNicol), pp. 331, Academic Press, London (1984); R. J. Bergeron: *Inclusion Compounds* (Volume 3, Ed. J. L. Atwood, J. E. D. Davies, and D. D. MacNicol), pp. 391, Academic Press, London (1984); I. Tabushi: *Inclusion Compounds* (Volume 3, Ed. J. L. Atwood, J. E. D. Davies, and D. D. MacNicol), pp. 445, Academic Press, London (1984); R. Breslow: *Inclusion Compounds* (Volume 3, Ed. J. L. Atwood, J. E. D. Davies, and D. D. MacNicol), pp. 473, Academic Press, London (1984); M. Komiyama and M. L. Bender: *The Chemistry of Enzyme Action* (Ed. M. I. Page), pp. 505, Elsevier, Amsterdam (1984); M. L. Bender; *Enzyme Mechanisms* (Eds. M. I. Page and A. Williams), pp. 56, The Royal Society of Chemistry, London (1987).
2. W. A. Freeman, W. L. Mock, and N.-Y. Shih: *J. Am. Chem. Soc.* **103**, 7367 (1981); W. L. Mock and N.-Y. Shih: *J. Org. Chem.* **48**, 3618 (1983); **51**, 4440 (1986); W. L. Mock, T. A. Mirra, J. P. Wepsiec, and T. L. Manimaran: *J. Org. Chem.* **48**, 3619 (1983).
3. D. J. Cram: *Science* **219**, 1177 (1983); D. J. Cram, K. D. Stewart, I. Goldberg, and K. N. Trueblood: *J. Am. Chem. Soc.* **107**, 2574 (1985).
4. A. P. Croft and R. A. Bartsch: *Tetrahedron* **39**, 1417 (1983); C. M. Spencer, J. F. Stoddart, and R. Zarzycki: *J. Chem. Soc., Perkin Trans.* 2, 1323 (1987).
5. P. Garratt: *Endeavour* **11**, 36 (1987).
6. F. Diederich and H. A. Staab: *Angew. Chem. Int. Ed. Engl.* **17**, 372 (1978); H. A. Staab and F. Diederich: *Chem. Ber.* **116**, 3487 (1983); H. A. Staab, F. Diederich, C. Krieger, and D. Schweitzer: *Chem. Ber.* **116**, 3504 (1983).
7. E. Clar: *Polycyclic Hydrocarbons* (Volume 1), Academic Press, New York (1964); W. J. Bailey and C.-W. Liao: *J. Am. Chem. Soc.* **77**, 992 (1955).
8. J. Luo and H. Hart: *J. Org. Chem.* **52**, 4833 (1987).
9. J. E. Rice, T. J. Lee, R. B. Remington, W. D. Allen, D. A. Clabo, Jr., and H. F. Schaefer III: *J. Am. Chem. Soc.* **109**, 2902 (1987).
10. S. Kivelson and O. L. Chapman: *Phys. Rev.* **B28**, 7236 (1983).
11. R. W. Alder and R. B. Sessions: *J. Chem. Soc., Perkin Trans.* 2, 1849 (1985); A. Nickon and E. F. Silversmith: *Organic Chemistry: The Name Game*, pp. 110, Pergamon Press, New York (1987).
12. A. Wasserman: *Diels Alder Reactions: Organic Background and Photochemical Aspects*, Elsevier, Amsterdam (1965).
13. R. Breslow and U. Maitra: *Tetrahedron Lett.* **25**, 1239 (1984) and references therein.
14. N. S. Isaacs and A. V. George: *Chem. Brit.* **23**, 47 (1987) and references therein.
15. M. Bednarski and S. Danishefsky: *J. Am. Chem. Soc.* **105**, 3716, 6968 (1983) and references therein.
16. I. Fleming: *Frontier Orbitals and Organic Chemical Reactions*, Wiley, Chichester (1976).
17. F. H. Kohnke and J. F. Stoddart: *Abstracts of 194th ACS National Meeting*, New Orleans, 30 Aug–4 Sept 1987, CARB 32.
18. F. H. Kohnke, A. M. Z. Slawin, J. F. Stoddart, and D. J. Williams: *Angew. Chem. Int. Ed. Engl.* **26**, 892 (1987).
19. P. Vogel and A. Florey: *Helv. Chim. Acta* **57**, 200 (1974).
20. H. Hart, N. Raja, M. A. Meador, and D. L. Ward: *J. Org. Chem.* **48**, 4357 (1983).

21. C. Mahaim, P.-A. Carrupt, J.-P. Hagenbuch, A. Florey, and P. Vogel: *Helv. Chim. Acta* **63**, 1149 (1980).
22. P. -A. Carrupt and P. Vogel: *Tetrahedron Lett.* 4537 (1979); Y. Bessiére and P. Vogel: *Helv. Chim. Acta* **63**, 232 (1980).
23. A. A. Pinkerton, D. Schwarzenbach, J. H. A. Stibbard, P.-A. Carrupt, and P. Vogel: *J. Am. Chem. Soc.* **103**, 2095 (1983); J.-M. Tornare, P. Vogel, A. A. Pinkerton, and D. Schwarzenbach: *Helv. Chim. Acta* **68**, 2195 (1985).
24. F. H. Kohnke, A. M. Z. Slawin, J. F. Stoddart, and D. J. Williams: *Acta Cryst. C.* **44**, 736, 738, 740, 742 (1988).
25. P. G. Jones: *Chem. Brit.* **17**, 222 (1981).
26. J. Rebek, Jr., B. Askew, P. Ballester, C. Buhr, S. Jones, D. Nemeth, and K. Williams: *J. Am. Chem. Soc.* **109**, 5033 (1987); J. Rebek, Jr., K. Williams, K. Parris, P. Ballester, and K.-S. Jeong: *Angew. Chem. Int. Ed. Engl.* **26**, 1244 (1987).
27. L. A. Paquette: *Stereochemistry and Reactivity of Systems Containing π-Electrons* (Ed. W. H. Watson), pp. 41, Verlag Chemie International, Deerfield Beech, FL (USA) (1983).
28. R. O. Gould, A. M. Gray, P. Taylor, and M. D. Walkinshaw: *J. Am. Chem. Soc.* **107**, 5921 (1985); S. K. Burley and G. A. Petsko: *Science* **229**, 23 (1985); *J. Am. Chem. Soc.* **108**, 7995 (1986); *FEBS Lett.* **203**, 139 (1986).
29. A. M. Z. Slawin, N. Spencer, J. F. Stoddart, and D. J. Williams: *J. Chem. Soc., Chem. Commun.* 1070 (1987); D. R. Alston, A. M. Z. Slawin, J. F. Stoddart, D. J. Williams, and R. Zarzycki, *Angew. Chem. Int. Ed. Engl.* **26**, 692 (1987); G. J. Moody, R. K. Owusu, A. M. Z. Slawin, N. Spencer, J. F. Stoddart, J. D. R. Thomas, and D. J. Williams: *Angew. Chem. Int. Ed. Engl.* **26**, 890 (1987); P. R. Ashton, E. J. T. Chrystal, J. P. Mathias, K. P. Parry, A. M. Z. Slawin, J. F. Stoddart, and D. J. Williams: *Tetrahedron Lett.* **28**, 6367 (1987).
30. M. A. F. Elmosalmy, G. J. Moody, J. D. R. Thomas, F. H. Kohnke, and J. F. Stoddart: *Anal. Proc.* in press.
31. P. R. Ashton, N. S. Isaacs, F. H. Kohnke, A. M. Z. Slawin, C. M. Spencer, J. F. Stoddart, and D. J. Williams: *Angew. Chem. Int. Ed. Engl.* **27**, 966 (1988).
32. L. L. Miller, A. D. Thomas, C. L. Wilkins, and D. A. Weil: *J. Chem. Soc., Chem. Commun.* 661 (1986); A. D. Thomas and L. L. Miller: *J. Org. Chem.* **51**, 4160 (1986); W. C. Christopfel and L. L. Miller: *J. Org. Chem.* **51**, 4169 (1986); P. W. Kenny and L. L. Miller: *J. Chem. Soc., Chem. Commun.* 84 (1988).
33. J. F. Stoddart: *Nature* **334**, 10 (1988).
34. J. F. Stoddart: *Chem. Brit.* **24**, 1203 (1988).
35. L. Milgrom: *New Scientist* No. 1641, 3 December 1988, p. 61.
36. P. Ellwood, J. P. Mathias, J. F. Stoddart, and F. H. Kohnke, *Bull. Soc. Chem. Belg.* **97**, 669 (1988).

Journal of Inclusion Phenomena and Molecular Recognition in Chemistry 7 (1989), 247–256.

Reaction Control by a Host–Guest Complexation Method

FUMIO TODA
Department of Industrial Chemistry, Faculty of Engineering, Ehime University, Matsuyama 790, Japan

(Received: 1 February 1988)

Abstract. Very successful stereo-, regio-, and enantio-controls of the photoreaction of a guest compound were achieved by irradiation of the host–guest complex in the solid state. In order to discover the reason for the successful control, the X-ray crystal structure of the complex was studied. The complexation method was also effective to freeze the equilibrium in solution and isolate the labile tautomeric isomer as an inclusion complex.

Key words. Irradiation, solid state, enantiocontrol, organic reaction.

1. Introduction

Since guest molecules are arranged close together and in one conformer in a host–guest complex, inter- and intramolecular photoreactions of the guest compound would proceed regio- and stereo-selectively and efficiently by irradiation of the complex in the solid state. When an optically active host compound is used, enantio-control of the reaction is expected. When the complexation is applied to an equilibrium mixture of tautomers, one labile tautomer can be isolated as a host–guest complex. By an irradiation of the complex in the solid state, stereo- and enantio-controls of the reaction of the labile tautomer can be achieved.

2. Regio- and Stereo-controls

Photodimerization of chalcone (**1a**) is not easy either in solution or in the solid state. For example, irradiation of **1a** in solution gives a mixture of **1a**, its *cis*-isomer, and polymer. Irradiation of **1a** in the solid state gives a complex mixture of all possible stereoisomeric photodimers in low yields. X-ray crystal structural studies of two dimorphs of **1a** showed that the distance between the double bond center is greater than the limit of an intermolecular reaction (4.2 Å). However, irradiation for 6 h in the solid state of a 1 : 2 complex of 1,1,6,6-tetraphenylhexa-2,4-diyne-1,6-diol (**2**) and **1a** which had been prepared by recrystallization of the two components from solvent gave the *syn*-head-to-tail dimer (**3a**) selectively [1, 2]. Similar irradiation of 1 : 2 complexes of **2** and **1b–d** gave **3b–d**, respectively, in the yields shown in Table I. The reason for the well controlled reaction can be interpreted as follows: in the complex with **2**, two molecules of **1** are packed close together in the positions which give *syn*-head-to-tail dimer **3**. In order to clarify the reason, X-ray crystal analysis of a 1 : 2 complex of **2** and **1a** was carried out. The X-ray analysis disclosed that the above assumption is correct and the distance between the double bond centers of **1a** is short enough (3.682 Å) to react easily [3].

Benzylideneacetone (**4**) is also photoinactive in the solid state. However, irradiation of a 1 : 2 complex of **4** with 2,6-diphenylhydroquinone (**5**) in the solid state gave the *syn*-head-to-tail dimer (**6**) in 70% yield. An X-ray crystal structural study of the complex showed that a hydrogen bond between the components makes the packing tight and the distance between the double bonds shorter (3.787 Å) [2, 3].

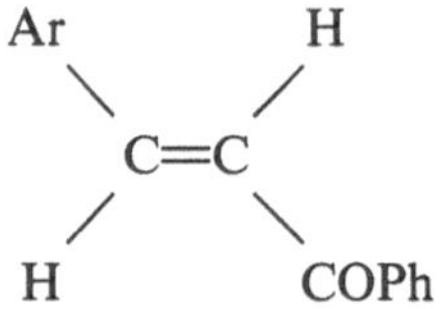

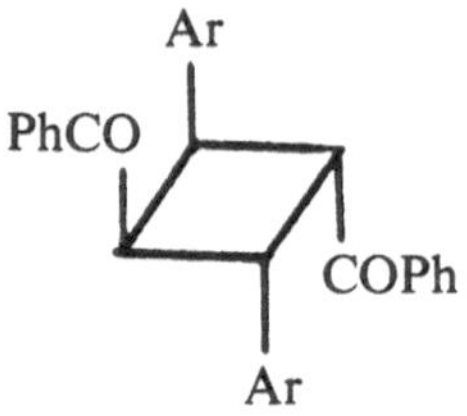

a: Ar = Ph
b: Ar = *o*-Me-C$_6$H$_4$—
c: Ar = *o*-MeO—C$_6$H$_4$—
d: Ar = 2-Naphthyl

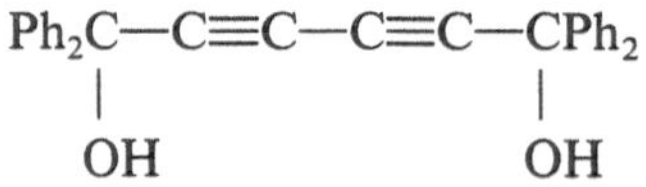

Ph$_2$C—C≡C—C≡C—CPh$_2$
   | |
   OH   OH
          2

Table I. Reaction time, products, and yields of photocycloaddition reaction of **1a–d** in the complex with **2**, in the solid state.

Chalcone	Reaction time (h)	Product	
		No.	Yield (%)
1a	6	**3a**	90
1b	6	**3b**	85
1c	1.5	**3c**	88
1d	1	**3d**	82

PhCh=CH—CO—CH=CHPh
          4

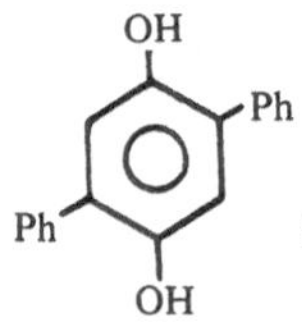

PhCH=CHCO

COCH=CHPh
        6

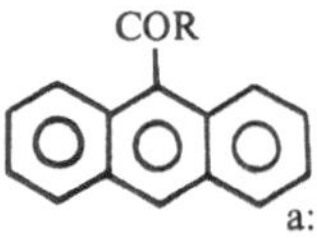

a: R = H
b: R = Me
      7

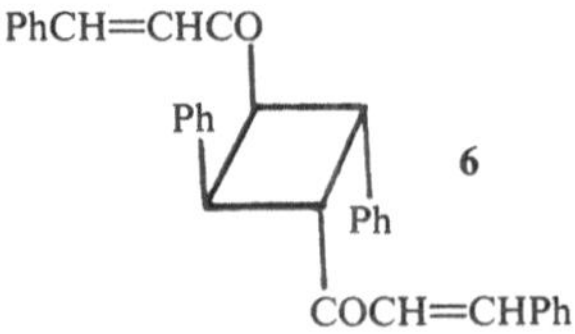

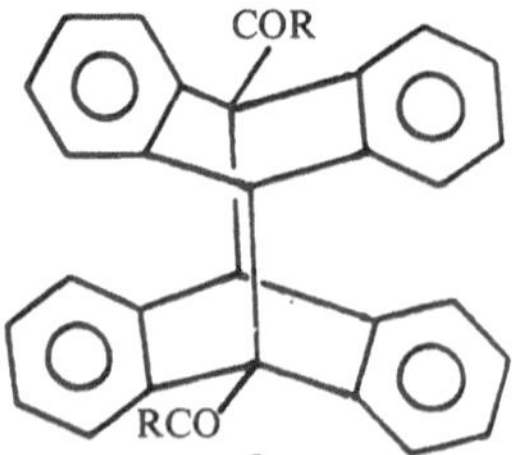

Although irradiation of 9-formylanthracene (**7a**) in solution for 24 h gives the *anti*-photocycloaddition product (**8a**) in a low yield [4], irradiation of a 1:2 complex of **2** and **7a** in the solid state for 8 h gave **8a** in 86% yield [2]. An X-ray crystal structural study of the complex showed that two molecules of **7a** are arranged between two host molecules by forming a hydrogen bond in the direction which gives the *anti*-dimer (**8a**) by the photodimerization and that the distance between the two reaction centers of **7a** is short enough to react readily (4.042 Å) [3]. Similar irradiation of a 1:2 complex of **2** and 9-acetylanthracene (**7b**) in the solid state for 1.5 h gave **8b** in 87% yield.

3. Control and Acceleration of Photoreaction by a Combination of Freezing of Equilibrium by Complexation and Irradiation of the Complex in the Solid State

Since 2-pyridone (**9**) exists as an equilibrium mixture with 2-hydroxypyridine (**10**), it is difficult to isolate **9** in a pure state. However, the complexation method is applicable for an isolation of the *keto*-form (**9**) in a pure state. For example, **9** was isolated by complexation with **2** and **5** as 1:2 complexes with **2** and with **5**, respectively. The structure of the former complex was studied by an X-ray crystal analysis [5]. Inclusion of the *keto*-form is reasonable, because **2** and **5** form stronger hydrogen bonds with the carbonyl oxygen of the *keto*-form than with the hydroxyl oxygen of the *enol*-form.

The selective inclusion of the *keto*-form (**9**) can be used to effect an efficient dimerization of **9**. Moreover, since molecules of **9** are packed regularly in a host–guest complex, regio- and stereo-selective intermolecular reaction of **9** is expected. Irradiation of a 1:2 complex of **2** and **9** in the solid state for 6 h gave the *trans-anti*-dimer (**11**) in 76% yield [2]. This efficient reaction is in contrast to that in solution which gives **10** in 40% yield after irradiation for 72 h [6]. An X-ray crystal structural study of a 1:2 complex of **2** and **9** showed that two molecules of **9** are arranged between two molecules of **2** in the positions which give the *trans-anti*-dimer (**11**) by dimerization and the distance between the reaction centers is very short (3.837 Å) [5].

Freezing of the equilibrium by the complexation method is applicable to some other compounds. 2-Mercapto substituted tropone (**12**) has been reported to exist as an equilibrium mixture of 2-mercapto-tropone (**12a**) and 2-hydroxytropothione

(**12b**), and the latter is predominant both in solution [7] and in the solid state [8]. The equilibrium is frozen and the former was isolated by complexation with **2**. When a solution of **2** and **12** in petroleum ether was kept at room temperature, a 1:1 complex of **2** and **12a** was obtained in 90% yield as orange prisms of mp 101–103°C. The structure of **12a** in the complex with **2** was elucidated by an X-ray crystal structural study, and the study disclosed that **12a** has a delocalized structure (Figure 1). This is a very interesting result because the delocalized structure has never been reported for either tropolone or **12**. As an exception, ferric troponate has been reported to have a delocalized structure [9].

$$
\underset{\textbf{12a}}{\text{O}\diagup\text{SH}} \quad\rightleftharpoons\quad \underset{\textbf{12b}}{\text{HO}\diagdown\text{S}} \qquad\qquad \underset{\textbf{13}}{\text{S—S}}
$$

A packing diagram of the complex shows that two molecules of **12a** are packed in the *anti*-positions to each other and the distance between the two seven-membered rings is short enough (3.49 Å) to cause a π–π interaction. The stacking of **12a** in the complex is comparable to that of graphite (3.5 Å). Although **12** is thermally labile and gives dimer **13** easily, **12a** in the complex is stable. The stability is probably due to the packing of **12a** in the *anti*-direction.

1,2,4-Triazole (**14**) also exists as an equilibrium mixture of the two tautomers, 1,2,4-triazacylopenta-3,5-diene (**14a**) and 1,2,4-triazacyclopenta-2,5-diene (**14b**), and it is difficult to isolate one tautomer in a pure state. The former was isolated in a pure state as a 1:1 complex with 1,1-bis(2,4-dimethylphenyl)but-2-yn-1-ol (**15**). The complex was obtained from a solution of **14** and **15** in MeOH as colorless prisms of mp 101–102 °C, in 88% yield [10]. An X-ray analysis of the complex showed that **14a** is included and the closest contact is a hydrogen bond between the hydroxyl group of **15** and the nitrogen at N4. The distance of the hydrogen bond was found to be 2.697 Å [10].

Cycloocta-2,4,6-trien-1-one exists as an equilibrium mixture of a stable form (**16a** and **16b**) and an unstable form (**17**) in a 95:5 ratio at 20°C [11]. Conversion

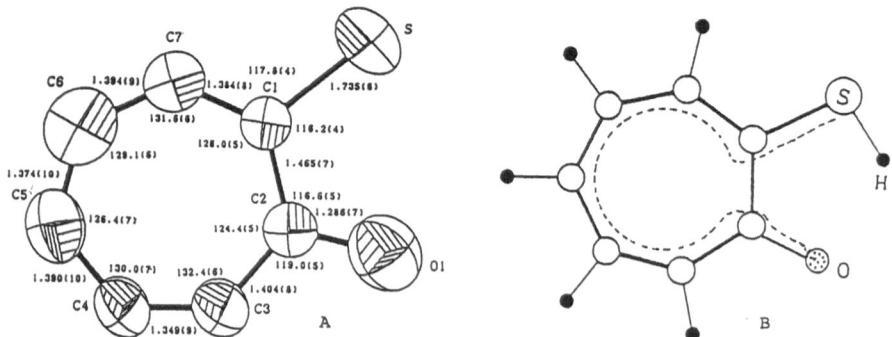

Fig. 1. Bond lengths and angles with numbering of the atoms (A) and a delocalized structure (B) of **12a** in the complex with **2**.

between the two optical conformers **16a** and **16b** is rapid around room temperature with the activation energy of 11.9 kcal/mol [11]. By complexing with optically active 1,6-di(o-chlorophenyl)-1,6-diphenylhexa-2,4-diyne-1,6-diol (**18**), the flipping equilibrium of **16** was frozen in one conformer as a 1:2 complex. The IR spectrum of the complex did not show any carbonyl absorption which is assignable to **17**. In order to know whether one enantiomer is included selectively or not, photoconversion of **16** in the complex to bicyclo[4.2.0]octa-4,7-dien-2-one (**19**) was carried out. When a 1:2 complex of **18a** and **16** was irradiated in the solid state for 168 h, 50% conversion occurred and (−)-**19** was obtained in 28% yield ($[\alpha]_D - 69.0°$ (c 0.12, CHCl$_3$)) [12]. Although the optical purity of **19** was not determined, it is clear that one of the two enantiomers **16a** and **16b** is included selectively or at least predominantly.

Photoreaction of **16** in pentane has been reported to give racemic **19** in 30% yield after irradiation for 21 days [11]. Therefore, it is obvious that the photoreaction of **16** in the complex proceeds not only enantioselectively but also much more efficiently.

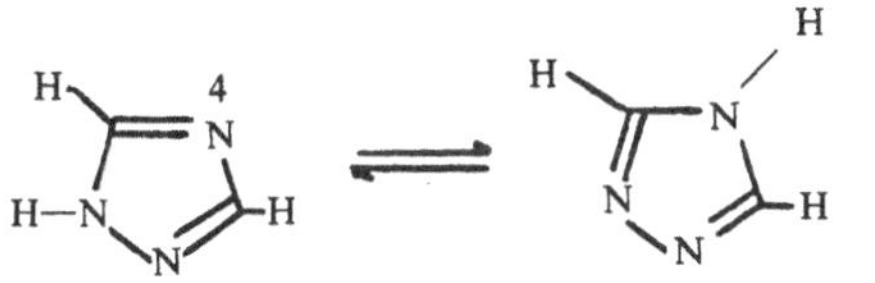

14a   14b 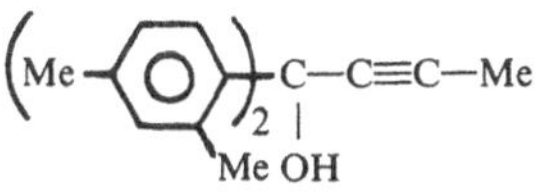   15

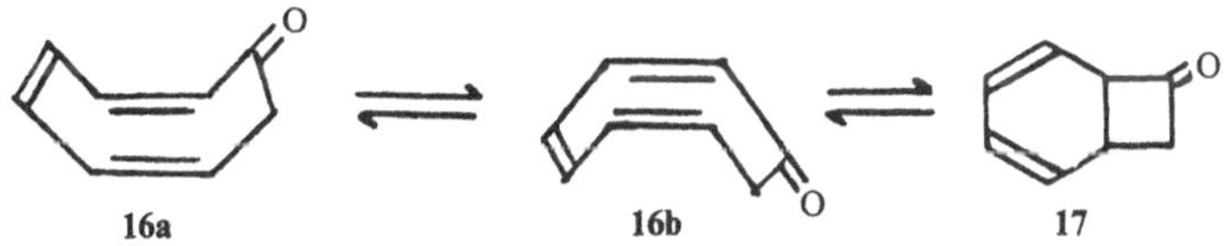

16a   16b   17 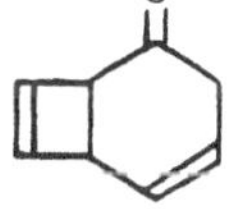   19

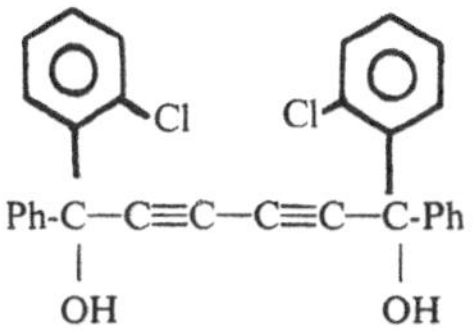

18

4. Control and Acceleration of Photoreaction by a Combination of Host–Guest Complex Formation by Solid–Solid Reaction and Irradiation in the Solid State

We found that the host–guest complex is formed by a solid–solid reaction of host and guest compounds [13]. The solid–solid reaction can be done either by grinding both components using an agate mortar and pestle or agitating using a test-tube shaker, and in some cases just by keeping a mixture of host and guest compounds at room temperature [13]. For example, when a mixture of finely powdered **2** and an equimolar amount of finely powdered benzophenone is agitated using a test-tube shaker for 0.2 h at room temperature, a 1:1 complex of both components was obtained in quantitative yield. The IR spectrum of the product was identical to that of an authentic sample prepared by complexation in solution. The complexation method for the solid–solid reaction is applicable to a variety of host and guest compounds. Interestingly, freezing of equilibrium of **9** and **10** occurs even by complexation with **2** in the solid state.

When the complex formation by solid–solid reaction is combined with the photoreaction in the solid state, a stereoselective photoreaction can be carried out continuously. In other words, the host compound can be used catalytically. Upon irradiation of 1:1 and 1:2 mixtures of **2** and chalcone (**1a**) in the solid state with occasional mixing by an agate mortar and pestle for 10 and 40 h, respectively, **3a** was obtained selectively in 80 and 82% yields, respectively. These results show the formation of a 1:2 complex of **2** and **1** before the photodimerization reaction, since **1a** itself does not give **3a** upon irradiation in the solid state. Furthermore, irradiation of a 1:4 mixture of **2** and **1a** under the same conditions for 72 h gave **3a** in 87% yield. This result shows that host compound **2** was used almost twice like a catalyst. This is illustrated in Scheme I. By mixing **2** and **1a**, their 1:2 complex is formed, and the irradiation to the complex gives **2** and **3a**. Further mixing forms new complexes which upon irradiation gives **3a**. Finally, all **1a** is converted to **3a**.

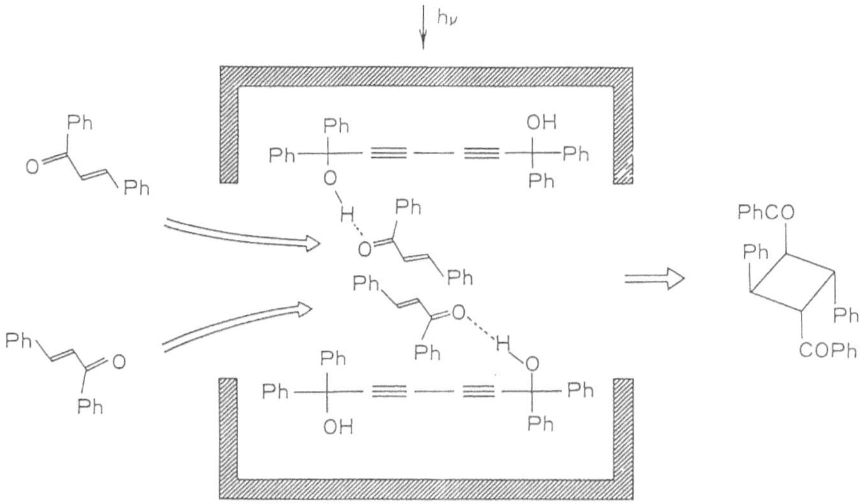

Scheme I

5. Enantiocontrol

Irradiation of a 1:1 complex of **18a** and 2-methoxytropone (**20a**) in the solid state for 72 h gave (1S, 5R)-(−)-1-methoxybicyclo[3.2.0]hepta-3,6-dien-2-one (**21a**) of 100% ee in 11% yield together with (S)-(+)-methyl-4-oxocyclopent-2-ene-1-acetate (**22a**) of 91% ee (26% yield) [14, 15]. Similar irradiation of a 1:1 complex of **18a** and 2-ethoxytropone (**20b**) gave **21b** of 100% ee (12%) and **22b** of 72% ee (14% yield).

The enantioselective photoreaction can be interpreted as follows: a disrotatory [2 + 2]-photoreaction of **20** in the complex with **18a** occurs only in the A direction according to a steric hindrance of the o-chlorophenyl group of **18a** (Scheme II). This interpretation was shown to be reasonable by an X-ray crystal structural study of the complex [16]. Formation of **22** is due to the side-reaction in the presence of water (Scheme III) [17]. According to a photochemical enolization, **22** is partly racemized [17].

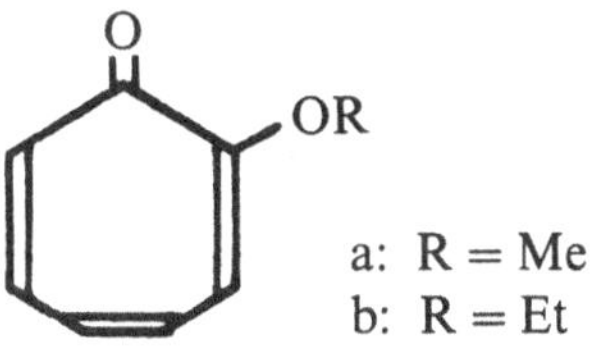

a: R = Me
b: R = Et

20 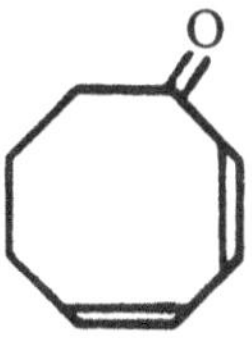   **23** 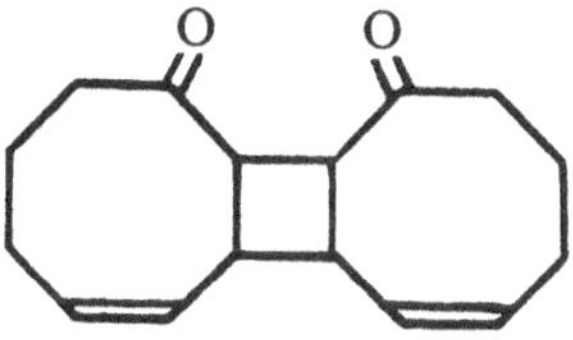   **24**

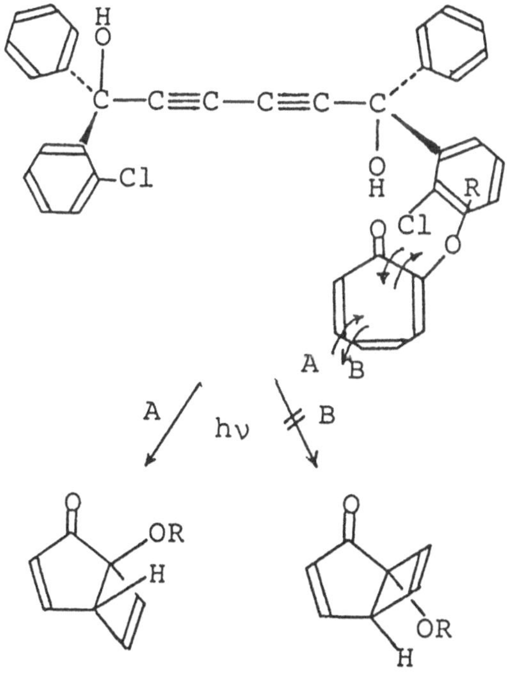

(1S, 5R)-(−)-   (1R, 5S)-(+)-

21

Scheme II

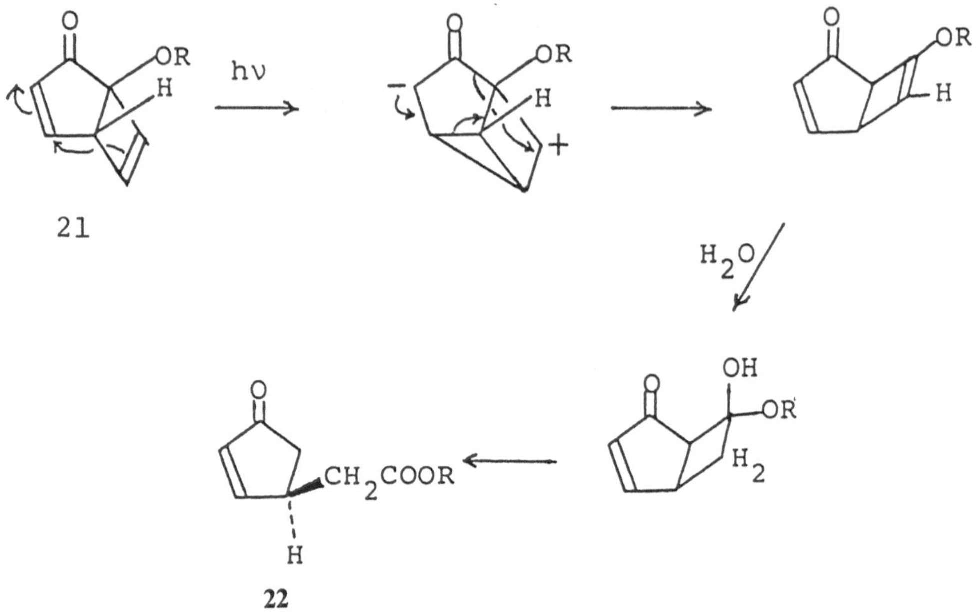

21

22

Scheme III

Irradiation of cycloocta-2,4-dien-1-one (23) in pentane gives a racemic photodimer, *anti*-tricyclo[8.6.0.0^{2,9}]hexadeca-7,11-diene-3,16-dione (24) in 10% yield along with polymeric materials [18]. Efficient and enantioselective photodimerization of 23 was achieved by irradiation of a 2:1 complex of 18a and 23. Irradiation of the complex in the solid state for 48 h gave (−)-24 of 78% ee in 55% yield [12]. Since one unit of the complex contains one molecule of 23 and two molecules of 18a, at least two of these units should take part in the photodimerization of 23. Interestingly, irradiation for 6 h of a 1:1 complex of 23 and optically active host (25) [19] in the solid state gave (−)-bicyclo[4.2.0]oct-7-en-2-one (26) ([α]$_D$ − 60.6°(*c* 0.18, CHCl$_3$)) in quantitative yield.

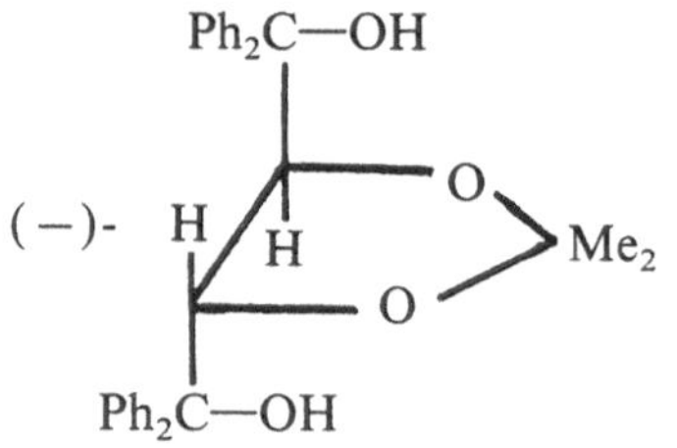

25 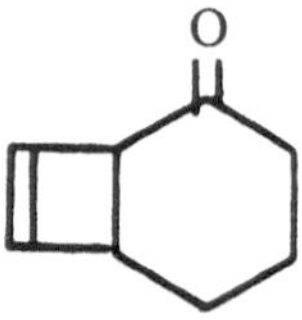   26

Although optically active oxaziridines are useful reagents for enantioselective oxidation of olefins, those of more than 30% ee have not been obtained by any enantioselective synthetic method. Very efficient enantioselective photocyclization of nitrones (27) in the complex with 18 into oxaziridines (28) of high optical purity was found (Table II) [20]. Enantioselectivity in the formation of 28b, 28d, and 28e is high. In the case of **27e**, optically pure **27e** is included in the complex with **18**, and irradiation of the complex gives **28e** in which all three chiral centers are 100% ee [20].

Table II. Melting points of complexes of **27** and **18**, irradiation time, yields, and optical purities of the product (**28**)

Complex			Irradiation time (h)	Product (**28**)	
Ar	R	mp (°C)		Yield (%)	Optical Purity (% ee)
a p-Cl—C$_6$H$_4$—	t-Bu	112–115	32	74	30
b o-Cl—C$_6$H$_4$—	t-Bu	108–110	12	51	100
c 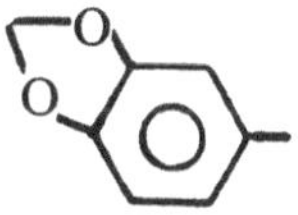	i-Pr	95–103	16	63	28
d 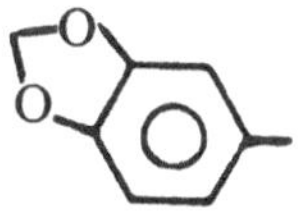	t-Bu	104–111	21	52	94
e Ph	i-PrMeCH	126–128	24	40	100

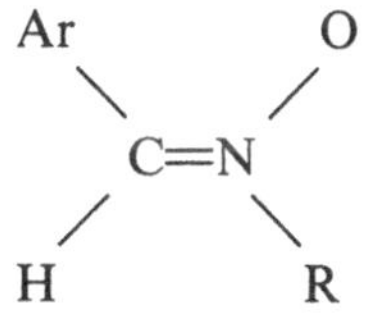

Enantioselective photocyclization of oxoamides to β-lactams has been achieved by irradiation of the complex of oxoamide and an optically active host compound [15, 16]. For example, irradiation of a 1:1 complex of *N,N'*-dimethylbenzoylformamide (**29**) and **18a** in the solid state for 8 h gave ($-$)-**30** of 100% ee in almost quantitative yield. An X-ray crystal structural study of the complex showed that a molecule of **29** is held in a fixed conformation determined by two hydrogen bonds and by neighboring host molecules which prevent free rotation about the CO—CO single bond in the complex (Scheme IV). Free rotation about this bond would enable the production of the two possible enantiomers. The fixed conformation of the guest molecule by the chiral host molecule causes the least molecular motion during the photocyclization and the high enantioselectivity [16].

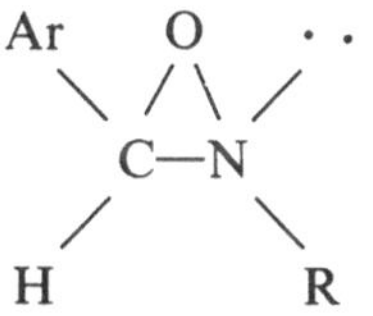

The most exciting enantioselective photocyclization of an oxoamide was found in the case of *N,N*-diisopropylbenzoylformamide (**31**). **31** forms chiral crystals which upon irradiation in the solid state gives 93% ee β-lactam (**32**) in 75% yield [21]. This does not need any chiral source.

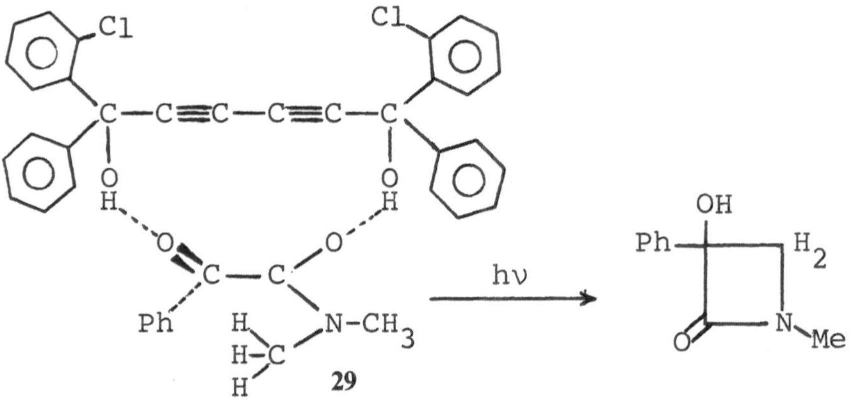

Scheme IV

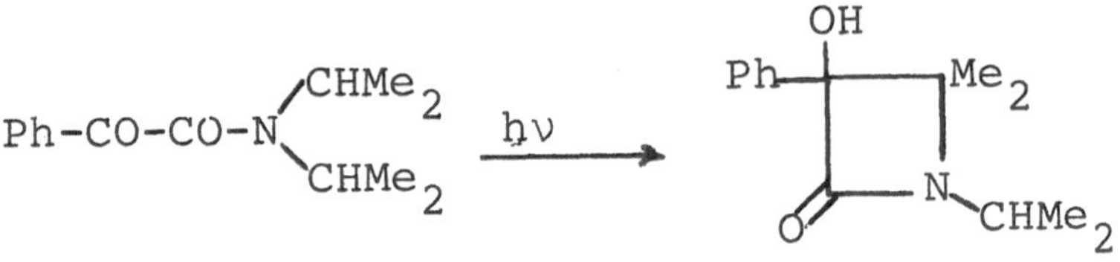

References

1. K. Tanaka and F. Toda: *J. Chem. Soc., Chem. Commun.* 593 (1983).
2. K. Tanaka and F. Toda: *Nippon Kagaku Kaishi*, 141 (1984).
3. M. Kaftory, K. Tanaka, and F. Toda: *J. Org. Chem.* **50**, 2154 (1985).
4. F. D. Greene, S. L. Misrock, and J. R. Wolf, Jr.: *J. Am. Chem. Soc.* **77**, 3852 (1955).
5. M. Kaftory: *Tetrahedron* **43**, 1503 (1987).
6. E. C. Taylor and R. O. Kan: *J. Am. Chem. Soc.* **85**, 776 (1963).
7. T. Nozoe, M. Sato, and K. Matsui: *Proc. Jpn. Acad.* **28**, 407 (1953).
8. Y. Ikegami: "Infrared Absorption Spectra," Nankōdō, Tokyo (1959), Vol. 8, p. 33.
9. T. A. Hamor and D. J. Watkin: *J. Chem. Soc., Chem. Commun.* 440 (1969).
10. F. Toda, K. Tanaka, J. Elguero, L. Nassimbeni, and M. Niven: *Chem. Lett.* 2317 (1987).
11. C. Ganter, S. M. Pokras, and J. D. Roberts: *J. Am. Chem. Soc.* **88**, 4235 (1966).
12. F. Toda, K. Tanaka, and M. Oda: *Tetrahedron Lett.* **29**, 653 (1988).
13. F. Toda, K. Tanaka, and A. Sekikawa: *J. Chem. Soc., Chem. Commun.* 279 (1987).
14. F. Toda and K. Tanaka: *J. Chem. Soc., Chem. Commun.* 1429 (1986).
15. F. Toda, K. Tanaka, and M. Yagi; *Tetrahedron* **43**, 1495 (1987).
16. M. Kaftory, K. Tanaka, and F. Toda: *J. Org. Chem.* **53**, 4391 (1988).
17. W. G. Dauben, K. Koch, S. L. Smith, and O. L. Chapman: *J. Am. Chem. Soc.* **85**, 2616 (1963).
18. T. S. Cantrell and J. S. Solomon: *J. Am. Chem. Soc.* **92**, 4656 (1970).
19. F. Toda and K. Tanaka: *Tetrahedron Lett.* **29**, 4299 (1988).
20. F. Toda and K. Tanaka: *Chem. Lett.* 2283 (1987).
21. F. Toda and M. Yagi: *J. Chem. Soc., Chem. Commun.* 1413 (1987).

Journal of Inclusion Phenomena and Molecular Recognition in Chemistry 7 (1989), 257–266.

Electrochemical Switching in Reducible Lariat Ethers: from Cation Binding Enhancements to Electrochemically-Mediated Transport

LUIS ECHEGOYEN*, GEORGE W. GOKEL, LOURDES E. ECHEGOYEN, ZHI-HONG CHEN, and HYUNSOOK YOO
Department of Chemistry, University of Miami, Coral Gables, FL 33124, U.S.A.

(Received: 1 February 1988)

Abstract. The evolution of lariat ethers from relatively simple, substituted crown ethers into electrochemically sensitive ligands is presented. Although nitrogen-pivot lariats were observed to be better binders than the corresponding parent crowns and to retain considerable flexibility after complexation, overall stability constants were not favorable for cation transport applications. This led to the syntheses of nitrobenzene- and anthraquinone-substituted systems capable of reversible redox behavior and drastically enhanced cation binding abilities when reduced. Application of these in enhanced cation transport processes has been demonstrated.

Key words. Cation transport, electrochemical reduction, EPR, lariat ethers.

1. Introduction

The 1987 Nobel prize in chemistry was awarded to Charles J. Pedersen, Donald J. Cram, and Jean-Marie Lehn not for a single discovery, but for the invention of macrocyclic 'crown' polyethers [1] and their elaboration into cryptands [2], spherands, cavitands [3], and other, related macrocycles. Although very closely related, each of these structures has a unique character. The original monocyclic crown ethers are moderately strong binders of alkali and alkaline earth metal cations and they exhibit a modest cation selectivity. Binding dynamics in crown ether complexation reactions is relatively high. Evolution along the cryptand line gave spherands which, like their predecessors, are very strong and selective cation binders, but the rates of their decomplexation reactions are generally slow.

Cryptands are thus relatively rigid, three-dimensional structures while monocyclic crown ethers are generally flexible, two-dimensional structures [4]. The former exhibit high binding constants and the latter more modest ones under similar circumstances. These considerations led us to develop a class of compounds we call the 'lariat ethers' [5] that were designed to have high binding while retaining considerable flexibility. Structural comparisons of a crown, a cryptand, and a lariat ether are shown in Figure 1 below. In order to emulate the properties of the cryptands, however, the lariat ethers had to be elaborated into systems that could exhibit enhanced cation binding and increased rigidity [6]. This was done by incorporating an electrochemically reducible function [7].

---

*Author for correspondence.

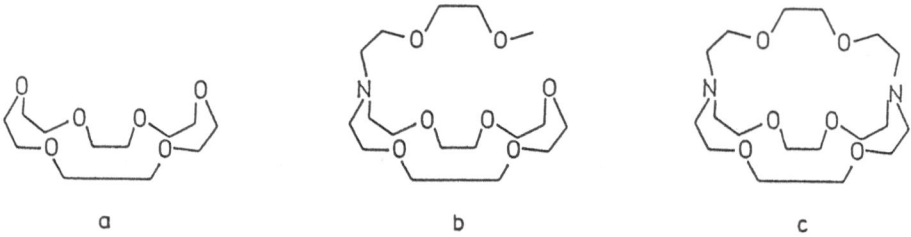

Fig. 1.   Structural comparisons between (a) crown ether, (b) lariat ether, and (c) cryptand.

Based on these considerations, we have developed, in an evolutionary fashion reminiscent of the crown, cryptand, spherand series, compounds that have become increasingly more specialized in their structural variations and corresponding functions. One of the important considerations guiding this development has been the potential application of these systems in membrane transport. At the beginning of a transport step, strong binding is required. At the end of the transport step, week binding and high dynamics are required. Throughout, good binding strength is needed and this is often accompanied by rigidity and slow kinetics.

The stages of this development are described below along with an assessment of the directions still to be followed.

2. Results and Discussion

2.1. STRUCTURE AND FLEXIBILITY

Initially, two classes of lariat ethers were prepared, the carbon-pivot [5, 8] and the nitrogen-pivot [9] compounds. These differ by the atom to which the donor-group-bearing sidearm is attached to the macroring. Although the cation binding strengths for the carbon-pivot compounds were generally disappointing, the nitrogen-pivot lariat ethers exhibited significant enhancements over structures such as 15-crown-5 or 18-crown-6. For example, aza-15-crown-5 having a $CH_3OCH_2CH_2OCH_2CH_2$ sidearm attached to the macroring nitrogen, binds Na^+ in anhydrous methanol with an equilibrium constant (expressed hereinafter as log K_S) of 4.54. This compares to a binding constant of 3.29 [10] for Na^+ with 15-crown-5 and to a value of 4.34 for Na^+ with 18-crown-6 in methanol. Of course, this is several powers of ten lower than the value of 7.8–8.0 reported for 2.2.2.[cryptand] binding Na^+ [11]. Nevertheless there is a clear binding enhancement when compared to the parent crown ethers. Additional evidence for this was obtained from the solution of several solid state structures of lariat ether-cation complexes showing clear sidearm participation in binding to the cation [12].

Our belief that dynamics and flexibility are both retained in N-pivot lariat ether complexes derives from two sources. First, we have studied the relaxation times (T_1) for a variety of N-pivot lariat ethers using ^{13}C-NMR [6]. The ^{13}C-NMR results show essentially no change in the mobility of detectable nuclei whether or not the lariat ether is complexed by sodium or potassium cations. Second, ultrasound relaxation time studies conducted collaboratively with Petrucci and Eyring [13] have confirmed high binding rates for a two-step complexation mechanism.

From these lines of evidence, it is clear that lariat ethers of this general type do offer enhanced cation binding capability while retaining much of the fast cation exchange dynamics inherent to the monocyclic crown ethers.

2.2. ELECTROCHEMICAL PROPERTIES AND ENHANCED CATION BINDING: THE PRINCIPLE OF SWITCHABILITY

Once the essential structures and dynamics of the lariat ether interactions were established, we began our quest for enhanced cation binding strengths. Our approach was to replace the polyethyleneoxy donor groups in the lariat ether sidearms by systems capable of reversible, electrochemical reduction. The first series of compounds were based on the nitrobenzene system. In the neutral state, the nitrobenzene oxygens served as neutral donors for a ring-bound cation. After electrochemical reduction, a nitrobenzene radical anion was formed, substantially increasing the donicity of the sidearm. As a consequence, the overall binding constant was increased dramatically.

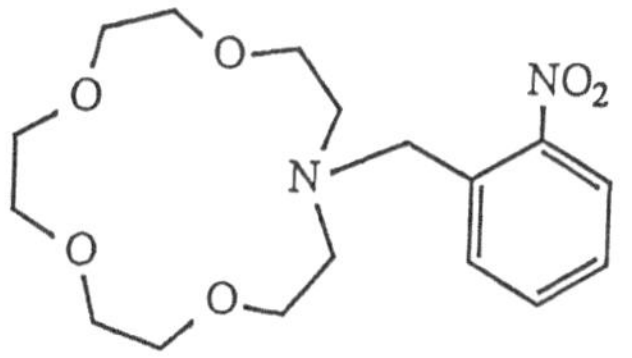

Nitrobenzene-substituted lariat ether

The cyclic voltammogram for the compound illustrated above appeared essentially identical to that observed for 2-nitrotoluene in the absence of any cation. When Na^+ was added to the solution, a significant change in the voltammogram was observed. This change was not observed when the nitro group was in the *para*-position, indicating the importance of proper geometry. The new cyclic voltammogram exhibited a new, quasi-reversible redox couple at a more positive potential than the original one. This was surprising to us since a shift of the existing peak was expected rather than the appearance of a new peak, the intensity of which depended upon the cation concentration (see Figure 2).

The appearance of simultaneous redox pairs, a rather unusual observation, instead of a shift of the existing peak, has not been fully rationalized and is under current investigation in our laboratories. At least part of this behavior can be attributed to the numerical value of K_1 for the neutral ligand. When K_1 is larger than 10 000, two waves are always observed. When K_1 is less than 1, a single peak is observed which appears to be the original peak shifted to a new position as a result of added cation. At K_1 values in the range 1–10 000, the behavior, as expected, is intermediate between the two extremes noted above and depends on the ratio K_2/K_1. Computer simulations based purely on thermodynamic considerations (no kinetic parameters) permit us to simulate the behavior described above. In order to further evaluate all of the chemical and structural contributions to this behavior, we are currently synthesizing the carboxylic acid lariat ether shown below.

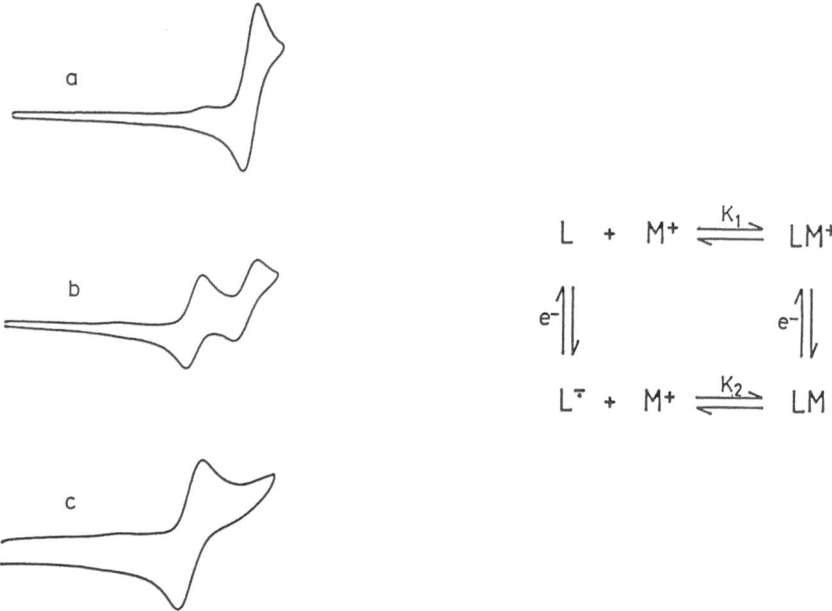

Fig. 2. *Left*: Cyclic voltammograms for *N*-(2-nitrobenzyl)aza-15-crown-5 in the (a) absence, (b) presence of 0.5 equiv., and (c) presence of 1 equiv. of NaClO$_4$. *Right*: Corresponding electrochemical cycle.

This compound can form an anion by deprotonation and its complexation can be followed in solution using NMR techniques [6]. We believe this compound is the ideal diamagnetic analog of the nitrobenzene lariat ether in its radical ion form. It should provide information on the effect of rigidity on the overall complexation process.

Benzoic acid-substituted lariat ether

Despite these mechanistic intricacies and uncertainties, it is possible to extract considerable information from the electrochemical behavior. When two separate redox couples are observed, it is possible to determine the ratio K_2/K_1, the factor of electrochemically-enhanced cation binding that occurs upon electrochemical switching. The enhancements observed always follow the trend $Li^+ > Na^+ > K^+$. This result is expected in view of coulombic considerations. Factors as high as 10^6 were observed for Li^+ cation and lower values, typically 10^2 were observed for K^+ [7]. Although these enhancement factors exhibit a clear selectivity, the overall binding ability is actually leveled. The latter is due to the fact that K_1 for typical, neutral crowns and lariat ethers is in the order $K^+ > Na^+ > Li^+$, the reverse of the order observed for electrochemical enhancement [14]. This levelling effect can be overcome by imposing a more rigid binding structure on the reduced ligand. This was demonstrated in a reducible aza-cryptand [15].

At this stage it was demonstrated that reduced lariat ethers could bind cations more strongly than the neutral, parent compounds. These structures, when reduced, exhibit cation binding affinities much closer to those of cryptands than do the neutral lariat ethers and behave as if they are more rigid, cryptand-like structures as well. In contrast to the cryptands, they can be switched back to their low-binding, flexible structures on demand using electrochemical or chemical oxidation.

2.3. ANTHRAQUINONE SUBSTITUENTS FOR ONE AND TWO ELECTRON REDUCTION IN AQUEOUS MEDIA: AN ELEMENT OF TUNABILITY

Before attempting to apply these structures directly in cation transport experiments across lipid membranes, it was necessary to replace the nitrobenzene group by one that would afford a water-stable anion radical. After careful consideration of synthetic possibilities, the anthraquinone residue was chosen. This selection afforded an additional flexibility in the switching scheme, namely the possibility of one- and/or two-electron reduction. The latter can be regarded as a tunability feature of the overall system. Several of the anthraquinone derivatives that were prepared as part of this study are shown below.

As anticipated, two reversible redox waves were observed for all of these compounds in the absence of a bound cation [16]. Addition of Li^+, Na^+, or K^+, results in most cases in the appearance of additional redox waves. All electrochemical observations are consistent with those presented in the previous section for nitrobenzenes except that they contain an additional redox wave. By controlling the potential, it is possible to observe binding enhancement in two distinct stages: K_2/K_1 and K_3/K_2 (see scheme below).

In contrast to the nitrobenzene derivative results, the observed enhancements are typically lower for the anthraquinone cases. This was expected in view of the fact that charge and spin are much more delocalized in the latter case. Typical enhancement values for Li^+ are 10^3–10^4 for one electron reduction in the case of anthraquinones and 10^2 for Na^+ or K^+. The second electron reduction results in a further enhancement (K_3/K_2). The binding enhancement of the anthraquinone anion is reduced by the more diffuse charge relative to nitrobenzene, but this is compensated by addition of a second electron, a step not easily accessible in the nitrobenzene system.

We have outlined the electrochemical and electron spin resonance properties of several of these systems in a recent series of papers [17]. Results with the anthraquinone lariat ether were successful but unremarkable. Likewise, the single-armed anthraquinone podands behaved as expected. Unexpected behavior was observed for the anthraquinone bis(podands) having the arms arranged either *syn* or *anti*. The anion radical of the *anti* compound bound two cations as expected but the *syn* two-armed anthraquinone podand radical anion did so as well. This behavior was not anticipated and is surprising. The capability of these anion

n = 1,2   *n = 0–5*   *n = 1,4*

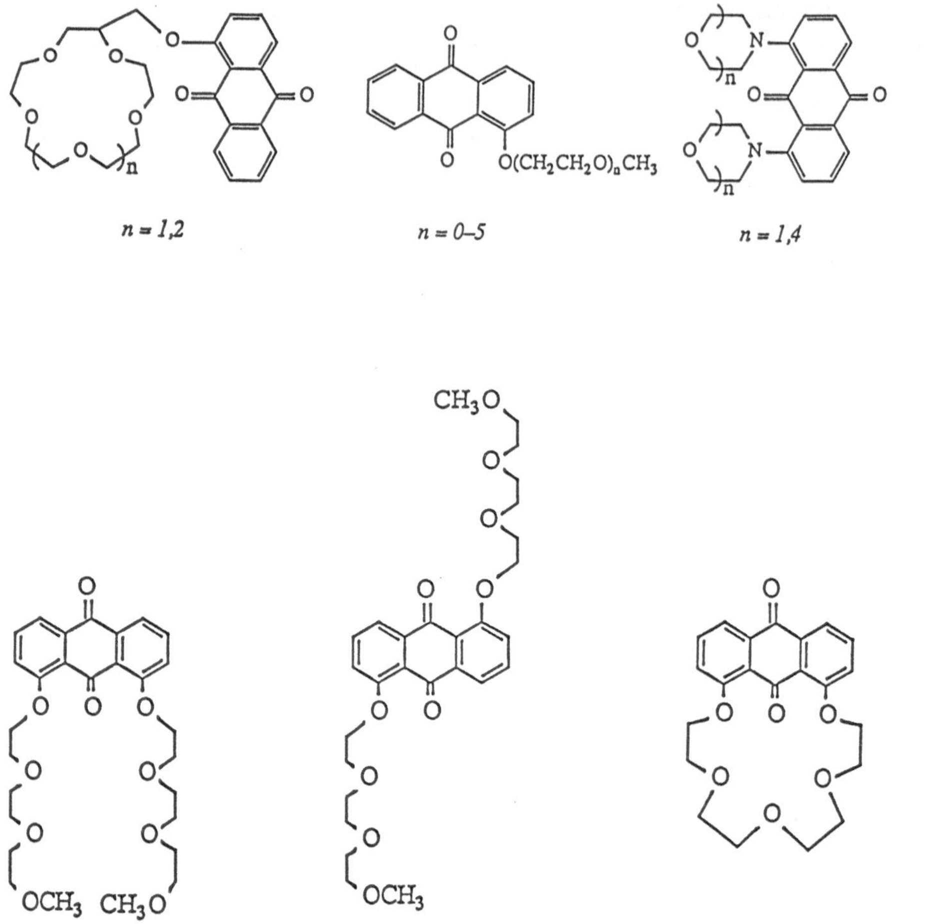

Anthraquinone derivatives

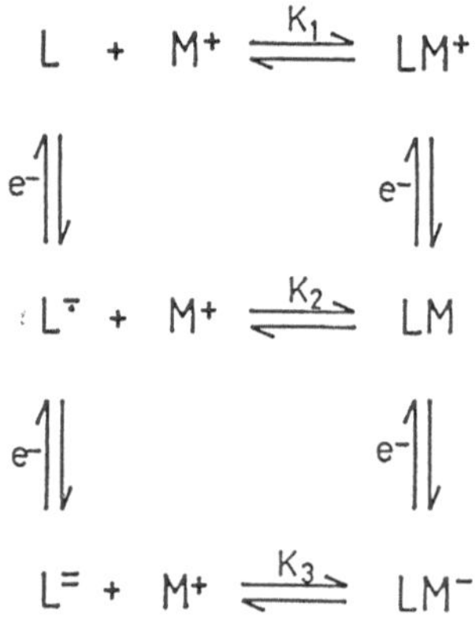

Scheme 1.   Electrochemical cycle for anthraquinone derivatives.

radicals to bind one and/or two cations adds an extra dimension in the degree of flexibility, if not tunability of these systems [18]. EPR results also indicate that the cyclic anthraquinone compound possesses an exceptionally high affinity for Na^+ when reduced.

Preliminary results with the morpholino and aza-15-crown-5 compounds illustrated above have shown interesting cation selectivities upon electrochemical reduction. It appears, for example, that the bis-crown is unaffected by K^+ although the electrochemical behavior is more normal in the presence either of Li^+ or Na^+. The morpholino case exhibits a gradual disappearance of the second redox couple as the cation concentration is increased. This apparently indicates the formation of a strong 2 : 1 ligand-to-metal complex. Further synthetic, voltammetric, and EPR experiments are underway.

2.4. APPLICATIONS IN CATION TRANSPORT

An important advantage of the anthraquinone system is the stability of its radical anion in the presence of water at neutral pH. When an anthraquinone radical anion podand was added to a two-phase, CH_2Cl_2–water system containing Li^+, the cation was transported into the organic phase. This was demonstrated unequivocally using electron spin resonance spectroscopy by detection of a 0.33 G Li-hyperfine splitting.

When an H-shaped cell was constructed, the neutral anthraquinone podand failed to transport Li^+. Upon coulometric reduction to the corresponding anion radical, the podand was able to transport Li^+ at a rate of 2.2×10^{-7} moles/hour [19]. This electrochemically-enhanced transport scheme, clearly shows that it is possible to

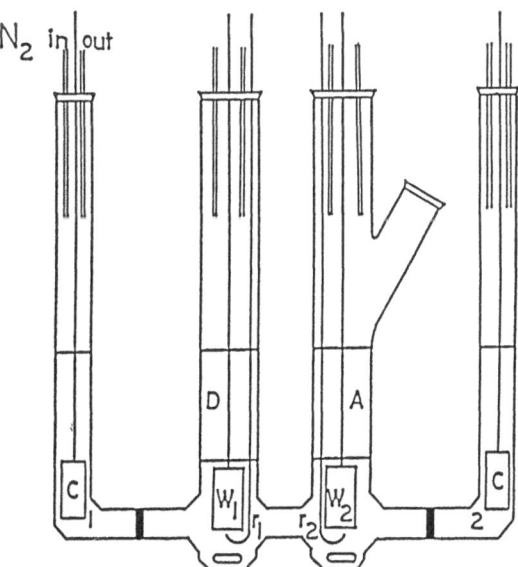

Fig. 3.   Diagram of modified transport cell: (D) aqueous donor phase; (A) aqueous receiving phase; (W₁) and (W₂) working electrodes for reduction and oxidation respectively; (r₁) and (r₂) reference electrodes; (C₁) and (C₂) counter electrodes.

attain both enhanced cation binding and enhanced cation transport ability upon reduction.

More recent results have been obtained using a modified version of the previously reported H cell (see Figure 3 below). Two potentiostats are used simultaneously, one to reduce the anthraquinone ligand near the donor organic interface and the other to oxidize the complexed, neutral species near the receiving interface. Such an experimental arrangement has been reported by Saji [20] and is clearly not without its problems. Nevertheless, we have recently shown further cation transport rate enhancements using this approach. Refinements in this approach are still required.

2.5. THE FUTURE FOR CHEMICAL REDOX-DRIVEN SYNTHETIC ION PUMPS

Our efforts to prepare suitable switchable carriers have suffered from several difficulties that did not become apparent until actual transport experiments were attempted. It now appears that the combination of a macroring or podand, an anthraquinone, and a lipophilic residue are all required. The lipophilic residue must be attached through a non-hydrolytically-sensitive linkage and best results seem likely when the lipophilic residue and cation binding portions of the molecule are non-adjacent. We have already begun studies of switchable transport in lipid bilayers but for these experiments, highly lipophilic carriers are imperative. Furthermore, reduction and oxidation must be accomplished by chemical rather than electrochemical means. Suitable redox agents are currently under evaluation. Another important but currently remote goal is to develop ion channel structures which would incorporate reducible groups capable of providing on–off gating in these channels.

3. Experimental

3.1. GENERAL METHODS

Preparations for the compounds described herein are given in the references cited above. All new compounds were characterized by NMR, IR, and combustion analysis as a minimum as described in those references.

3.2. REAGENTS FOR ELECTRON SPIN RESONANCE AND CYCLIC VOLTAMMETRY

Tetrahydrofuran (THF, Aldrich) was flask-to-flask distilled from the vacuum line immediately prior to use. Dichloromethane (Aldrich) was distilled from CaH_2. All solutions were prepared either under vacuum or under an atmosphere of dry N_2. Tetrabutylammonium perchlorate (TBAP, Fluka) was twice recrystallized from EtOAc and stored in a desiccator. Alkali metal perchlorate salts (Aldrich) were recrystallized from deionized water and dried in a vacuum oven at 100°C for 24 h.

3.3. ELECTRON PARAMAGNETIC RESONANCE

EPR spectra were recorded for dry THF solutions by using the X-band of an IBM ER-200D SRC spectrometer. Samples were prepared under vacuum (10^{-3} mm) by reaction of a 10^{-3} M solution of the compound with either sodium or potassium metal. Spectra in the presence of Li^+ were obtained in purified CH_2Cl_2 by controlled-potential electrolysis by using TBAP as the supporting electrolyte and 1 M $LiClO_4$ aqueous solution directly in the ESR cavity.

3.4. CYCLIC VOLTAMMETRY EXPERIMENTS

The electrochemical experiments were performed at 25°C under N_2 in MeCN 0.1 M in TBAP. The electroactive species was present in millimolar concentrations. Glassy carbon was used as the working electrode and a Pt wire as the counter electrode. E^0 values are reported vs. saturated calomel electrode (SCE). The measurements were done on a Bioanalytical Systems (Model 100) electrochemical analyzer, equipped with IR compensation.

Acknowledgement

We warmly thank the NIH for a grant (GM 33940, jointly to LE and GWG) that supported this work.

References

1. C. J. Pedersen: *J. Am. Chem. Soc.* **87**, 7017 (1967).
2. J.-M. Lehn: *Science* **227**, 849 (1985).
3. D. J. Cram and K. Trueblood: 'Concept, Structure and Binding in Complexation', in F. Vögtle and E. Weber (Eds.) *Host–Guest Complex Chemistry: Macrocycles*, Springer Verlag, 1985, 125–188.
4. G. W. Liesegang and E. M. Eyring: 'Kinetic Studies of Synthetic Multidentate Macrocyclic Ligands', in *Synthetic Multidentate Macrocyclic Compounds*, R. M. Izatt and J. J. Christensen, Eds.; Academic Press, New York, 1978, p. 245.
5. (a) G. W. Gokel, D. M. Dishong, and C. J. Diamond: *J. Chem. Soc., Chem. Commun.* 1053 (1980); (b) D. M. Dishong, C. J. Diamond, and G. W. Gokel: *Tetrahedron Lett.* 1663 (1981); (c) R. A. Schultz, D. M. Dishong, and G. W. Gokel: *Tetrahedron Lett.* 2623 (1981); (d) R. A. Schultz, D. M. Dishong, and G. W. Gokel: *J. Am. Chem. Soc.* **104**, 625 (1982); (e) R. A. Schultz, E. Schlegel, D. M. Dishong, and G. W. Gokel: *J. Chem. Soc., Chem. Commun.* 242 (1982); (f) Y. Nakatsuji, T. Nakamura, M. Okahara, D. M. Dishong, and G. W. Gokel: *Tetrahedron Lett.* 1351 (1982).
6. (a) A. Kaifer, H. D. Durst, L. Echegoyen, D. M. Dishong, R. A. Schultz, and G. W. Gokel: *J. Org. Chem.* **47**, 3195 (1982); (b) H. D. Durst, L. Echegoyen, G. W. Gokel, and A. Kaifer: *Tetrahedron Lett.* 4449 (1982); (b) L. Echegoyen, A. Kaifer, H. D. Durst, and G. W. Gokel: *J. Org. Chem.* **49**, 688 (1984); (c) A. Kaifer, L. Echegoyen, H. Durst, R. A. Schultz, D. M. Dishong, D. M. Goli, G. W. Gokel: *J. Am. Chem. Soc.* **106**, 5100 (1984).
7. (a) A. Kaifer, L. Echegoyen, D. Gustowski, D. M. Goli, and G. W. Gokel: *J. Am. Chem. Soc.* **105**, 7168 (1983); (b) D. A. Gustowski, L. Echegoyen, D. M. Goli, A. Kaifer, R. A. Schultz, and G. W. Gokel, *J. Am. Chem. Soc.* **106**, 1633 (1984).
8. D. M. Dishong, C. J. Diamond, M. I. Cinoman, and G. W. Gokel: *J. Am. Chem. Soc.* **105**, 586 (1983).
9. R. A. Schultz, B. D. White, D. M. Dishong, K. A. Arnold, and G. W. Gokel: *J. Am. Chem. Soc.* **107**, 6659 (1985).
10. G. W. Gokel, D. M. Goli, C. Minganti, and L. Echegoyen: *J. Am. Chem. Soc.* **105**, 6786 (1985).
11. (a) B. G. Cox, H. Schneider, J. Stroka: *J. Am. Chem. Soc.* **100**, 4746 (1978); (b) M.-F. Lejaille, M.-H. Livertoux, C. Guidon, and J. Bessiere: *Bull. Soc. Chim. Fr.* 1373 (1978); (c) B. G. Cox, J. Garcia-Rosas, and H. Schneider; *J. Am. Chem. Soc.* **103**, 1384 (1981).

12. (a) F. R. Fronczek, V. J. Gatto, R. A. Schultz, S. J. Jungk, W. J. Colucci, R. D. Gandour, and G.
    W. Gokel: *J. Am. Chem. Soc.* **105**, 6717 (1983); (b) F. R. Fronczek, V. J. Gatto, C. Minganti, R.
    A. Schultz, R. D. Gandour, and G. W. Gokel: *J. Am. Chem. Soc.* **106**, 7244 (1984); (c) B. D. White,
    K. A. Arnold, F. R. Fronczek, R. D. Gandour, and G. W. Gokel: *Tetrahedron Lett.* **26**, 4035 (1985);
    (d) R. D. Gandour, F. R. Fronczek, V. J. Gatto, C. Minganti, R. A. Schultz, B. D. White, K. A.
    Arnold, D. Mazzocchi, S. R. Miller, and G. W. Gokel: *J. Am. Chem. Soc.* **108**, 4078 (1986).
13. (a) L. Echegoyen, G. W. Gokel, M. S. Kin, E. M. Eyring, and S. Petrucci: *J. Phys. Chem.* 3854
    (1987); (b) G. W. Gokel, L. Echegoyen, M. S. Kim, E. M. Eyring, and S. Petrucci: *Biophys. Chem.*
    26, 225 (1987).
14. A. Kaifer, D. A. Gustowski, L. Echegoyen, V. J. Gatto, R. A. Schultz, T. P. Cleary, C. R. Morgan,
    A. M. Rios, and G. W. Gokel: *J. Am. Chem. Soc.* **107**, 1958 (1985).
15. D. A. Gustowski, V. J. Gatto, A. Kaifer, L. Echegoyen, R. E. Godt, and G. W. Gokel: *J. Chem.
    Soc., Chem. Commun.* 923 (1984).
16. L. Echegoyen, D. A. Gustowski, V. J. Gatto, and G. W. Gokel: *J. Chem. Soc., Chem. Commun.* **220**
    (1986).
17. (a) M. Delgado, L. Echegoyen, V. J. Gatto, Vincent, D. A. Gustowski, and G. W. Gokel: *J. Am.
    Chem. Soc.* **108**, 4135 (1986); (b) D. A. Gustowski, M. Delgado, V. J. Gatto, L. Echegoyen, and G.
    W. Gokel: *Tetrahedron Lett.* 3487 (1986); (c) D. A. Gustowski, M. Delgado, V. J. Gatto, L.
    Echegoyen, and G. W. Gokel: *J. Am. Chem. Soc.* **108**, 7553 (1986).
18. M. Delgado, D. A. Gustowski, H. K. Yoo, V. J. Gatto, G. W. Gokel, and L. Echegoyen: *J. Am.
    Chem. Soc.* **110**, 119 (1988).
19. L. Echeverria, M. Delgado, V. J. Gatto, G. W. Gokel and L. Echegoyen: *J. Am. Chem. Soc.* **108**,
    6825 (1986).
20. T. Saji: *J. Chem. Soc., Chem. Commun.* 716 (1986).

Journal of Inclusion Phenomena and Molecular Recognition in Chemistry **7** (1989), 267–276.

Multi-Functionalized Chiral Crown Ethers as Enzyme Models for the Synthesis of Peptides. Multiple Chiral Recognition in the Enzyme Model

SHIGEKI SASAKI** and KENJI KOGA
Faculty of Pharmaceutical Sciences, University of Tokyo, 7-3-1 Hongo, Bunkyo-ku, Tokyo 113, Japan

(Received: 1 February 1988)

Abstract. A novel approach to the enzyme model for the synthesis of peptides has been established by using multi-functionalized chiral crown ethers as hosts. The new strategy consists of three key steps as follows. (1) Guest assembly: the host having one free thiol and one thioester with N-protected α-amino acid or peptide proceeds via rapid intra-complex thiolysis of α-amino acid ester salts to form the dithioester, and assembles two guests. (2) Amide formation: the intramolecular aminolysis occurs between the bound guests to form the amide bond. (3) Peptide chain elongation: as the thiol reactive group is regenerated, the above two reactions are repeated to elongate the peptide chain. In the present paper, we describe the multiple chiral recognition that could be achieved by the chiral crown ether in both the intra-complex thiolysis and the intramolecular aminolysis. For explanation of the chiral recognition, we propose a likely structure for the intermediate of the aminolysis.

Key words. Chiral, crown ether, enzyme model, peptide synthesis, chiral recognition, molecular recognition, thiolysis, aminolysis, thiol, thioester.

1. Introduction

One of the most important studies in the area of molecular recognition is the design and synthesis of organic molecules as models of enzymes which catalyze useful synthetic reactions. Artificial macrocycles have been expected to work as hosts to assemble guests into the host reactive sites by forming non-covalent host–guest complexes [1]. Recently, much attention has been paid to a new type of host which can assemble plural guests in a host cavity, since such a host may effect the mutual proximity between the bound guests and accelerate the synthetic reaction between them. Thus, hosts bearing multiple binding sites for plural guests have been designed [2], but few of them have been successfully applied to useful synthetic reactions.

We recently reported a novel enzyme model for the synthesis of peptides by using the multi-functionalized chiral 18-crown-6 derivatives [3]. The new hosts have achieved the assembly of plural guests by covalent bonds formed through non-covalent complexes between the host and the guest, and then enhanced the bond formation between the bound guests. This enzyme model has mimicked the general concept of enzyme catalysis, in which the reactive enzyme-substrate covalent intermediate ($E \sim S_1$) is formed from the noncovalent complex ($E \cdot S_1$), and then reacts with the second substrate (S_2) to give the product ($S_1 - S_2$) as shown in Equation (1) [4].

---

* Author for correspondence.

2. Design

2.1. STRATEGY

It was already reported that thiol-bearing chiral crown ethers of type **1** showed rate enhancements in the thiolysis of α-amino acid p-nitrophenyl ester salts, due to the intra-complex nature of the reaction, forming the corresponding thioester as shown in Scheme 1 [5]. Evaluating the thioester as the reactive intermediate for nucleophile

$$(1) \quad E + S_1 \rightleftharpoons E \cdot S_1 \longrightarrow E \sim S_1 \xrightarrow{\;S_2\;} E + S_1 \text{-} S_2$$

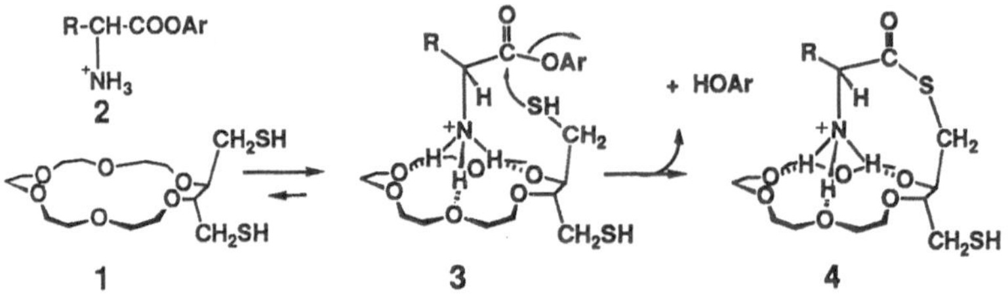

Scheme 1.   Thiol protease model.

[6], a strategy for enhancement of the reaction with the second substrate was designed as shown in Scheme 2. This enzyme model contains three key steps as follows. (1) The host (**5**) having one free thiol and one thioester with N-protected α-amino acid or peptide can form a non-covalent complex with the guest (**2**) and undergoes intra-complex thiolysis to form another thioester (**7**). (2) After neutralization of **7**, intramolecular aminolysis occurs in the host **8** to form the amide bond (**9**). (3) As the thiol reactive group is regenerated in **9**, the above two reactions can be repeated to elongate the peptide chain. We have directed this study toward achievement of actual catlaytic activity for peptide elongation, as a final goal for the enzyme model.

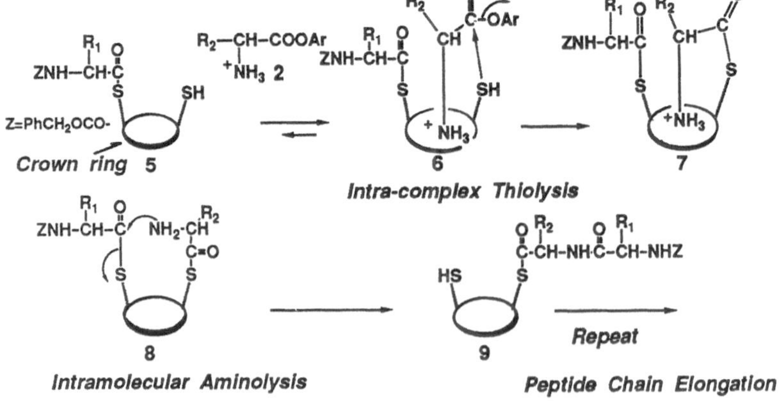

Scheme 2.   Enzyme model for peptide synthesis.

2.2. PREVIOUS RESULTS

In an earlier study, we pointed out some important aspects of aminolysis of thioester in our enzyme model. First, the fastest rate for aminolysis of thioester was obtained in the presence of equimolar amounts of acid and base catalysts. Second, the reaction proceeded in aprotic nonpolar solvents such as benzene, ethyl acetate, dichloromethane, and so on [7]. Thus, the peptide syntheses by the enzyme model have been performed in benzene buffered with equimolar amounts of pivalic acid and triethylamine as acid and base catalysts, respectively. Third, the superiority of intramolecular aminolysis over an intermolecular one was clearly demonstrated, despite the large membered cyclic intermediate expected for the intramolecular reaction. The host 10 could achieve the synthesis of the tetrapeptide derivative (11) by formal turnover of the intra-complex thiolysis and the intramolecular aminolysis, but its efficiency as an enzyme model has remained to be improved [3].

2.3. DESIGN OF THE NEW HOSTS

The *syn*-type host 12 used in the present study was designed so as to improve host efficiency, to have reactive groups on the same face (*syn*-orientation of the reactive groups), and opposite side of the crown ring. The *anti*-type host (13), having reactive groups on the opposite face (*anti*-orientation of the reactive groups), was used for comparison with 12. These hosts were synthesized by an unambiguous method starting from L-tartaric acid. We expected the following two improvements for 12. (1) In the intra-complex thiolysis step, 12 was expected to exhibit a larger rate acceleration than 10, since the closer proximity between the ester of the guest and the thiol of the host was expected because of the short arm of the reactive group as shown in 14 vs. 15. (2) In the intramolecular aminolysis step, 12 might afford better proximity between nucleophilic amine and electrophilic thioester, as shown in 16 vs. 17. Substituents, R = o-MeOC$_6$H$_4$OCH$_2$, were introduced in order to keep the thiomethyl reactive groups perpendicular to the crown plane in the host–guest complex [5c].

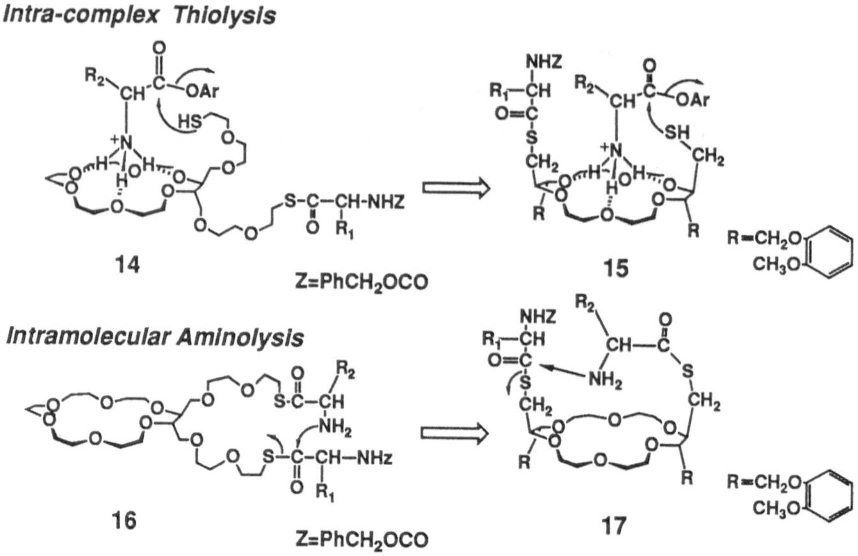

Intra-complex Thiolysis

Fig. 1.   Design of new host.

3. Intra-complex Thiolysis

3.1. RATE ENHANCEMENT OF THIOLYSIS BY CROWN HOSTS

Rate enhancements of thiolysis by thiol-bearing crown ethers were compared by kinetic experiments, using 18 [5a] as the reference host with long side arms, and the results are summarized in Table I. As shown in the column of enhancement ratio, large rate enhancements by the crown ethers have been clearly demonstrated in comparison with the rate constants by the acyclic dithiol (19), (entry 2~4 vs. 1). Particular advantages of the mercaptomethyl reactive group of 12 and 13 over the reactive group with the long side arm of 18 were shown by the fact that an enhancement ratio of only 5 was obtained by 18 but rates of more than 150 were obtained by the hosts 12 and 13 in the thiolysis of D-phenylalanine guest (entry 6~12).

3.2. CHIRAL RECOGNITION BY CROWN HOSTS

As shown in the column of D/L ratio in Table I, 12 and 13 showed chiral recognition by the factors of $k_D/k_L = 4.5$, 4.0, and 5.2 for phenylalanine guest by 12, 13, and 20, respectively. The reaction by 20, which has a free thiol and a thioester with Z-glycine, exhibited high chiral recognition with the ratio of 10 for valine guest. In contrast, 18 showed no chiral recognition. Here, superiority of the structure of the thiomethyl reactive group of 12 over that of 18 has been clearly demonstrated in the intra-complex thiolysis. It seems reasonable that the *syn*-type 12 and the *anti*-type 13 showed the same preference for D-guests, because both hosts have the same chirality of the thiomethyl reactive groups.

   The previous study showed that chiral recognition by the chiral crown ethers occurs not in the complex formation but in the intra-complex thiolysis, and that

$$R_1-\underset{\underset{NH_3Br}{|}}{CH}-\overset{\overset{O}{\|}}{C}-O-\underset{}{\bigcirc}-NO_2$$

$$R-SH \xrightarrow{\hspace{3cm}} R_1-\underset{\underset{NH_3Br}{|}}{CH}-\overset{\overset{O}{\|}}{C}-SR \quad + \quad HO-\bigcirc-NO_2$$

Table I. Chiral recognition in the intra-complex thiolysis[a]

Entry	Host	Guest (R₁=)	k_ψ, 10^{-4} s⁻¹	Enhancement ratio[b]	D/L Ratio
1	19	H	5.2	1	
2	18	H	340	65	
3	12	H	244	47	
4	13	H	198	38	
5	19	L-CH₂Ph	0.65		
6	19	D-CH₂Ph	0.73	1	1.2
7	18	L-CH₂Ph	3.9		
8	18	D-CH₂Ph	3.4	5.2	1.2
9	12	D-CH₂Ph	25		
10	12	D-CH₂Ph	114	156	4.5
11	13	L-CH₂Ph	29		
12	13	D-CH₂Ph	115	158	4.0
13	20	H	170		
14	20	L-CH₃	260		
15	20	D-CH₃	780		3.0
16	20	L-CH₂Ph	30		
17	20	D-CH₂Ph	160		5.2
19	20	L-CH₂(CH₃)₂	2.4		
20	20	D-CH₂(CH₃)₂	24		10

[a] Intra-complex thiolysis was carried out in CH₂Cl₂-EtOH (95:5) buffered with 0.02 M pyridine and 0.01 M AcOH (pH = 5.4 in H₂O) at 25°C, using 5 mM of host (10 mM of 20, entry 13–20) and 0.1 mM of guest, and released p-nitrophenol was followed photometrically at 320 nm.
[b] Enhancement ratios were calculated using the rate constants by 19 for glycine guest (entry 2–4) and phenylalanine guest (entry 7–12) as standard values.

18

19

20

conformational restriction of the thiomethyl reactive group in the tetrahedral intermediates seems to play an important role in the chiral recognition [5c]. Such plausible tetrahedral intermediates of the intra-complex thiolysis are depicted in Figure 2, where the conformation of the thiomethyl was assumed to be *anti* to the adjacent C—C bond of the crown ring [7]. In the intermediate with the L-guest, there seems to be large steric repulsion between the hydrogen of the thiomethyl group and the α-substituent of the guest as depicted by the arrow in **21**. By contrast, in the intermediate with the D-guest, such steric repulsion between two hydrogens seems to be much smaller as depicted in **22**. Thus, the faster thiolysis of D-guests may be explained by the existence of a more stable tetrahedral intermediate with D-guests.

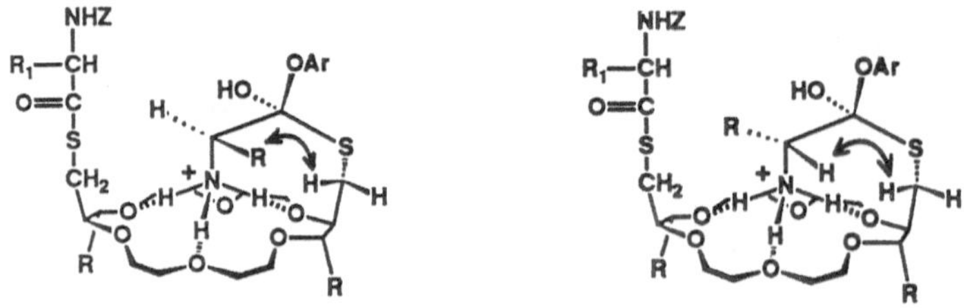

21 With L-guest (Less stable)

22 With D-guest (More stable)

Fig. 2. Tetrahedral intermediates in thiolysis.

3.3. SELECTIVE MONOACYLATION OF THE HOSTS THROUGH THE INTRA-COMPLEX THIOLYSIS

The enhanced thiolysis has been applied to the selective monoacylation of the hosts. The thiolyses of *p*-nitrophenyl ester salts of α-amino acids were performed in slightly acidic solution (CH_2Cl_2 buffered with pyridine–acetic acid, pH = 5.0) to form the monoacylated host selectively. Protection of the amine could be achieved cleanly and rapidly by using carbobenzyloxychloride in the presence of pyridine. The results are summarized in Table II. The intra-complex thiolysis could be also applied to the formation of dithioester under the similar thiolysis condition, to give the desired dithioesters rapidly as shown in Table III. It should be noted that **20** exhibited the largest reaction rate for D-alanine guest and almost the same rates for glycine and D-phenylalanine guests, in both kinetic and preparative experiments.

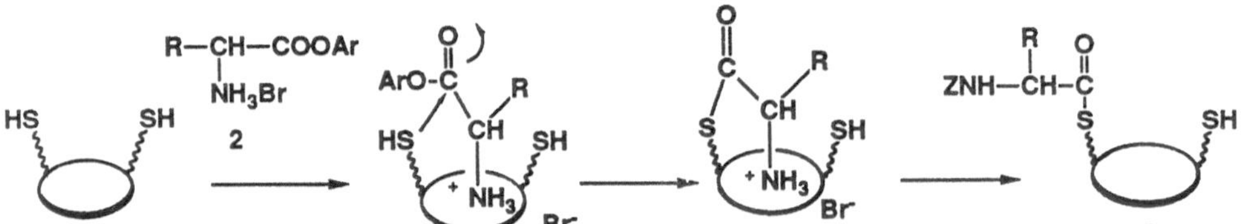

23 24 25 5

Table II. Monoacylation of the hosts through the intra-complex thiolysis

Parent Host	Guest (R=)	Thiolysis Time (min)	Yield of 5 (%)
12	H	60	76
13	H	60	71
12	L-CH_3	<10	81
12	D-CH_3	<10	68

Thiolysis was performed in CH_2Cl_2 buffered with 10 mM each of pyridine and acetic acid at room temperature using 2 mM each of host and guest. Protection was carried out in CH_2Cl_2 in the presence of 0.03 M each of pyridine and carbobenzyloxy chrolide and 0.01 M of host at room temperature for 1 hour.

Table III. Dithioester formation by thiolysis $(5 \rightarrow 6 \rightarrow 7)$

Parent Host	Guest		Time (min)	Yield of 7 (%)
	$R_1 =$	$R_2 =$		
12	H	D-CH_3	10	81
12	H	L-CH_3	10	83
12	H	D-CH_2Ph	60	80
12	H	L-CH_2Ph	60	82

The thiolysis was performed in CH_2Cl buffered with 20 mM of pyridine and 10 mM of acetic acid using 2 mM each of host and guest at room temperature

4. Intramolecular Aminolysis

4.1. CHIRAL RECOGNITION IN THE INTRAMOLECULAR AMINOLYSIS

The intramolecular aminolysis was carried out in benzene buffered with pivalic acid and triethylamine, which function as general acid and base catalysts in aminolysis. The reaction rates were examined by quantitative analysis of the disappearance of the starting host (**8**), and results are compared in Table IV. The aminolysis by using **12** as a parent host, Z-glycine as the electrophilic guest, and D-alanine as the nucleophilic guest showed the fastest reaction rate so far obtained. The same reaction using L-alanine as the nucleophilic guest gave a much smaller reaction rate by the factor of $k_D/k_L = 4.8$. In the case of the reaction using phenylalanine as the nucleophilic guest, a chiral recognition factor of $k_D/k_L = 4.2$ was obtained. It is interesting that *syn*-type **12** showed the common preference for D-guest in both the intra-complex thiolysis and intramolecular aminolysis. Although *anti*-type **13** showed chiral selection in the thiolysis, chiral recognition was not observed in the aminolysis.

| 8 | $R_1 = H$ | 9 |

Table IV. Chiral recognition in intramolecular aminolysis

Parent Host	R_2	k, 10^{-3} min^{-1}	$t_{1/2}$ (hr)	D/L Ratio
12	D-CH_3	11.3	1.0	4.8
12	L-CH_3	2.4	4.8	
12	D-CH_2Ph	4.2	2.8	4.2
12	L-CH_2Ph	1.0	11.6	
13	D-CH_3	4.9	2.4	
13	L-CH_3	4.7	2.4	1.0

The intramolecular aminolysis was carried out in benzene buffered with 0.15 M each of triethyl amine and pivalic acid at 27°C using 1 mM of host, and the disappearance of 8 was quantitatively followed by HPTLC.

4.2. LIKELY STRUCTURE FOR THE INTERMEDIATE OF AMINOLYSIS

For consideration of chiral recognition by the *syn*-type **12**, the stability of the dipolar tetrahedral structure (**26**) was considered as the rate-determining intermediate for aminolysis of thioester [8]. The tetrahedral intermediate (**26**) formed across

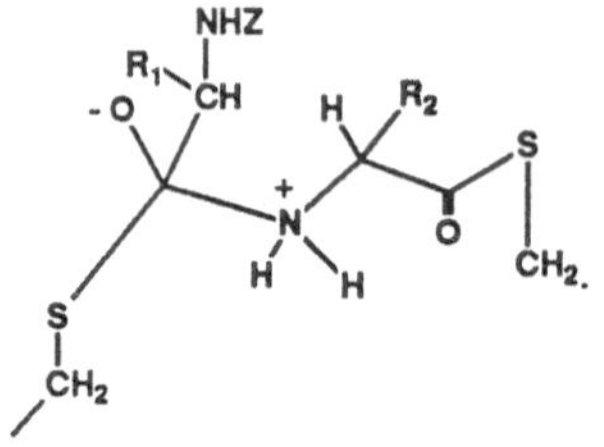

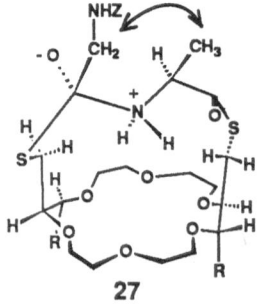

26

Dipolar tetrahedral intermediate of aminolysis

the crown ring seems to have significant conformational restriction, first by the tetrahedral nature of the ammonium cation, and second as a result of the conformational preference of thiomethyl groups, such as is discussed in the section on intra-complex thiolysis. Taking account of such conformational restriction, **27** and **28**, which have a common structure except the chirality of the nucleophilic guests, were proposed as likely structures for tetrahedral intermediates from investigation using molecular models. In these structures, all tetrahedral centers are thought to have a *gauche* relationship with the adjacent center. A structure, in which the residue of the electrophilic thioester is supposed to take a position close to the methylene of the thiomethyl group, should suffer larger steric repulsion than **27** and **28**. Using **27** and **28** as likely structures for intermediates, differences in stability between the two intermediates formed from L- and D-guests may be explained as follows. In the intermediate with L-guest (**27**), there seems to exist steric repulsion between the α-substituent of the nucleophilic guests and the methylene of the glycine unit. In contrast, such steric repulsion seems to be smaller in the intermediate with a D-guest (**28**) as depicted by the arrow.

27

The syn type host with L-guest

(Less stable)

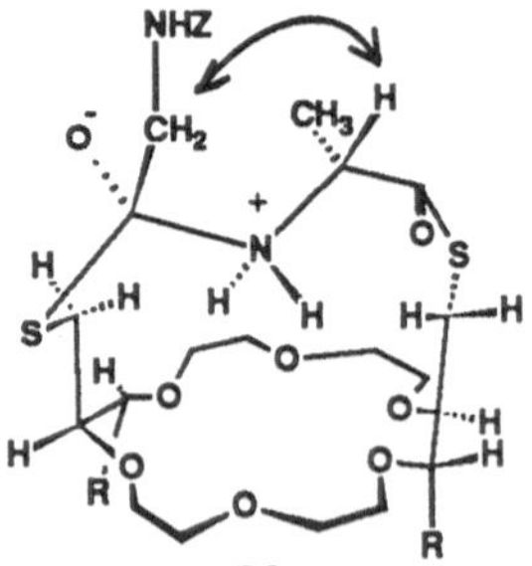

28

The syn type host with D-guest

(More stable)

The same assumption on the conformation of the dipolar intermediates was also applied to the reaction of the *anti*-type host. Rotation around a C—C bond of the crown ring is needed so that the amine nucleophile can attack the thioester as shown in **29**. In the case of the aminolysis by L-alanine as the nucleophilic guest, rotations around the C—C bonds of the crown ring as depicted by **a** or **b** give two different intermediates such as **30** and **31**, respectively. The similar steric restriction of the tetrahedral intermediate as discussed above was applied to that of the *anti*-type host. In **31** there seems to be steric repulsion, as depicted by the arrow, whereas such repulsion seems to be much smaller in **30**. In the case of the intermediates with a D-guest, similarly, both less and the more stable structures may be formed. Thus, the *anti*-type host may form stable intermediates with both D- and L-guests, resulting in no chiral recognition.

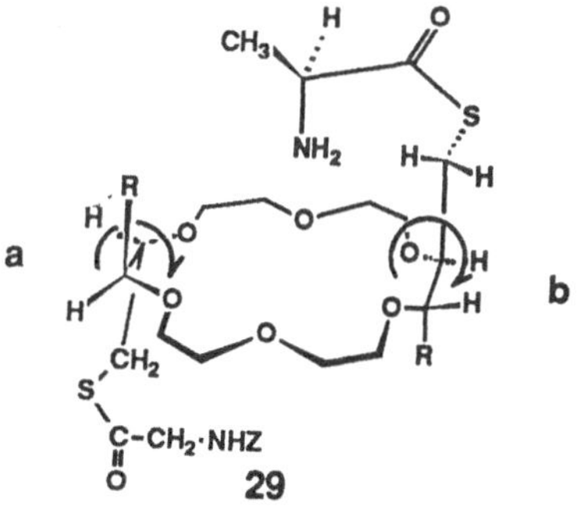

29

The anti-type host with L-guest

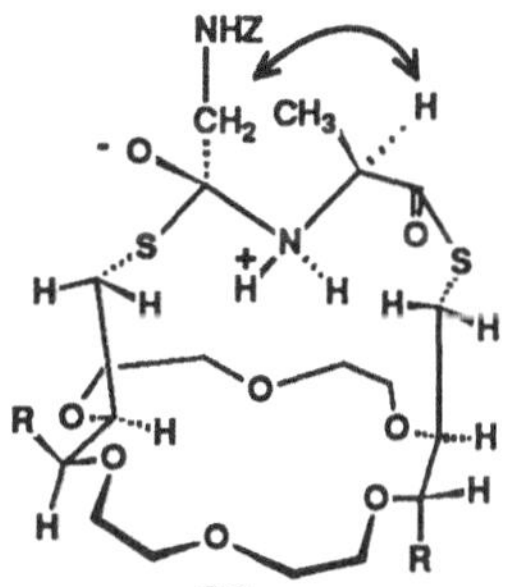

30

More stable intermediate

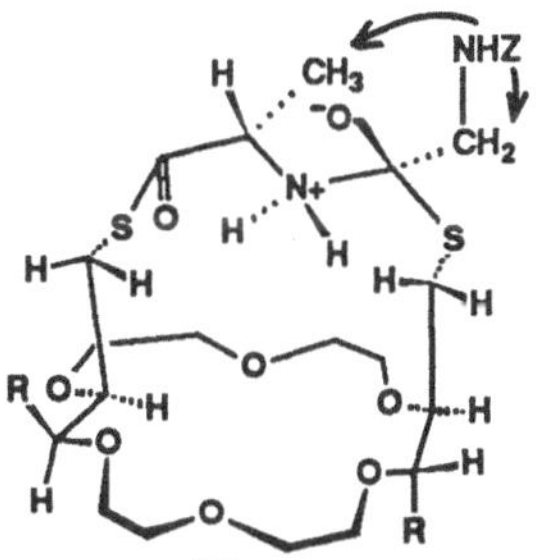

31

Less stable intermediate

The fact that *syn*-**12** gave only double the reaction rate compared with *anti*-**13** showed that the conformational orientation of the thiomethyl reactive groups, that was expected to be advantageous in the intramolecular aminolysis for *syn*-type host, was not efficient enough to assemble the two reactive groups into proximity. We expect that design of a more efficient host may be facilitated from the insight into the structures for the intermediates of the aminolysis.

5. Conclusion

As an approach to the enzyme model for a synthetic reaction, we have established a new strategy by using multi-functionalized chiral crown ethers. Our enzyme model has the characteristics of a model for catalysis of a synthetic reaction in that the host has a single binding site and two reactive sites for plural guests and assembles the guests by covalent bonds formed through the intra-complex reaction. The present study has revealed that the chiral host could achieve multiple chiral recognition in both the intra-complex thiolysis and the intramolecular aminolysis, and that we could assume likely structures for the intra-complex thiolysis and intramolecular aminolysis as well. The achievement of the multiple chiral recognition has suggested the possibility of constructing a new type of enzyme model which catalyzes the synthesis of optically active peptide by using racemic guests.

Acknowledgements

We are grateful to our coworkers: Dr. Motoji Kawasaki, Mr. Mitsuhiko Shionoya, Mr. Hisaaki Chaki, and Mr. Kenji Ohta for their great contribution. The Science and Technology Agency is also acknowledged for financial support.

References

1. For example, F. Vögtle and E. Weber (ed.): *Host Guest Complex Chemistry I, II, and III*, Springer-Verlag, Berlin, 1984.
2. For example, (a) Y. Murakami, K. Aoyama, K. Dobashi, and M. Kida: *Bull. Chem. Soc. Jpn.* **49**, 3633 (1976), (b) I. Tabushi, K. Yamamura, K. Fujita, and H. Kawakubo: *J. Am. Chem. Soc.* **101**, 1019 (1979), (c) F. Kotzyba-Hilbert, J. M. Lehn, and P. Viering: *Tetrahedron Lett.* **21**, 741 (1980), (d) I. Tabushi, Y. Kuroda, M. Yamada, H. Higashimura, and R. Breslow: *J. Am. Chem. Soc.* **107**, 5545 (1985), (e) C. Rai, K. Odashima, and K. Koga: *Tetrahedron Lett.* **26**, 5197 (1985).
3. (a) S. Sasaki, M. Shionoya, and K. Koga: *J. Am. Chem. Soc.* **107**, 3371 (1985), (b) K. Koga and S. Sasaki: *Pure and Appl. Chem.* **60**, 539 (1988).
4. W. P. Jencks: *Catalysis in Chemistry and Enzymology*, McGraw-Hill, New York, 1969, Chapter 2.
5. (a) T. Matsui and K. Koga: *Tetrahedron Lett.* **19**, 1115 (1978), (b) T. Matsui and K. Koga: *Chem. Pharm. Bull.* **27**, 2295 (1979), (c) S. Sasaki, M. Kawasaki, and K. Koga: *Chem. Pharm. Bull.* **33**, 4247 (1985). Other thiol protease models, (d) Y. Chao and D. J. Cram: *J. Am. Chem. Soc.* **98**, 1025 (1976), (e) Y. Chao, G. R. Weissman, G. D. Sogah, and D. J. Cram: *J. Am. Chem. Soc.* **101**, 4948 (1979), (f) J. M. Lehn and C. Sirlin: *J. Chem. Soc., Chem. Commun.* 949 (1978).
6. S. Yamada, Y. Yokoyama, and T. Shioiri: *Experientia* **32**, 967 (1976).
7. S. Sasaki and K. Koga: unpublished data.
8. W. P. Jencks: in Ref. 4, Chapter 10.

Author Index

Subject Index

ADVANCES IN INCLUSION SCIENCE